Biology of Proteoglycans

Biology of Extracellular Matrix:
A Series

Editor
ROBERT P. MECHAM

A list of books in this series is available from the publisher on request.

BIOLOGY OF PROTEOGLYCANS

Edited by **THOMAS N. WIGHT**

Department of Pathology
School of Medicine
University of Washington
Seattle, Washington

ROBERT P. MECHAM

Pulmonary Disease and Critical Care Division
Department of Cell Biology and Physiology
Washington University School of Medicine
Jewish Hospital
St. Louis, Missouri

1987

ACADEMIC PRESS, INC.
Harcourt Brace Jovanovich, Publishers
Orlando   San Diego   New York   Austin
Boston   London   Sydney   Tokyo   Toronto

ACADEMIC PRESS, INC.
Orlando, Florida 32887

United Kingdom Edition published by
ACADEMIC PRESS INC. (LONDON) LTD.
24–28 Oval Road, London NW1 7DX

Library of Congress Cataloging in Publication Data

Biology of proteoglycans.

(Biology of extracellular matrix)
Includes bibliographies and index.
1. Proteoglycans.   I. Wight, Thomas N.   II. Mecham,
Robert P.   III. Series.   [DNLM: 1. Proteophycans.
QU 55 B6179]
QP552.P73B56 1987   599'.019'2454   86-26476
ISBN 0–12–750650–0 (alk. paper)

PRINTED IN THE UNITED STATES OF AMERICA

87 88 89 90   9 8 7 6 5 4 3 2 1

Contents

BIOSYNTHESIS OF HEPARIN AND HEPARAN SULFATE

ULF LINDAHL and LENA KJELLÉN

PROTEOGLYCAN METABOLISM BY RAT OVARIAN GRANULOSA CELLS *IN VITRO*

MASAKI YANAGISHITA and VINCENT C. HASCALL

INTEGRAL MEMBRANE PROTEOGLYCANS AS MATRIX RECEPTORS: ROLE IN CYTOSKELETON AND MATRIX ASSEMBLY AT THE EPITHELIAL CELL SURFACE

ALAN RAPRAEGER, MARKKU JALKANEN,
and MERTON BERNFIELD

STRUCTURAL ORGANIZATION OF PROTEOGLYCANS
IN CARTILAGE

ERNST B. HUNZIKER and ROBERT K. SCHENK

PROTEOGLYCANS, CHONDROCALCIN, AND THE CALCIFICATION
OF CARTILAGE MATRIX IN ENDOCHONDRIAL OSSIFICATION

A. ROBIN POOLE and LAWRENCE C. ROSENBERG

BIOCHEMICAL BASIS OF AGE-RELATED CHANGES
IN PROTEOGLYCANS

EUGENE J-M. A. THONAR and KLAUS E. KUETTNER

Preface

Proteoglycans are a group of extracellular matrix molecules that have "come of age." Recognized in the early 1960s as important structural components of cartilage, proteoglycans were once thought to be specific to that tissue. We know that they are found in the matrices of all tissues and function not only as structural elements supporting cells and providing tissue turgor but also as mediators of events that characterize development and a variety of diseases. It is the goal of this volume to present a representative, but by no means inclusive, sample of current research on the role of proteoglycans in the cell biology of the extracellular matrix. The book is organized into four areas: methodological developments; proteoglycan metabolism; proteoglycans in cartilage; and proteoglycans in "soft" tissues.

The first section (Chapters 1 and 2) deals with current methodological developments which have had enormous impact on our understanding of the complexity of proteoglycan structure. In Chapter 1, Dr. Caterson and his colleagues describe the use of monoclonal antibodies to probe the structure of the protein and carbohydrate portions of proteoglycans. This chapter illustrates how the use of these antibodies has added to our appreciation of the widespread occurrence of proteoglycans in a variety of tissues. Dr. Sandell (Chapter 2) reviews current work regarding the genes that code for the proteoglycan protein cores and associated proteins. From this chapter, the reader will gain insight into how the "tools" of the molecular biologist can be used to solve questions not only of primary structure and homology between proteoglycan families but also of regulation of proteoglycan biosynthesis.

Chapters 3 through 5 deal with various aspects of proteoglycan metabolism. Drs. Lindahl and Kjellén (Chapter 3) discuss the intricate synthetic pathway used by cells to synthesize proteoglycans that contain heparin and heparan sulfate. This chapter stresses the importance of post-translational modification in the assembly of these proteoglycans. Drs. Yanagishita and Hascall (Chapter 4) discuss the intracellular pathways taken by proteoglycans once they are synthesized. It has now become clear that not only do most, if not all, cells synthesize proteoglycans, but many synthesize more than one family. This chapter reviews a series of experiments which show that different proteoglycans synthesized by a cell are "processed" differently. Chapter 5 (Drs. Rapraeger, Jalkanen, and Bernfield) describes the characteristics of a specific class of proteoglycan which is processed for insertion into the plasma membrane of epithelial cells. This chapter illustrates that proteoglycans participate in the cell adhesion process and that this class of molecule may be capable of "communicating" with both intracellular and extracellular compartments. Such a strategic location implicates proteoglycans as potential "information transducers."

Chapters 6 through 8 focus on proteoglycans in cartilage. Drs. Hunziker and Schenk (Chapter 6) provide insight into the molecular organization of different proteoglycans in cartilage using updated methodology. Their work describing increased preservation of proteoglycans in cartilage challenges some of our current concepts as to how proteoglycans are organized in this tissue. Drs. Poole and Rosenberg (Chapter 7) next focus on a specific aspect of proteoglycans in cartilage—the role of proteoglycans and associated proteins in the calcification process in growth plate. As in the preceding chapter, the authors demonstrate how "state of the art" techniques allow us to question some of the dogma regarding the role of proteoglycans in this process. Chapter 8 by Drs. Thonar and Kuettner considers age-related changes in the structure of cartilage proteoglycans. These authors review experiments which show marked changes in the metabolism of proteoglycans in cartilage as a function of age and present evidence that such changes may predispose this tissue to degenerative disease such as osteoarthritis.

The last section of this volume (Chapters 9–12) deals with the emerging field of proteoglycans in "soft" or noncartilagenous tissue. Drs. Kelly, Carlson, and Caroni emphasize that proteoglycans may have a key role in the nervous system. Evidence is reviewed to indicate that proteoglycans are found within the nerve–muscle synapse and may be important in nerve transmission and synaptic regeneration. Over the years there has been considerable interest in the role of proteoglycans in the pathophysiology of blood vessels. My colleagues and I (Chapter 10) review current work in this area and illustrate how studying vascular cells in culture may contribute to our understanding of the role of proteoglycans in events associated with vascular wall development and disease. Chapter 11 by Drs. Marcum, Reilly, and

Rosenberg focuses on a specific class of proteoglycan in blood vessel. These authors review the evidence that heparin and heparan sulfate are not only critical to the maintenance of the nonthrombogenic blood–tissue interface but also may be important regulators of vascular cell growth. Such a role for these proteoglycans emphasizes their importance in key events associated with atherosclerosis. Chapter 12 by Drs. Harper and Reisfeld discusses proteoglycan changes that occur in malignancy. These authors illustrate that some cells such as human malignant melanoma cells possess a specific class of cell surface proteoglycan which may play a role in key events associated with the metastatic processes such as regulation of cell proliferation and migration. The last chapter addresses a rather unexplored area—the role of proteoglycans inside the cell. Dr. Stevens describes studies which indicate that proteoglycans are present within secretory granules of cells of the immune system and presents an interesting hypothesis which implicates these macromolecules as important mediators in the inflammatory response.

The editors would like to express a sincere appreciation to Dr. Vincent C. Hascall, who unselfishly spent many hours discussing ideas for this volume. We thank Academic Press for their insight, patience, and tolerance. It is hoped that this book will serve to inspire both "young and old" to continue to probe the molecular and cell biology of these marvelous molecules.

THOMAS N. WIGHT

Monoclonal Antibodies as Probes for Elucidating Proteoglycan Structure and Function

Bruce Caterson, Tony Calabro, and Anne Hampton

Departments of Biochemistry and Orthopedic Surgery, West Virginia University Medical Center, Morgantown, West Virginia 26506

I. Introduction

Proteoglycans are found in all connective tissues throughout the mammalian body. Together with the collagens and/or elastin, they form the major structural elements of the extracellular matrix of connective tissues, and the various interactions that occur between these macromolecules endow the tissue with its specific biomechanical or physiological properties. The relative proportion of proteoglycan and collagen in tissues changes during normal development, in aging, and with the onset of pathological conditions. At present we know little about the biological mechanisms controlling the organization of these macromolecules within the extracellular matrix and, unlike the collagens or elastin, we also know little about the primary structural features of the numerous different classes of proteoglycans found in connective tissues. The major reason for this dilemma is that proteoglycans are inherently heterogeneous macromolecules. The heterogeneity of these molecules is derived from the occurrence of different classes of glycosaminoglycan and N- and O-linked oligosaccharide carbohydrate moieties that are differentially substituted, on possibly both common or different polypeptide backbones. The presence of these large numbers of carbohydrate substitutions on the core protein of the proteoglycan has hampered the determination of the primary amino acid sequence of these molecules, although the recent use of recombinant DNA technology (1) has offered a potential solution to this problem. In this chapter we will review how we have used mono-

1

clonal antibody technology to study the structure, localization, and metabolism of proteoglycans in various connective tissues. The recent introduction of monoclonal antibody technology has offered researchers the potential of elucidating characteristic structural features of the different classes of proteoglycans independent of their inherent heterogeneity and polydispersity. This technique, possibly in combination with recombinant DNA technology, offers great potential for future studies on proteoglycan structure, function, and metabolism.

II. Production, Screening, and Characterization of Monoclonal Antibodies Directed Against Epitopes Present on Proteoglycan Substructures

In this laboratory our major experience has been in the production and characterization of monoclonal antibodies directed against hyaline cartilage proteoglycans, and thus, the majority of the examples cited in this chapter will apply to monoclonal antibodies recognizing epitopes present on these subclasses of proteoglycan molecules. However, problems that we have encountered in these studies and our suggested solutions to some of these problems can be applied to related studies that use different classes of proteoglycans.

A. *Immunization and Fusion*

When planning the production of monoclonal antibodies to an antigen a major consideration has to be the amount of purified (or partially purified) antigen that is available for the several different phases of the project—that is, immunization, screening, cloning, and characterization of antibody specificity. Often each of these procedures can consume considerable quantities of "valuable" antigen, and thus it is important to consider the general availability of antigen for each of these procedures before commencing the initial immunization of animals.

There are several immunization protocols described for the preparation of activated lymphocytes for hybridoma production. The most common method utilizes the spleen of immunized animals as the source of lymphocytes for fusion and hybridoma production. In this laboratory we use a somewhat less common procedure utilizing the draining lymph nodes as the source of activated lymphocytes after immunization. Procedures outlining the variety of different immunization protocols available for monoclonal antibody production have been excellently reviewed by Kearney (2). The procedure most com-

monly used in this laboratory for the production of antibodies to proteoglycans is briefly described below.

Young 4- to 6-week-old BALB/c mice are injected subcutaneously
with proteoglycan antigen (200–400 μg/ml) at six sites (Fig. 1) in the
hind footpads and in the inguinal and lateral thoracic regions of the
mouse. Five injections of antigen are administered 3 days apart over a
2-week period. For the first injection (day 1) the antigen (400 μg/ml,
50 μl per injection site) was mixed 1:1 with Freund's complete adjuvant, the second (day 3) Freund's incomplete adjuvant, and the subsequent three injections (days 6, 9, and 12) isotonic saline (0.15 M NaCl).
Approximately 50 μl of the antigen mixture is injected into each of the
six regions of the mouse (Fig. 1). Hybridoma fusion is performed 1 or 2
days after the last injection (day 14), and draining lymph nodes from
the regions nearest the sites of injection of the mouse (axillary, brachial, inguinal, and popliteal lymph nodes; see Fig. 1) are dissected and

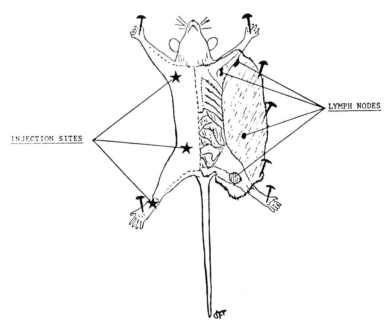

FIG. 1. Injection sites and areas used as the primary source of draining lymph nodes
for the isolation of proliferating lymphocytes prior to hybridoma production. Antigen
was administered to six injection sites subcutaneously in the footpads and in the inguinal and lateral thoracic regions of the mouse. Five series of injections are administered over a 2-week period: Draining lymph nodes are obtained from the axillary,
brachial, inguinal, and popliteal lymph nodes.

used as the source of proliferating lymphocytes for hybridoma production. Preparation of the lymphocytes and details of the fusion protocol utilizing the X 63-Ag 8.653 myeloma cell line have been described elsewhere (3). A potential problem with this particular immunization protocol is that it requires considerably more antigen for immunization than alternative methods utilizing the spleen as the source of lymphocytes. More recently we have found that immunization of one mouse (instead of three, as indicated in Ref. 3) provides sufficient lymphocytes for adequate hybridoma production. The hyperimmunization protocol described above has been particularly successful for production of monoclonal antibodies to both carbohydrate and protein epitopes on proteoglycans. It is possible that this procedure preferentially induces the proliferation of initial lymphocyte subpopulations that are not necessarily activated in immunization protocols involving subsequent booster injections.

B.   Screening Procedures

Perhaps one of the biggest problems we have encountered relating to the production of monoclonal antibodies to proteoglycans has come from the screening procedures we used in our initial detection of antibody-producing hybridomas. The most common procedure used in the detection of antibody-producing hybridomas is the enzymeimmunoassay or enzyme-linked immunoabsorbent assay (ELISA). This procedure relies on the passive absorption of antigen to plastic microtiter plates. Details of this procedure have been described elsewhere (4). The problem that arises in this assay procedure comes from the fact that proteoglycan antigens often do not adsorb very well to the plastic surfaces of the microtiter plates that are used in enzymeimmunoassay screening procedures. Thus the ability to detect hybridomas producing antibodies to epitopes present on proteoglycan antigens is markedly reduced because of the small or negligible amounts of antigen present for primary antibody binding. In order to circumvent this problem we have developed new methods and/or modified existing immunoassay procedures such that potential hybridomas synthesizing antibodies directed against particular epitope(s) on proteoglycan antigens can be initially detected. Examples using these procedures are described below.

C.   Dot–Blot Enzymeimmunoassay

This procedure utilizes nitrocellulose paper as a means of binding or immobilizing proteoglycan antigens for the subsequent detection and screening of antibody-producing hybridomas. Nitrocellulose paper is applied to a Schleicher and Scheull Minifold apparatus. (This apparatus is a suction manifold with 96 wells arranged in a format similar to an ELISA plate.) Antigen (1–20 μg in 200 μl PBS-azide) is applied to each well and vacuum applied to suck through the antigen solution and thus coat the nitrocellulose paper with antigen in a 96-well "dot–blot" template. The nitrocellulose paper is then removed from the Minifold apparatus and the sheet "blocked" with a solution of 5% BSA in Tris–saline buffer (0.05 M NaCl, 0.1 M Tris, pH 7.3). The sheet of nitrocellulose is then replaced in the Minifold apparatus and 200 μl of each of the hybridoma supernatants from culture dishes are added to appropriate wells of the 96-well Minifold apparatus and incubated with the coated antigen for 1–2 hr at 37°C. After incubation the hybridoma supernatants are removed from the wells of the Minifold apparatus and the wells are washed successively three times with Tris–saline buffer. The nitrocellulose sheet is removed from the Minifold apparatus and incubated for 1 hr at 37°C with 20 ml of an appropriate enzyme-linked second antibody (e.g., peroxidase-conjugated goat anti-mouse immunoglobulin diluted in 5% BSA, Tris–saline buffer) in 250-mm Petri dishes. The second antibody is removed and the nitrocellulose sheet is washed successively five times for 10 min with 20 ml of Tris–saline buffer. After washing, the nitrocellulose sheet is incubated with a solution containing substrate for the enzyme conjugated to the second antibody (e.g., H_2O_2 and 4-chloro-1-naphthol that produces a precipitating product after the peroxidase reaction) until appropriate color development occurs. Antibody-secreting hybridomas are detected by the presence of colored spots (precipitates) on the nitrocellulose sheets. An example of this procedure, comparing it to conventional ELISA analyses, is given in Fig. 2.

The antigen used for injection and also coated to the nitrocellulose sheet or ELISA plate was rat chondrosarcoma proteoglycan core protein produced by chondroitinase ABC digestion of rat chondrosarcoma proteoglycan monomer (5). Analyses of hybridoma supernatants by conventional ELISA assay detected only five antibody-secreting hybridoma wells (indicated by stars in Fig. 2), whereas at least 25 strong positive reactions were detected using the nitrocellulose "dot–blot" screening analyses. In conventional ELISA analyses the more hydrophobic (protein-rich) proteoglycan fragments preferentially adsorb to the surface of the plastic microtiter plates, resulting in a greater likelihood of detecting antibody-secreting hybridomas recognizing epitopes in these regions of the proteoglycan molecule. The increased sensitivity of the nitrocellulose "dot–blot" detection method results from (a) the increased amount of antigen coated on the nitrocellulose

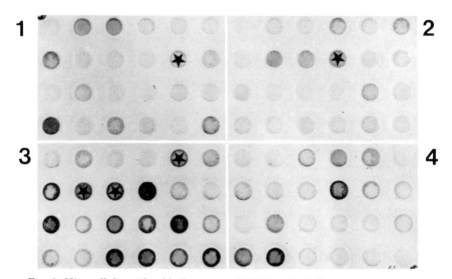

FIG. 2. Nitrocellulose "dot–blot" assay and ELISA of hybridoma supernatants containing antibodies directed against epitopes present on the core protein of rat chondrosarcoma proteoglycan. The antigen coated on nitrocellulose or plastic microtiter plates
was rat chondrosarcoma proteoglycan that had been predigested with chondroitinase
ABC. Ten micrograms of antigen in 200 µl PBS-azide was applied to each well.
Hybridoma supernatants (200 µl) were tested for the presence of antibody-recognizing
antigens. Wells containing a star indicate the five positive results that were obtained
using conventional ELISA immunoassay procedures with alkaline phosphatase-conju-
gated second antibodies. Dark spots (precipitates) indicate positive reactions detected
with the nitrocellulose "dot–blot" screening procedure using horseradish peroxidase-
conjugated second antibodies.

spots and (b) the fact that a representative sample of the different
proteoglycan subpopulations present in the antigen mix are immobilized on the nitrocellullose.

A variation of this "dot–blot" enzymeimmunoassay procedure can
also be used in the preliminary screening and characterization of
monoclonal antibody specificity. Once again, advantage is taken of the
fact that large amounts of proteoglycan antigen (up to 40 µg per well)
can be immobilized on the nitrocellulose sheets. An example of the use
of this procedure in the preliminary screening analyses of monoclonal
antibodies recognizing epitopes present in the hyaluronic acid-binding
region (HABR) of hyaline cartilage proteoglycan monomer is given in
Fig. 3.

Four monoclonal antibodies (12/21/1-C-6, 5/6/4-A-3, 6/1/7-D-1, and
6/1/8-C-1) recognizing epitopes present in the HABR of proteoglycan
were tested against proteoglycans and proteoglycan substructures iso-

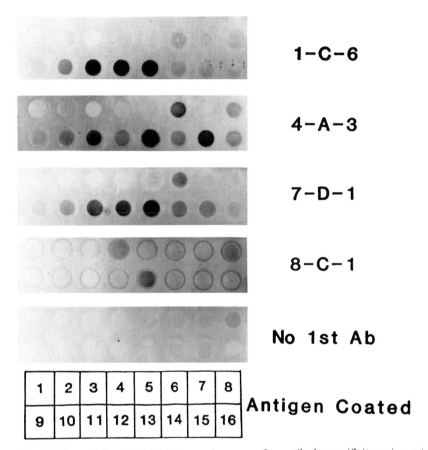

FIG. 3. Nitrocellulose "dot–blot" screening assay for antibody specificity using rat chondrosarcoma (RC) and bovine nasal cartilage (BNC). Antigens coated in positions 1–16 were as follows: 1. RC-A1, proteoglycan aggregate; 2. RC-A1D1, proteoglycan monomer; 3. BNC proteoglycan aggregate; 4. BNC proteoglycan monomer; 5. no antigen; 6. RC alkali-core protein; 7. no antigen; 8. RC-A1D4, protein-rich proteoglycan and link protein; 9. RC-A1, reduced and alkylated proteoglycan aggregate; 10. RC-A1D1, reduced and alkylated proteoglycan monomer; 11. BNC-A1, reduced and alkylated proteoglycan aggregate; 12. BNC-A1D1, reduced and alkylated proteoglycan monomer; 13. RC-hyaluronic acid-binding region (HABR) produced by clostripain digestion of RC proteoglycan aggregate; 14. RC alkali-core, reduced and alkylated; 15. RC ternary complex produced after clostripain treatment of RC proteoglycan aggregate (this antigen contains both HABR and link protein); 16. BNC-A1D4, protein-rich proteoglycan and link protein. Monoclonal antibody supernatants tested against the different antigens were 12/21/1-C-6, 5/20/4-A-3, 6/1/7-D-1, and 6/1/8-C-1. A "no first antibody" control was included to test for nonspecific binding by the horseradish peroxidase-conjugated second antibody.

lated from rat chondrosarcoma and bovine nasal cartilage. For this assay antigen solutions were applied to appropriate wells of the Schleicher and Scheull Minifold apparatus such that replicate strips containing each of the individual antigens could be obtained (see grid pattern in Fig. 3). After coating with antigen, the nitrocellulose sheet was blocked and replicate strips, containing individual spots of the test antigens, were incubated with 15 ml of different monoclonal antibody hybridoma supernatants. A control strip incubated without primary antibody must also be included in each of these screening assays. Each strip is then incubated with peroxidase-conjugated second antibody (e.g., goat anti-mouse Ig), washed ($\times 5$) with Tris–saline buffer, and incubated with peroxidase substrate (4-chloro-1-naphthol and H_2O_2). Antigens recognized by individual monoclonal antibody supernatants are detected by the presence of purple spots (precipitates) present in different locations on the nitrocellulose strip. Relative color intensities can be used as a qualitative estimate of degree of antigenicity of different antigens recognized by different monoclonal antibody supernatants (see Fig. 3). All of the monoclonal antibodies recognize an epitope present on HABR of proteoglycan monomer prepared after clostripain digestion (6) of rat chondrosarcoma proteoglycan aggregate. The 5/6/4-A-3, 6/1/7-D-1, and 12/21/1-C-6 monoclonal antibodies preferentially recognized epitopes present on reduced and alkylated proteoglycan preparations. Both 5/6/4-A-3 and 6/1/7-D-1 recognize a wide range of reduced and alkylated antigens coated on the nitrocellulose strips; 12/21/1-C-6 shows a more limited range of reactivity. The 6/1/8-C-1 antibody recognizes only native HABR, indicating that the epitope recognized by this antibody is conformation dependent. This 6/1/8-C-1 antibody appears to recognize only native HABR in proteoglycan preparations.

D. *Antibody-Bound Radioimmunoassay*

Enzymeimmunoassays using plastic microtiter plates or nitrocellulose sheets require considerable amounts of antigen for coating surfaces for monoclonal antibody detection. Often only small amounts of antigen are available for both immunization and detection of antibody-secreting hybridomas. In some instances only radiochemical amounts of *in vitro*-labeled antigen are available for screening antibody specificity. For these situations we use a solid-phase "antibody-bound" radioimmunoassay procedure (AB-RIA) for detecting potential antibody-secreting hybridomas. The basic principles of this technique are described below.

The AB-RIA procedure uses nitrocellulose sheets as a means of binding monoclonal antibodies (and other proteins present in hybridoma supernatants) to be tested for reactivity against radiolabeled antigens. Hybridoma supernatants (200–300 μl) are applied to separate wells of a Schleicher and Scheull Minifold apparatus containing a sheet of nitrocellulose. The hybridoma supernatants are sucked through the Minifold apparatus, thus binding maximal amounts of antibody present in each supernatant sample. The nitrocellulose sheet is removed from the Minifold apparatus and blocked with 5% BSA in PBS-azide buffer. After washing, the nitrocellulose sheet is incubated for 1–4 hr with 15–20 ml radiolabeled antigen (100,000–500,000 cpm/ml) in 5% BSA PBS-azide buffer. The nitrocellulose sheet is washed for 10 min successively with 5% BSA PBS-azide (×3) and then air-dried. Wells containing monoclonal antibodies recognizing the radiolabeled antigen are identified by autoradiography ([125I]-labeled antigens) or after cutting out appropriate spots on the nitrocellulose plate and determining radioactivity on a scintillation counter. The sensitivity of this assay is directly proportional to the specific activity of the radiolabeled antigen. The use of [125I]-labeled antigens is recommended; however, [3H]- and [14C]-labeled antigens can be adequately detected. This method is extremely sensitive and fast, and modification of this assay can be used in all stages of antibody production and characterization. An example of the use of this method for detecting antibodies that recognize the HABR of cartilage proteoglycan is given in Fig. 4.

Hybridoma supernatants were obtained from the wells of four 24-cell culture dishes and tested for the presence of monoclonal antibodies recognizing epitopes present in the HABR of rat chondrosarcoma proteoglycan. The antigen used for injection was rat chondrosarcoma proteoglycan core protein, and in this experiment we wanted specifically to isolate hybridomas producing antibodies recognizing epitopes present in the proteoglycan HABR and not those present in the polysaccharide attachment region of the proteoglycan monomer. HABR was radiolabeled with [125I]iodine (∼1000 cpm/ng protein) and used in the AB-RIA described above. After 16 hr of autoradiography, 10–20 strong positive supernatants were detected (Fig. 4).

III. Monoclonal Antibodies to Epitopes Present on Proteoglycan Monomers

A. Antibodies to Chondroitin Sulfate Isomers

In several publications (5, 7–9) we have described the production and characterization of both polyclonal and monoclonal antibodies

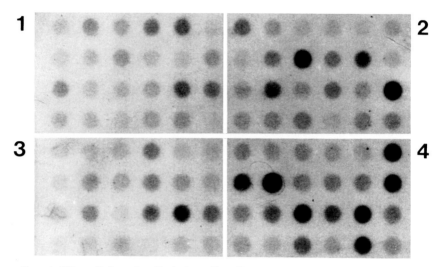

Fɪɢ. 4. Nitrocellulose "antibody-bound" radioimmunoassay. Proteins (monoclonal antibodies) from test supernatants (200 μl) of four 24-well hybridoma tissue culture dishes were bound to nitrocellulose using a Schleicher and Scheull Minifold apparatus. The nitrocellulose sheet was "blocked" with 5% BSA Tris–saline, washed, and incubated for 2 hr with ^{125}I-labeled HABR (2 × 10^5 cpm/ml in 5% BSA Tris–saline) that had been isolated from rat chondrosarcoma proteoglycan. The nitrocellulose sheet was washed, air-dried, and subjected to autoradiography. Wells containing hybridomas secreting antibodies recognizing epitopes on the HABR of proteoglycans are indicated by dark spots on the developed X-ray film.

that specifically recognize epitopes present on proteoglycans containing the different isomers of chondroitin sulfate (and dermatan sulfate). These monoclonal antibodies can distinguish between glycosaminoglycan chains containing unsulfated chondroitin sulfate, chondroitin 4-sulfate, dermatan sulfate, and chondroitin 6-sulfate attached to the polypeptide backbones of proteoglycans (Fig. 5). The specificity of the antibody recognition relies on the prior digestion of antigens with either chondroitinase ABC or chondroitinase AC II. Digestion with chondroitinase involves an elimination reaction that introduces a delta-4,5-unsaturation in the glucuronic acid moiety of the disaccharides released during enzymatic digestion (10). In addition, small oligosaccharide stubs containing at least one characteristic unsaturated disaccharide of the chondroitin sulfate isomer (0, 4, or 6 sulfated), and the linkage region of the glycosaminoglycan chain, remain attached to the protein core of the proteoglycan (9). The structure of the epitope recognized by each of these monoclonal antibodies is given in Fig. 5. The specificity of these antibodies was determined by

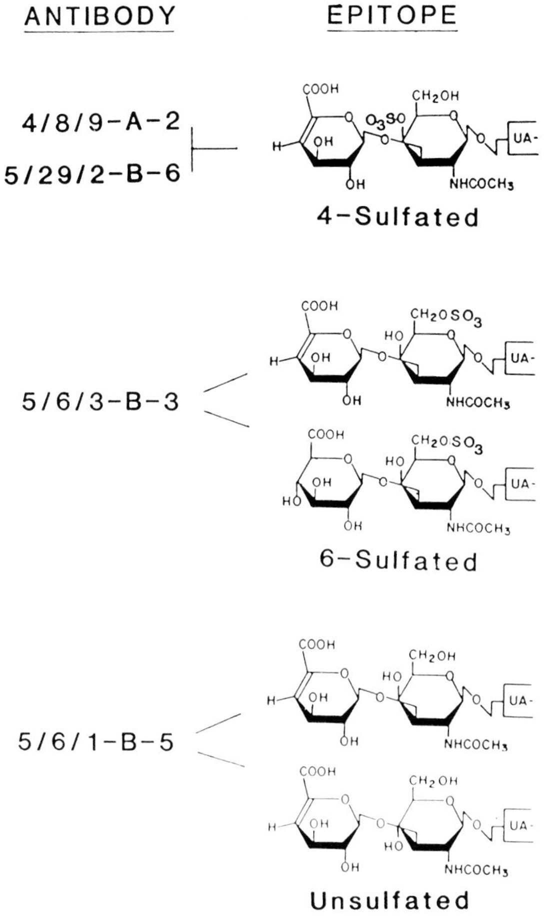

ANTIBODY   EPITOPE

FIG. 5. Structures of the various epitopes that are recognized by monoclonal antibodies directed against chondroitin sulfate isomers. Monoclonal antibody specificities are as follows: 4/8/9-A-2 and 5/29/2-B-6, anti-chondroitin 4-sulfate (dermatan sulfate); 5/6/3-B-3, anti-chondroitin 6-sulfate; 5/6/1-B-5, anti-unsulfated chondroitin sulfate (hyaluronic acid). See text for details of specificities.

competitive radioimmunoassay analyses using characteristic unsaturated disaccharide obtained from chondroitinase digestion of chondroitin 4-sulfate, chondroitin 6-sulfate, and chondroitin as competing antigens (8). All of the monoclonal antibodies recognize an epitope containing a 4,5-delta unsaturation in the glucuronic acid moiety of the oligosaccharide stub attached to the proteoglycan core protein. However, one of the monoclonal antibodies (5/6/3-B-3) reacts with both unsaturated and saturated oligosaccharides that are generated after either chondroitinase or testicular hyaluronidase digestion of the chondroitin 6-sulfated glycosaminoglycan. This finding indicates that the presence of a 4,5-unsaturation in the glucuronic acid moiety is not needed for antibody binding. The results also suggest that glucuronic acid residues are not present as the nonreducing terminal of 6-sulfated glycosaminoglycan chains because untreated proteoglycan preparations (no chondroitinase or hyaluronidase treatment) do not react with this antibody. The specificity of the 5/6/3-B-3 monoclonal antibody is different to that described by Jenkins *et al.* (11) that also recognizes oligosaccharides of chondroitin 6-sulfate after hyaluronidase digestion.

Competitive inhibition radioimmunoassays also indicated that one of the antibodies (5/6/1-B-5) with a specificity for unsulfated chondroitin sulfate could also be inhibited by oligosaccharides derived after *Streptomyces* hyaluronidase digestion of hyaluronic acid (J. E. Christner, unpublished result). This finding indicates that the 5/6/1-B-5 antibody does not distinguish between the occurrence of either *N*-acetylglucosamine or *N*-acetylgalactosamine in the hexosamine residue adjacent to the nonreducing unsaturated glucuronic acid moiety (Fig. 5). Thus digestion with *Streptomyces* hyaluronidase prior to immunoassay (e.g., immunolocation) can potentially be used as a means of specifically detecting hyaluronic acid in tissue preparations.

Another of these antibodies, 4/8/9-A-2, can be used in immunoassay analyses to distinguish between the presence of either dermatan sulfate (4-sulfated) or chondroitin 4-sulfate glycosaminoglycans substituted on proteoglycan core proteins. This specificity is introduced by the use of either chondroitinase ABC or chondroitinase AC II prior to immunoassay analyses (10). Chondroitinase ABC digestion of either dermatan sulfate or chondroitin sulfate produces the same nonreducing unsaturated terminal disaccharide containing an unsaturated glucuronic acid residue and 4-sulfated *N*-acetylgalactosamine on the oligosaccharide stub attached to the proteoglycan core protein. In contrast, chondroitinase AC II will only digest chondroitin sulfate glycosaminoglycans. Thus differential digestion utilizing the two different chondroitinases can be used in conjunction with the 4/8/9-A-2 monoclonal antibody for immunolocation analyses identifying both chondroitin-4-sulfate proteoglycan and dermatan sulfate proteoglycan in different connective tissues (8). Another monoclonal antibody (5/29/2-B-6) has similar specificity to the 4/8/9-A-2 monoclonal antibody.

Several of these monoclonal antibodies have been used for immunohistochemical localization of 4-, 6-, and 0-sulfated chondroitin sulfate and dermatan sulfate proteoglycans in rat ear connective tissues. An example depicting their use in immunolocation analyses is given in Fig. 6. Part of these data have been presented in preliminary form elsewhere (9). An inherent advantage in these studies came from the requirement that the tissue sections needed to be pretreated with chondroitinase in order to generate the antigenic determinant recognized by each specific monoclonal antibody. Thus, non-enzyme-treated sections could be used as controls for nonspecific staining (see Fig. 6b). With the exception of the 5/6/3-B-3 monoclonal antibody, testicular hyaluronidase-treated sections were also used as additional controls to test for nonspecific immunofluorescence in tissue sections.

Figure 6a shows a hematoxlyn–eosin stained section of rat ear. In the center of the section is the rat ear cartilage (C). The large vacuoles in the cartilage apparently contain lipid (12) and give the cartilage a "honeycomb" appearance. Adjacent to the cartilage are muscle fiber bundles (M), nerve fiber bundles (N), and loose connective tissue. Also indicated in the section is the dermis (D), the epidermis (E), and a hair follicle (HF). Figure 6b shows a typical control section which has been subjected to immunohistochemical analyses. Controls were (1) no enzyme digestion, (2) hyaluronidase digestion (exception being the 5/6/3-B-3 studies), (3) no first antibody, and (4) no second antibody. No autofluorescence or nonspecific immunofluorescence was observed in these control sections. Figure 6c shows the results of chondroitinase ABC digestion of a tissue section and subsequent immunohistochemical staining using the 5/6/3-B-3 monoclonal antibody. The results show strong fluorescence in the cartilage and in regions surrounding the basement membrane of muscle fiber bundles. Figure 6d shows a nerve fiber bundle from the same section. Strong immunofluorescence was also observed in the fibrous sheath surrounding nerve fibers. Preliminary conclusions drawn from these analyses indicate that, in addition to cartilage, chondroitin 6-sulfated proteoglycans are located in the fibrous sheath surrounding muscle fibers and nerve fibers. Figure 6e shows the results of immunohistochemical staining using the 5/6/1-B-5 monoclonal antibody after chondroitinase ABC digestion of a tissue section. The results indicate that unsulfated chondroitin sulfate proteoglycans only occur in the cartilage. Immunofluorescence was absent in all other connective tissues of the rat ear. The result supports the notion that the fluorescence seen surrounding the muscle and nerve fiber was due to the presence of chondroitin-6-sulfated proteoglycans and not unsulfated proteoglycans. (RIA analyses of the 5/6/3-B-3 monoclonal antibody indicated that it also weakly recog-

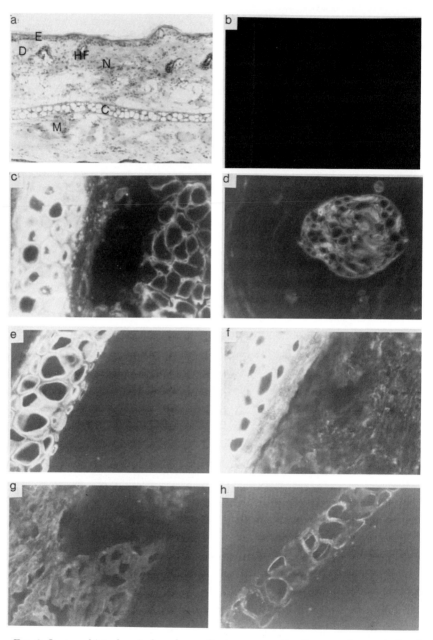

FIG. 6. Immunohistochemical analyses of rat ear connective tissues using monoclonal antibodies directed against different chondroitin sulfate isomers. Sections c/g had been digested with chondroitinase ABC prior to immunolocation with monoclonal antibodies. Section h was pretreated with chondroitinase AC II. (a) Hematoxylin–eosin stain of rat ear tissues (C, cartilage; D, dermis; E, epidermis; HF, hair follicle; M, muscle; N, nerve fibers in loose connective tissue); (b) typical control obtained where no first antibody or

nized unsulfated chondroitin sulfate proteoglycans after chondroitinase digestion—see Ref. 8.) Figure 6f and g show the results obtained using 4/8/9-A-2 monoclonal antibody in immunohistochemical analyses of chondroitinase ABC-treated rat ear connective tissue sections. After chondroitinase ABC digestion, cartilage and loose connective tissue adjacent to the cartilage show positive immunofluorescence (Fig. 6f). Similarly, Fig. 6g shows positive immunofluorescence in the dermis of the ear. However, no fluorescence was found in the epidermis or with a hair follicle shown in this tissue section.

It is of interest to note that all the above sections had been treated with chondroitinase ABC prior to immunohistochemical staining. However, if the sections were treated with chondroitinase AC II (which only digests chondroitin sulfates but not dermatan sulfate) prior to staining with the 4/8/9-A-2 antibody, then immunofluorescence is only observed in the rat ear cartilage (Fig. 6h). This result indicates that the fluorescence observed in the loose connective tissue and dermis of Fig. 6f and g was due to the presence of dermatan sulfate proteoglycan within these tissues.

B. Antibodies to Keratan Sulfate

Two types of keratan sulfate have been identified in the extracellular matrices of connective tissues (13). One of these, keratan sulfate type I, has been exclusively found in the cornea; the other keratan sulfate, type II, is located in a wide variety of skeletal tissues (cartilage, bone, and intervertebral disc). In addition, recent reports have indicated that keratan sulfate-like molecules may also occur in proteoglycans isolated from fish skin (14) and human aorta (B. Caterson, S. Jagannathan, and W. Wagner, unpublished result).

N-Acetyllactosamine is the repeating disaccharide unit of keratan sulfate glycoaminoglycans. The sulfate groups are predominantly found in the C-6 position of the glucosamine unit of the N-acetyllactosamine disaccharide, although they may also occur in some of the galactose residues at the C-6 position. Disulfated disaccharides are more commonly found in skeletal (type II) keratan sulfate (15, 16). N-Acetyllactosamine is also a commonly occurring disaccharide component of oligosaccharides in many glycoproteins, glycolipids, and

---

when no enzymatic digestion had been performed; (c, d) staining observed with monoclonal antibody 5/6/3-B-3 (anti-chondroitin 6-sulfate); (e) staining observed with monoclonal antibody 5/6/1-B-5 (anti-unsulfated chondroitin sulfate); (f/h) staining observed with monoclonal antibody 4/8/9-A-2.

mitogens that are used to stimulate lymphocyte proliferation in mice. Structures that contain *N*-acetyllactosamine appear to be very antigenic to mice, and thus monoclonal antibodies generated against glycoproteins containing this structure are often directed against this carbohydrate moiety (17). Similarly, it has been our experience that whenever we immunize BALB/c mice with proteoglycan antigens containing keratan sulfate, the majority of the resultant hybridomas appear to be synthesizing antibodies with specificities directed against epitopes present in the keratan sulfate glycosaminoglycan chains.

Monoclonal antibodies to keratan sulfate have now been produced by several laboratories (3, 18–20). Preliminary results from this laboratory suggest that there are several antigenic determinants present in the keratan sulfate glycosaminoglycan and that many of the hybridomas resulting from a given fusion are secreting antibodies that recognize subtle differences in the sulfated poly *N*-acetyllactosamine structure. An example of this is illustrated in Fig. 7. Competitive-inhibition radioimmunoassays using two different monoclonal antibodies recognizing epitopes present on keratan sulfate (types I and II) were compared. Competing antigens tested in this radioimmunoassay analysis were corneal keratan sulfate (type I) and skeletal keratan sulfate (type II); the radiolabeled antigen was bovine nasal cartilage core protein containing skeletal keratan sulfate. The binding of one of the antibodies (1/20/5-D-4) was equally inhibited by both corneal and skeletal keratan sulfate preparations, whereas the other (4/8/1-B-4) was inhibited by corneal keratan sulfate to a much greater extent (50% inhibition for type I and type II keratan sulfate were 40 ng and 270 ng, respectively). These results indicate that, on a dry-weight basis, the epitope recognized by the 1/20/5-D-4 monoclonal antibody is represented equally on both corneal and keratan sulfate preparations. However, there is approximately seven times more of the 4/8/1-B-4 epitope present on corneal keratan sulfate compared to skeletal keratan sulfate.

Recent studies performed in the laboratory of Dr. Ten Feizi (17) have compared the relative reactivity of different oligosaccharides of keratan sulfate (derived by keratanase digestion of corneal keratan sulfate) in competition radioimmunoassays using monoclonal antibodies provided by this laboratory (1/20/5-D-4 and 4/8/1-B-4) and other laboratories (20). Their results show that the 1/20/5-D-4 and 4/8/1-B-4 monoclonal antibodies recognize different oligosaccharide structures within the keratan sulfate glycosaminoglycan. The epitopes recognized by both of these antibodies contain a minimum of six sugar units containing disulfated *N*-acetyllactosamine residues (20). The rel-

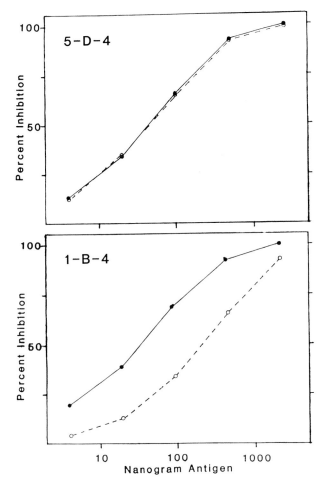

FIG. 7. Competitive radioimmunoassay using corneal keratan sulfate (●) and skeletal keratan sulfate (○) as competing antigens. The differential inhibition with these two antigens and two different monoclonal antibodies 1/20/5-D-4 and 4/8/1-B-4 that recognize epitopes in the keratan sulfate glycosaminoglycan are shown. The radioimmunoassay procedure is described in Ref. 4.

ative occurrence of these different oligosaccharide moieties (epitopes) is different in both skeletal and corneal keratan sulfates and may vary during tissue development, in aging, and in pathological conditions. Clearly, the use of monoclonal antibody technology for studies examining subtle differences in keratan sulfate structure look very promising.

Monoclonal antibodies to keratan sulfate (especially 1/20/5-D-4) have been used in a wide range of studies examining keratan sulfate proteoglycan biosynthesis and metabolism. Vertel *et al.* (21) have used the 1/20/5-D-4 monoclonal antibody in immunohistochemical analyses to demonstrate that the Golgi is the site of keratan sulfate substitution on the proteoglycan core protein of embryonic chick chondrocytes grown in tissue culture. More recently, Thonar *et al.* (22) have used 1/20/5-D-4 to quantify changes in the relative amounts of keratan sulfate fragments present in blood of normal and diseased patients, and Witter *et al.* (23) have used 1/20/5-D-4 to characterize proteoglycan fragments derived from human arthritic synovial fluid. Such assays may be useful for monitoring increases in cartilage metabolism in a wide variety of joint diseases.

C. Monoclonal Antibodies to the Hyaluronic Acid-Binding Region of Cartilage Proteoglycan

The HABR is an important functional domain of the proteoglycan monomer of aggregating proteoglycans (24, 25). Hyaluronic acid-binding proteoglycans were originally described in cartilagenous tissue (24, 25); however, related proteoglycan families have now been found in many other connective tissues: tendon, sclera, aorta, intervertebral disc, smooth muscle (26–28). It is not known whether this family of hyaluronic acid-binding proteoglycans share common HABRs or they result from separate gene products. In this laboratory we have produced and characterized several monoclonal antibodies recognizing epitopes present in the HABR of rat chondrosarcoma proteoglycan. These antibodies have been used in studies examining the primary structural characteristics of hyaluronic acid-binding proteoglycans from cartilage and other tissues.

Our approach to raising monoclonal antibodies to the HABR of proteoglycans has involved the use of chemically modified (reduced and alkylated) antigens for immunization of mice. Chemical modification of antigens essentially "haptenizes" the antigen and thus makes the antigen more immunogenic. Figure 8 depicts the product of reduction and alkylation of rat chondrosarcoma proteoglycans monomer. Cleavage of the disulfide bonds that predominantly occur in the globular HABR (25), and alkylation with iodoacetamide, introduces a foreign amino acid (S-carboxymethylcysteiamide) into the polypeptide backbone of the proteoglycan. This chemical modification and the unfolding of the proteoglycan polypeptide make this region of molecule particularly immunogenic. To date we have identified seven monoclonal

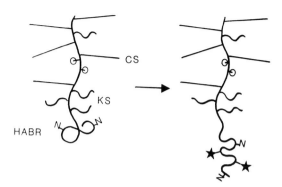

Fɪɢ. 8. Illustration depicting the results of reduction and alkylation of proteoglycan monomer. A foreign amino acid (S-carboxymethylcysteiamide, indicated by the star on the polypeptide) is introduced into the polypeptide backbone of the proteoglycan. These are confined to the hyaluronic acid-binding region (HABR) of the molecule. N- and O- on the polypeptide backbone depict the presence of N- and O-linked oligosaccharides, respectively, on the proteoglycan core protein. CS and KS indicate the presence of chondroitin sulfate and keratan sulfate glycosaminoglycans, respectively.

antibodies that recognize epitopes present in the HABR of rat chondrosarcoma. The methods we use to detect and characterize these monoclonal antibodies were described earlier in this chapter. The characteristics of some of these antibodies are described below.

Perhaps the best studied of these monoclonal antibodies is 12/21/1-C-6. This antibody has been used by Stevens et al. (29) in studies characterizing primary structural features of the core protein of aggregating proteoglycan from rat chondrosarcoma. This antibody recognizes an epitope common to many hyaluronic acid-binding proteoglycans from different animal species and tissues. Immunolocation analyses of three of these monoclonal antibodies (12/21/1-C-6, 12/21/2-A-5, and 12/21/5-C-4) are shown in Fig. 9. Ternary complex obtained after clostripain digestion of rat chondrosarcoma proteoglycan aggregate was subjected to sodium dodecyl sulfate–polyacrylamide gel electrophoresis (SDS–PAGE) under reducing conditions. This procedure separates the HABR (60 kDa) from link protein 3 (43 kDa). Immunolocation with each of the monoclonal antibodies indicates that the antibodies recognize epitopes on the HABR derived from rat chondrosarcoma proteoglycan monomer. One of the antibodies (12/21/5-C-4) reacts with both the HABR and link protein 3 (Fig. 9). This result suggests that there may be a common epitope shared by these two molecules. Recent studies by Neame et al. (30) have shown that there is a 14-amino acid polypeptide sequence in the

BRUCE CATERSON *ET AL.*

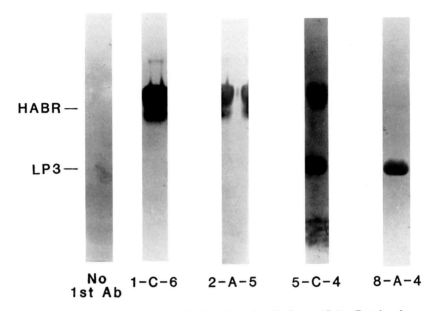

HABR—

LP3—

No   1-C-6   2-A-5   5-C-4   8-A-4
1st Ab

FIG. 9. Immunolocation analyses for detection of antibody specificity. Rat chondrosarcoma ternary complex containing proteoglycan hyaluronic acid-binding region (HABR), hyaluronic acid, and link protein (LP3) were subjected to SDS–PAGE in the presence of β-mercaptoethanol. The separation of HABR (~60 kDa) and LP3 (43 kDa) is indicated. Antibodies used in immunolocation analyses were 1-C-6, 2-A-5, and 5-C-4. These three monoclonal antibodies were produced from a fusion where reduced and alkylated rat chondrosarcoma proteoglycan (Fig. 8) was used as immunogen. Monoclonal antibody 8-A-4 was obtained from a fusion using rat chondrosarcoma LP2 as immunogen (see Ref. 34).

protein core of proteoglycan that shares a high degree of homology with sequences obtained from rat chondrosarcoma link protein. It is possible that the 12/21/5-C-4 monoclonal antibody is recognizing this common sequence.

It is of interest to note that all of the antibodies except 6/1/8-C-1 prefer that the proteoglycan antigen be reduced and alkylated for optimum antibody recognition (see Fig. 3). However, all of these antibodies also recognize "functional" HABR prepared after clostripain digestion of proteoglycan aggregate (Figs. 3 and 8). This finding indicates that preparation of HABR using the enzyme clostripain (which requires the presence of dithiothreitol for activation) probably causes some disulfide bond cleavage and rearrangement in the molecule, exposing epitopes normally masked in the native proteoglycan molecule.

IV. Monoclonal Antibodies to the Link Proteins
of Cartilage Proteoglycan Aggregate

The link proteins of cartilage proteoglycan are small glycoproteins that have important functions of maintaining and stabilizing the proteoglycan monomer–hyaluronic acid–link protein interaction that forms the proteoglycan aggregate structure (25, 31). The link protein(s) partakes in noncovalent interactions with both the proteoglycan monomer (32) and the hyaluronic acid (32, 33), essentially locking (31) all three components together in the proteoglycan aggregate structure. There have been several different link protein subpopulations isolated from hyaline cartilages of different animal species (34). Figure 10 shows the electrophoretic mobility of various link protein(s) isolated from Swarm rat chondrosarcoma, bovine nasal cartilage, human articular cartilage, and chick sternal cartilage. Three distinct link proteins (link protein 1, 48 kDa; link protein 2, 45 kDa; link protein 3, 43 kDa) have been isolated from bovine nasal cartilage (35). Similarly, three link protein subpopulations with comparable molecular weights have been identified in human articular cartilage (36). Only one link protein (link protein 2, 45 kDa) is present in proteoglycan aggregate from rat chondrosarcoma (37), and there appear to be only two major link proteins present in chick sternal cartilage.

The reason for the apparent microheterogeneity of the link proteins from different cartilages is not understood. Recent studies in this laboratory have used monoclonal antibody technology to study the microheterogeneity of link proteins in different cartilages from different sources. Monoclonal antibodies have been raised against link protein 2 (45 kDa) isolated from Swarm rat chondrosarcoma proteoglycan aggregate (34). These antibodies have been used in studies indicating that the major source of the microheterogeneity in the link proteins after SDS–PAGE, comes from the variable occurrence of N-linked oligosaccharide structures on the different link protein subpopulations (34). Removal of the high-mannose oligosaccharides that are present on link protein(s) from bovine nasal cartilage and rat chondrosarcoma alters their electrophoretic mobility to that of the smallest link protein (link protein 3, 43 kDa). One of these monoclonal antibodies (9/30/6-A-1) required the presence of high-mannose oligosaccharide on the link protein polypeptide for antibody binding, thus indicating that carbohydrate structures were part of the epitope recognized by one of these monoclonal antibodies.

Perhaps the best studied of these monoclonal antibodies is 9/30/8-A-4. This monoclonal antibody recognizes an epitope present on link

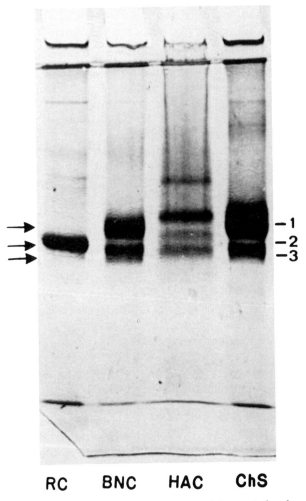

Fig. 10. Heterogeneity of the link proteins obtained from rat chondrosarcoma (RC), bovine nasal cartilage (BNC), human articular cartilage (HAC), and chick sternal cartilage (ChS). SDS–polyacrylamide gels (10%) were run in the absence of mercaptoethanol. The electrophoretic mobility of link proteins 1, 2, and 3 from bovine nasal cartilage are indicated.

proteins 1, 2, or 3 isolated from cartilage of a wide variety of animal species (rat, bovine, ovine, canine, rabbit, chicken, and human hyaline cartilages). The epitope(s) recognized by this antibody have been highly conserved during animal evolution and thus must be important for maintaining some structural or functional feature of the link proteins in proteoglycan aggregation. Recent studies by Neame *et al.* (30) have used the 9/30/8-A-4 monoclonal antibody to isolate two link protein peptides after trypsin or chymotrypsin digestion and perform amino acid sequence analyses. These two distinct peptides have very similar amino acid sequences and both contain an epitope recognized by the 9/30/8-A-4 monoclonal antibody. In addition, the amino acid sequence of these two link protein peptides show a high degree of homology with a small peptide sequence obtained from proteoglycan monomer (30). These data collectively suggest that the 9/30/8-A-4 antibody may be recognizing a region of the molecule (link protein and proteoglycan) involved in some common biological function, such as binding to hyaluronic acid.

The 9/30/8-A-4 monoclonal antibody has also been used in analyses showing the structural similarity and differences in peptide map patterns of link proteins from different animal species (34). Recent studies by Mort *et al.* (38) used 9/30/8-A-4 to identify naturally occurring link protein fragments in link proteins isolated from old human articular cartilage. Spitz-Fife *et al.* (39) have used 9/30/8-A-4 to identify for the first time link protein synthesized by canine synovium.

The 9/30/8-A-4 monoclonal antibody is a IgG_{2b} immunoglobulin with a κ light chain. This antibody binds strongly to *Staphylococcus aureus* protein A, and thus it can be used in solid-phase radioimmunoassay procedures (using heat- and formalin-treated *S. aureus*) to quantitate link protein in proteoglycan preparations. The epitope recognized by the 9/30/8-A-4 antibody is normally buried (or sequestered) in the proteoglycan aggregate. Thus, proteoglycan preparations need to be heat-denatured prior to immunoassay analyses in order to detect or quantitate the link protein(s) present in proteoglycan aggregates.

V. General Discussion

Our objectives for this chapter were 2-fold: first, to review methodological procedures commonly used in this laboratory that are particularly useful for production and characteristic of monoclonal antibodies to proteoglycan angtiens, and second, to review the specificities of several well-characterized monoclonal antibodies to epitopes present on different proteoglycan substructures. Methods have been described that provide a sensitive means of detecting hybridomas secreting antibodies directed against epitopes present or proteoglycans. Conventional immunoassay procedures using proteoglycan antigens often do not allow optimal detection of antibody-secreting hybridomas because proteoglycan antigens do not adsorb well to plastic microtiter plates or often do not bind well to *S. aureus* protein A. The use of monoclonal antibodies bound to nitrocellulose and radiolabeled antigens (AB-RIA) provides a rapid, sensitive means of detecting the presence of antibody-secreting hybridomas when only small amounts of antigen are available for screening and characterizing potential hybridoma specificities.

Well-characterized antibodies are now available that recognize epitopes present in carbohydrate and protein structures characteristic of different subtypes (or families) of proteoglycans. The inherent specificity of these antibodies can be used to distinguish subtle differences in the carbohydrate structures, for example, the isomers of chondroitin sulfate and differences in epitopes present in skeletal and corneal keratan sulfate. Monoclonal antibodies to protein determinants have also been used in primary structural analyses of the HABR (29) and link proteins (30, 34) of aggregating cartilage proteoglycans. It was of interest to note that many of our monoclonal antibodies cross-react with proteoglycans isolated from a variety of different animal tissues and species, suggesting the occurrence of highly conserved regions in proteoglycan structures throughout evolution.

Perhaps one of the most recent and exciting uses of monoclonal antibody technology has been in the area of recombinant DNA technology. Recent studies by Doege *et al.* (40) at the National Institute of Health have used well-characterized mixed monoclonal antibodies with specificities directed against rat chondrosarcoma link protein 2 to identify *Escherichia coli* recombinants, containing the λ gt phage expression system, secreting fusion proteins containing fragments of link protein. DNA sequence analyses has been performed on a DNA fragment accounting for approximately 50% of the link protein structure. Similar studies using mixed monoclonal antibodies directed against proteoglycan core protein determinants are under way. The use of monoclonal antibody methodology in conjunction with recombinant DNA technology offers great potential for the future in the determination of the structure–function relationships between the different families of connective tissue proteoglycans.

Acknowledgments

This work was supported by NIH Grants AM 32666, AM 32474, and AM 27127. The authors greatly appreciate the technical assistance of Ms. Dottie Heritage and Mr. James Spencer, and also Ms. Sally Bierer for typing this chapter. In addition, we would like to acknowledge the assistance of Dr. J. R. Couchman for the immunohistochemical studies and Ms. Irene Parrish for artwork in Fig. 1.

REFERENCES

1. Bourdon, M. A., Oldberg, A., Pierschbacher, M., and Ruoslahti, E. (1985). *Proc. Natl. Acad. Sci. U.S.A.* **82**, 1321–1325.
2. Kearney, J. F. (1984). In "Fundamental Immunology" (W. E. Paul, ed.), pp. 751–766. Raven, New York.
3. Caterson, B., Christner, J. E., and Baker, J. R. (1983). *J. Biol. Chem.* **258**, 8848–8854.
4. Baker, J. R., Caterson, B., and Christner, J. E. (1982). In "Methods in Enzymology" (V. Ginsberg, ed.), Vol. 83, pp. 216–235. Academic Press, New York.
5. Christner, J. E., Caterson, B., and Baker, J. R. (1980). *J. Biol. Chem.* **255**, 7102–7015.
6. Caputo, C. B., MacCallum, D. K., Kimura, J. H., Schrode, J., and Hascall, V. C. (1980). *Arch. Biochem. Biophys.* **204**, 220–233.
7. Caterson, B., Baker, J. R., Christner, J. E., and Couchman, J. R. (1982). *J. Invest. Dermatol.* **79**, 45s–50s.
8. Couchman, J. R., Caterson, B., Christner, J. E., and Baker, J. R. (1984). *Nature (London)* **307**, 650–652.
9. Caterson, B., Christner, J. E., Baker, J. R., and Couchman, J. R. (1985). *Fed. Proc. Fed. Am. Soc. Exp. Biol.* **44**, 386–393.
10. Yamagata, T., Saito, H., Habuchi, O., and Suzuki, S. (1968). *J. Biol. Chem.* **243**, 1523–1535.
11. Jenkins, R. B., Hall, T., and Dorfman, A. (1981). *J. Biol. Chem.* **256**, 8279–8282.
12. Kostovic-Knezevic, L., Bradamante, Z., and Svaiger, A. (1981). *Cell Tissue Res.* **218**, 149–160.
13. Meyer, K. (1970). In "Chemistry and Molecular Biology of the Intercellular Matrix" (E. A. Balazs, ed.), p. 5. Academic Press, New York.
14. Ito, M., Kitamikado, M., and Yamagata, T. (1984). *Biochim. Biophys. Acta* **797**, 221–230.
15. Seno, N., Meyer, K., Anderson, B., and Hoffman, P. (1965). *J. Biol. Chem.* **240**, 1005–1010.
16. Nakazawa, K., and Suzuki, S. (1975). *J. Biol. Chem.* **250**, 912–917.
17. Mehmet, H., Scudder, P., Tang, P. W., Hounsell, E. F., Caterson, B., and Feizi, T. (1986). *Eur. J. Biochem.* **157**, 385–391.
18. Funderburgh, J. L., Stenzel-Johnson, P. R., and Chandler, J. W. (1983). *Curr. Eye Res.* **2**, 769–776.
19. SunderRaj, N., Willson, T., Gregory, J. D., and Danle, S. P. (1985). *Curr. Eye Res.* **4**, 49–55.
20. Zanetti, M., Ratcliffe, A., and Watt, F. M. (1985). *J. Cell Biol.* **101**, 53–59.
21. Vertel, B. M., and Barkman, L. L. (1984). *Collagen Rel. Res.* **4**, 1–20.
22. Thonar, E. J-M. A., Lenz, M. E., Klintworth, G. K., Caterson, B., and Kuettner, K. E. (1985). *Arthritis and Rheumatism* **28**, 1367–1376.
23. Witter, J. P., Roughley, P. J., Caterson, B., and Poole, A. R. (1984). Trans. Orthop. Res. Soc. **9**, 313.
24. Hardingham, T. E., and Muir, H. (1973). *Biochem. J.* **135**, 905–908.
25. Heinegard, D., and Hascall, V. C. (1974). *J. Biol. Chem.* **254**, 927–934.
26. Wight, T. N., and Hascall, V. C. (1983). *J. Cell Biol.* **96**, 167–176.
27. Stevens, R. L., Ewins, R. J. E., Revell, P. A., and Muir, H. (1979). *Biochem. J.* **179**, 561–572.

28. Heinegard, D., Bjorne-Persson, A., Coster, L., Franzen, A., Gardell, S., Malmstrom, A., Paulsson, M., Sandfolk, R., and Vogel, K. (1985). *Biochem. J.* **230**, 181–194.
29. Stevens, J. W., Oike, Y., Handley, C. J., Hascall, V. C., Hampton, A., and Caterson, B. (1984). *J. Cell Biochem.* **26**, 247–259.
30. Neame, P. J., Perin, J-P., Bonnet, F., Christner, J. E., Jolles, P., and Baker, J. R. (1985). *J. Biol. Chem.* **260**, 12402–12404.
31. Hardingham, T. (1979). *Biochem. J.* **177**, 237–247.
32. Caterson, B., and Baker, J. R. (1978). *Biochem. Biophys. Res. Commun.* **80**, 496–503.
33. Oegema, T. R., Brown, M., and Dziewiatkowski, D. D. (1977). *J. Biol. Chem.* **252**, 6470–6477.
34. Caterson, B., Baker, J. R., Christner, J. E., Lee, Y., and Lentz, M. J. (1985). *Biol. Chem.* **260**, 11384–11356.
35. Baker, J. R., and Caterson, B. (1979). *J. Biol. Chem.* **254**, 2387–2393.
36. Roughley, P. J., Poole, A. R., and Mort, J. S. (1982). *J. Biol. Chem.* **257**, 11908–11914.
37. Oegema, T. R., Hascall, V. C., and Dziewiatkowski, D. D. (1975). *J. Biol. Chem.* **250**, 6151–6159.
38. Mort, J. S., Caterson, B., Poole, A. R., and Roughley, P. J. (1985). *Biochem. J.* **232**, 805–812.
39. Spitz-Fife, R., Caterson, B., and Myers, S. L. (1985). *J. Cell Biol.* **100**, 1050–1055.
40. Doege, K. J., Hassell, J. R., Caterson, B., and Yamada, Y. (1986). *Proc. Natl. Acad. Sci. USA* **83**, 3761–3765.

Molecular Biology of Proteoglycans and Link Proteins

Linda J. Sandell

Departments of Biochemistry and Orthopedic Surgery, Rush-Presbyterian–St. Luke's Medical Center, Chicago, Illinois 60612

I. Perspectives

The purpose of this chapter is to discuss the production of proteoglycans from the perspective of a molecular biologist and to review the initial studies on the nucleic acids that encode proteoglycan core proteins and link proteins. The contribution of molecular biology and recombinant DNA technology has been demonstrated by recent advances in our understanding of the biosynthesis and regulation of many complex proteins such as the immunoglobulins, globins, insulin, and collagens. Similar approaches will no doubt yield answers to many of the fundamental questions about proteoglycans, such as: Which chemically related proteoglycans represent different genetic types and which are posttranslational modifications of the same gene product; how has the diversity of proteoglycans evolved; and, does inappropriate gene expression result in development of specific diseases?

Proteoglycans are a heterogeneous group of glycosylated proteins whose members range from those having small core proteins substituted with one glycosaminoglycan chain to those with extremely large core proteins that contain >100 glycosaminoglycan chains. The large size and extent of glycosylation of many of the core proteins combined with the polydispersity of the extracellular matrix population and the small quantities synthesized by some tissues have made it difficult to determine whether genetically distinct core proteins actually exist. Even so, information gained from protein chemistry and immunology has already presented insights into the genetic relationships among core proteins and between core and link proteins. For example, common epitopes present among the large aggregating pro-

27

teoglycan core proteins from tissues such as cartilage, sclera, tendon, and aorta indicate that these proteins may be encoded by members of a gene family and that some of these proteins may in fact be encoded by the same gene.

Amino acid sequence data support the existence of separate mRNAs for two small proteoglycan core proteins. The amino acid sequence has been determined for the first 25 amino acids of a bovine skin dermatan sulfate proteoglycan (Pearson *et al.*, 1983) and the first 20 amino acids of human fetal membrane chondroitin sulfate–dermatan sulfate proteoglycan (Brennen *et al.*, 1984). An interesting relationship between the two small proteoglycans was uncovered: the first 10 amino acids are identical, after which they diverge for a few residues and then become similar again. These two proteins could be encoded by separate genes which evolved from a common ancestor; pressure to conserve the first region of amino acids (containing a site for glycosaminoglycan chain attachment) would have been high while the adjacent region was allowed to diverge with speciation or for a specific tissue function. Recent experiments discussed in Section II and III indicate that this core protein likely varies more between species rather than tissues.

In many tissues link proteins are associated with proteoglycan monomers and hyaluronic acid, forming a large proteoglycan aggregate. Link proteins derived from cartilage of many different species and from tissues other than cartilage have structural features in common as they cross-react with the monoclonal antibody, 8-A-4 (reviewed by Caterson *et al.*, this volume). The amino acid sequence of this epitope has been determined in the rat chondrosarcoma link protein. Two regions of this link protein contain the epitope which, judging from the high degree of homology within these regions, appear to have been generated by a tandem duplication of DNA within the gene (Neame *et al.*, 1986). Surprisingly, the amino acid sequence that contains this epitope is also found in the hyaluronic acid-binding region of the proteoglycan core protein (Neame *et al.*, 1985). This site may be important for a function common to the core protein and to link proteins, for example, binding to hyaluronic acid. Because of this common sequence, one can speculate that the link and core proteins have a common ancestral gene sequence.

In order to determine how these and other relationships arose and how the diversity of core and link proteins is generated, one must examine the nucleic acids encoding these proteins. This review will focus on recent data generated from cell-free (*in vitro*) translation of messenger RNA (mRNA) for core and link proteins and the cloning of complementary DNAs (cDNAs) for some of these proteins. Information will be presented on the core proteins for the large aggregating proteoglycan from cartilage, the L2 yolk sac tumor proteoglycan, the small proteoglycans from bone, cartilage, skin, and smooth muscle, and the polypeptide portion of the link proteins.

For more detailed information on the immunology, protein chemistry, and functions of proteoglycans, the reader is referred to the other chapters in this volume and to recent review articles by Heinegard and Paulsson (1984), Hassel *et al.* (1986), and Poole (1986).

II. Cell-Free Translation of mRNAs Encoding Proteoglycan Core Proteins and Link Proteins

Cell-free translation of mRNA serves two important functions: (1) to assay for the presence of a specific mRNA, and (2) to determine the size of the unmodified protein. The procedure includes purification of the mRNA population, translation in a cell-free protein-synthesizing system, isolation and identification of the protein precursor, and analysis by gel electrophoresis (SDS–PAGE). Systems generally used are the rabbit reticulocyte lysate and wheat germ lysate, from which the endoplasmic reticulum and enzymes necessary for cotranslational or posttranslational modifications of the protein have been removed. The lack of posttranslational processing allowed Milstein and colleagues (1972) to determine that secretory proteins were initially synthesized with a short sequence of amino acids, a "signal peptide," which provides the information necessary for transport of the nascent polypeptide into the lumen of the endoplasmic reticulum. *In vivo*, this peptide is removed during translation of the protein; however, in the products of cell-free translation, all proteins bound for the extracellular matrix will contain a signal sequence of 15–30 amino acids. A summary of translated core proteins is presented in Table I. The core proteins are classified as large and small based on the size of the proteoglycan as defined by Heinegard and Paulsson (1984).

A. Large Core Proteins

Translation of chondrocyte mRNA provided the first determination of the size of the unmodified core protein of the large aggregating proteoglycan. The products of translation of chicken sternal cartilage mRNA are shown in Fig. 1, lane A. Both polyclonal and monoclonal antibodies precipitated the chick sternal core protein as a single band, which migrated on SDS–PAGE at 340 kDa (Upholt *et al.*, 1979). This molecular size was in agreement with subsequent size estimates for the partially glycosylated intracellular core protein of 370 kDa for the major rat chondrosarcoma core protein (Kimura *et al.*, 1981) and 376

TABLE I

Primary Gene Product (from Translation of mRNA)

Protein	Size (Da)	Reference
Large core protein		
Chick sterna	340,000	Upholt *et al.* (1979)
Rat chondrosarcoma	330,000	Vertel *et al.* (1984)
Small core protein		
Monkey smooth muscle	40,500	Sandell *et al.* (1986)
Bovine smooth muscle	42,000	Sandell *et al.* (1986)
Human cartilage PGII	40,500	Sandell *et al.* (1986)
Bovine cartilage PGII	42,000	Sawhney and Sandell (1986)
Bovine bone	41,000 and 38,000	Day *et al.* (1986)
Link protein		
Rat chondrosarcoma	42,000	Hering and Sandell (1986a)
Bovine cartilage	41,000 and 27,000	Treadwell *et al.* (1980)
Bovine chondrocyte	42,000	Hering and Sandell (1986b)

kDa for a chick epiphyseal core protein (Geetha-Habib *et al.*, 1984). The intracellular core protein is considerably larger than the average core protein isolated from cartilage matrix, which has a continuous size distribution centered around 200–250 kDa (Hascall and Sajdera, 1970). Consequently, the size heterogeneity of the cartilage matrix proteoglycan is not due to the synthesis of core proteins of different sizes, but is likely due to proteolytic degradation in the extracellular matrix. mRNA from Swarm rat chondrosarcoma chondrocytes also has been translated (Vertel *et al.*, 1984). Like the sternal core protein, the core protein synthesized by the rat chondrosarcoma mRNA is a single species but has a molecular size approximately 8 to 10 kDa smaller than the sternal core protein (Fig. 2). Size fractionation of the sternal mRNA on sucrose gradients indicated that the mRNA encoding the core protein is >5 kilobases (kb) (Upholt *et al.*, 1981). These studies strongly support the existence of a single polypeptide as the core protein for the cartilage aggregating chondroitin sulfate proteoglycan.

The levels of translatable mRNA for the core protein have been determined during *in vitro* differentiation of mesenchyme to cartilage. mRNA was isolated from stage 24 chicken limb mesenchymal cells after culture at high density for 1–6 days (Vuorio *et al.*, 1982). Cell-free translation products corresponding to the core protein were not detected when RNAs from day 1 or day 2 cultures were translated (Fig. 3). The core protein was detected in the translation products of RNA

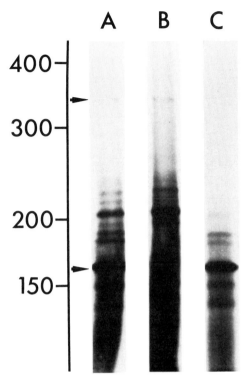

FIG. 1. Wheat germ cell-free translation of sternal RNA fractionated by Sepharose 4B column chromatography. Two milligrams of sternal RNA was applied to a 2.2 × 34 cm column in 650 mM NaCl–20 mM Tris-HCl, pH 7.5–2.5 mM EDTA–0.1% Sarkosyl according to the procedure of Frischauf *et al.* (1978). Bound RNA was eluted with 100 mM NaCl–20 mM Tris-HCl, pH 7.5–3.5 mM EDTA–0.1% Sarkosyl. RNA was translated in the presence of [^{35}S]methionine for 90 min and the products of translation were electrophoresed on a 5% SDS–polyacrylamide gel. Radioactive bands were observed by fluorography. Most of the proteins have been electrophoresed off the gel. Products of cell-free translation directed by: lane A, total sternal RNA; B, unbound RNA (enriched in core protein mRNA); C, bound RNA (enriched in type II collagen mRNA). The core protein (upper arrow) and type II collagen (lower arrow) translation products are indicated. Reproduced from Vuorio *et al.* (1982).

isolated from cultures on day 3 and continued to increase through day 6 (Fig. 3). The time course of appearance of core protein mRNAs closely paralleled the differentiation of the high-density mesenchymal cell cultures, as determined by a variety of techniques measuring the synthesis or accumulation of extracellular matrix products characteristic of cartilage (Levitt and Dorfman, 1974). It can be inferred that the increase in the production of proteoglycan in differentiating limb bud

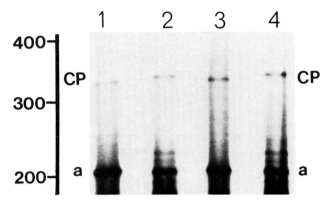

Fig. 2. RNA was translated in a wheat germ cell-free system in the presence of [³⁵S]methionine; products were electrophoresed on 5% SDS–PAGE and fluorographed. The core proteins (CP) are indicated. Only the upper portion of the fluorogram is shown. Lane 1, translation directed by 180 µg/ml rat chondrosarcoma mRNA; lane 3, 340 µg/ml rat chondrosarcoma mRNA; lanes 2 and 4, 160 µg/ml chicken sternal mRNA. Reproduced from Vertel *et al.* (1984).

cultures is due to an increase in the concentration of mRNA within the cells. In order to measure directly the amounts of specific mRNAs present in differentiating chondrocytes, it will be necessary to use DNA probes which can be hybridized to the core protein mRNA.

B. Small Core Proteins

Populations of small dermatan sulfate or chondroitin sulfate proteoglycans have been isolated from bone (Fisher *et al.*, 1983), articular cartilage (Rosenberg *et al.*, 1985), nasal cartilage (Heinegard *et al.*, 1981), cartilage chondrocytes (Sawhney and Sandell, 1986), skin fibroblasts (Pearson *et al.*, 1983), fetal membranes (Brennan *et al.*, 1984), tendon (Vogel and Heinegard, 1985), and smooth muscle cells (Chang *et al.*, 1983). Many of these proteoglycans are immunologically related, contain a core protein of approximately 45 kDa, 1–3 chondroitin sulfate or dermatan sulfate side chains, and 1–3 Asn-linked carbohydrate chains. Some of these proteoglycans can interact with type I collagen and fibronectin and may play an important role in extracellular matrix organization. It is not known whether the core proteins of these proteoglycans are the same or whether they are members of a family of closely related molecules. Rosenberg *et al.* (1985) have recently shown that in articular cartilage, at least two small proteoglycan spe-

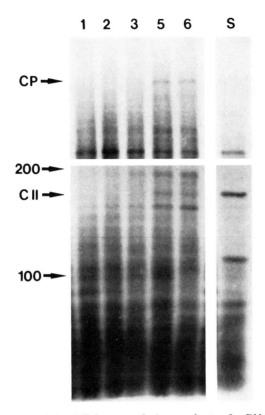

Fig. 3. Fluorogram of the cell-free translation products of mRNA prepared from differentiating limb bud cultures. Stage 24 chick limb mesenchymal cells were isolated and cultured at high density (Upholt *et al.*, 1979). RNA was prepared from a single series of differentiating plates after 1, 2, 3, 5, or 6 days in culture and translated in a wheat germ lysate in the presence of [^{35}S]methionine. Translation products of total sternal RNA (S) is included for comparison. A longer exposure of the portion of the gel containing polypeptides >200,000 is shown so that both the proteoglycan core protein (CP) and type II collagen (CII) translation products are visible. Arrows at 100 and 200 indicate the migration positions for polypeptides of 1×10^5 and 2×10^5 Da, respectively, as determined from molecular size standards. Reproduced from Vuorio *et al.* (1982).

cies can be separated which have core proteins that do not share immunologic identity.

The cell-free translation products of cartilage chondrocyte mRNA have been investigated with antisera specific for one of the cartilage small proteoglycan populations, called DS-PGII. Using this antisera to identify the primary translation product, it was found that the un-

1 2

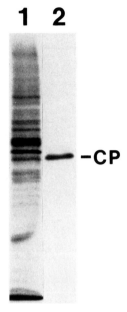

FIG. 4. RNA isolated from 5-day cultures of bovine metacarpophalangeal articular chondrocytes was translated in a reticulocyte lysate in the presence of [^{35}S]methionine. The products were immunoprecipiated with an antiserum specific for DS-PGII kindly provided by Dr. Robin Poole. Products were electrophoresed on a 10% SDS–polyacrylamide gel and a fluorograph was prepared. Lane 1, total translation products; lane 2, total translation products immunoprecipitated with antiserum.

modified protein is synthesized as a 42 kDa protein (Fig. 4). This is consistent with the estimated size of the unglycosylated human fibroblast core protein (Glossl *et al.*, 1984) and the unglycosylated bovine chondrocyte core protein (Sawhney and Sandell, 1987). The mRNA encoding this core protein is 1.5–2.5 kb. These results indicate that this core protein is a distinct gene product from the large core protein described in the previous section. In further studies the primary translation products of human and bovine chondrocytes, monkey and bovine smooth muscle cells, and human fibroblasts were compared (Sandell *et al.*, 1987). Core proteins were immunoprecipitated with antisera to both the predominant human fibroblast core protein, called proteodermata sulfate (provided by Dr. H. Kresse) and the bovine cartilage DS-PGII core protein (provided by Dr. R. Poole). The results show that these core proteins are immunologically related and that the bovine dermatan sulfate core proteins are the same size and the monkey and human core proteins are the same size, regardless of the tissue origin.

However, a size difference of 1000–1500 Da was observed between the primate and bovine core proteins: the bovine core protein being larger. These findings may help explain the amino acid sequence similarities and differences discussed in Section I between the small proteoglycan core proteins from human placenta and bovine skin. It is likely that the proteins have diverged between species and are encoded by a homologous gene.

In contrast to the tissues described above, cell-free translation of bone cell mRNA yielded two proteins migrating at approximately 38 and 40 kDa on SDS–PAGE, immunoprecipitated with a monospecific polyclonal antiserum to the small bone proteoglycan (A. Day, M. Young, and J. Termine, personal communication). The immunoprecipitation of more than one product of translation is quite unusual and the significance of these results is unclear at this time. Although the core proteins of these small proteoglycans have many characteristics in common, thev may not be identical. mRNAs encoding the bone and fibroblast core proteins have been cloned and will be discussed in Section III.

C. Link Proteins

Identification of link proteins in the products of cell-free translation has shown that the link protein is a distinct gene product and has opened the way for further studies on the biosynthesis of this group of proteins. The link protein precursor has been identified in translation products of rat chondrosarcoma mRNA (Hering and Sandell, 1986a) and bovine articular chondrocytes (Hering and Sandell, 1986b; 1987). From the products of translation of rat chondrosarcoma mRNA, one band was immunoprecipitated which migrated with an apparent molecular size of 42 kDa (Fig. 5). The size of the nascent protein is approximately 2.5 kDa less than the cellular link protein known to contain one asparagine-linked carbohydrate chain. Consistent with this data, Lohmander et al. (1983) have shown that when addition of this carbohydrate chain is inhibited, the size of link protein synthesized by rat chondrosarcoma chondrocytes is slightly smaller than the secreted link protein. As the size of the translated product is very similar in size to the unglycosylated mature link protein, it seems unlikely that link protein is biosynthesized as a much larger precursor.

To determine the size of the mRNA encoding the rat chondrosarcoma link protein, RNA was fractionated on sucrose density gradients and fractions were translated and immunoprecipitated (Hering and Sandell, 1986a). The result was unexpected in that most of the mRNA

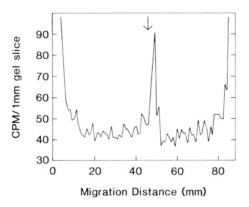

Migration Distance (mm)

FIG. 5. Cell-free translation products directed by rat chondrosarcoma RNA (1 mg/ml) were labeled by incorporation of [^{35}S]methionine, reduced and alkylated, and immunoprecipitated with a monoclonal antibody, 8-A-4 (kindly provided by Dr. B. Caterson). The immunoprecipitate was electrophoresed on a 10% acrylamide gel, cut into 1-mm sections, and radioactivity was determined by liquid scintillation counting. The migration of the link protein standard immunoprecipitated from [^{3}H]leucine-labeled rat chondrosarcoma culture medium is designated. The cell culture medium was kindly provided by Drs. J. Kimura and T. Shinomura.

was present in a larger size class (approximately 4–6 kb) than was predicted from the size of the translated protein (which would require a mRNA of approximately 1.0 to 2.0 kb). These data are consistent with a recent report of a cDNA clone encoding rat chondrosarcoma link protein which hybridized to multiple mRNAs on Northern blot analysis ranging in size from 1.5 to 5.5 kb (Doege *et al.*, 1985, 1986a) and indicates that only the larger mRNA(s) may be capable of translation into protein. These results will be discussed further in Section III.

In contrast to the rat chondrosarcoma chondrocytes, which produce only one link protein (Kimura *et al.*, 1980), two predominant link proteins have been isolated from cartilage of bovine (Tang *et al.*, 1979; Baker and Caterson, 1979), human (Roughley *et al.*, 1982), canine (Fife *et al.*, 1985), and chicken (McKeown-Longo *et al.*, 1982) origin. These link proteins have molecular sizes of approximately 48 and 45 kDa. A minor amount of a third smaller link protein has been observed in bovine, human, and avian cartilage. Are the multiple link proteins derived from the same precursor core protein? To begin to answer this question, RNA was isolated from cultured bovine articular cartilage chondrocytes. These cells synthesize and secrete link proteins of 49.5,

44.0, and 41.5 kDa (Fig. 6, lane 4). Cell-free translation of mRNA from these cultures with subsequent immunoprecipitation yielded a single band migrating at 42 kDa on SDS–PAGE (Fig. 6, lane 2). The size was approximately the same as that of the nascent rat chondrosarcoma link protein. To further investigate the origin of the multiple link proteins, cultured chondrocytes were incubated with tunicamycin (0.3–30 μg/ml) to inhibit the addition of asparagine-linked carbohydrate chains and labeled with [^{3}H]leucine. Link proteins were immunoprecipitated from cell lysates and medium and analyzed by SDS–PAGE. Tunicamycin treatment resulted in a dose-dependent increase in the intensity of a 41.5-kDa band and the elimination of the 49.5- and 44.0-kDa bands (Fig. 6, lane 3). Consequently, the 41.5-kDa band is likely the unglycosylated precursor for the two larger proteins. Taken together, these results agree with previous studies which conclude that the major difference between mature link proteins is the degree of glycosylation (Roughley et al., 1982, Le Gledic et al., 1983, Caterson et al., 1985; Caterson et al., this volume).

In a slightly different approach, Treadwell et al. have identified products of cell-free translation of cartilage mRNA which were immunoreactive with link protein antiserum. RNA was isolated from cartilage and translated into proteins. The translation products were then aggregated by the addition of hyaluronic acid and the aggregates purified by Sepharose 6B chromatography. Four proteins ranging in size from >300 kDa to 28 kDa were recovered, and the species migrating at 41 kDa and 28 kDa reacted with the antiserum. It is likely that the 41-kDa species represents the primary protein product of the link gene.

In experiments to investigate controls on the maintenance of chondrocyte phenotype, Adams and colleagues (1982) used cell-free translation of mRNA from chondroblasts infected with a temperature-sensitive mutant of Rous sarcoma virus (RSV). By translation of mRNA isolated from infected cells cultured at the permissive (36°C) and nonpermissive (41°C) temperatures for transformation, they determined that when the chondroblast phenotype is lost, the levels of mRNA available for translation of core protein, link protein, and type II collagen were reduced. This indicates that the oncogene product of RSV may affect matrix synthesis by decreasing the mRNA available for translation and that expression of proteoglycan core protein, link protein, and type II collagen are, in this case, coordinately regulated. This decrease in mRNA may be accomplished by reducing the rate of transcription, as has been reported for type I collagen in fibroblasts transformed with RSV (Sandmeyer et al., 1981).

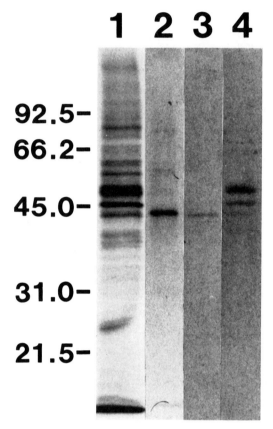

FIG. 6. Link proteins from cell-free translation of bovine chondrocyte mRNA and 5-day cell cultures treated with tunicamycin. Lane 1, total products of cell-free translation labeled with [35S]methionine; lane 2, link protein immunoprecipitated with a polyclonal antiserum, R13, provided by Dr. Bruce Caterson; lane 3, link protein immunoprecipitated from medium of cultures treated for 6 hr with 3 μg/ml tunicamycin, then labeled with [3H]leucine for 17 hr; lane 4, link proteins immunoprecipitated from medium of control cultures labeled with [3H]leucine for 17 hr. The migration of molecular size standards are shown on the left (× 10³). The molecular sizes of the link proteins are discussed in the text.

III. CLONING OF PROTEOGLYCAN CORE PROTEIN AND LINK PROTEIN cDNAs

Recombinant DNA technology can be used to determine the complete amino acid sequence of core and link protein precursors and provide probes for the detection of mRNA and isolation of genes. In order to assess rates of transcription and to investigate possible modifications of the gene involved in regulation, it is necessary to have cloned DNA sequences corresponding to the actual gene, including intervening sequences and the promoter region. There are technical and practical advantages to preparing complementary DNA (cDNA) rather than trying to isolate genes directly. First, mRNAs from tissues which synthesize the target protein are often far more abundant (many thousands of copies) than is the gene (one copy per haploid genome). Second, because the mRNA lacks introns, sequencing the cDNA will provide data about protein structure far more efficiently than having to locate exons within genomic fragments. The major disadvantage is that the tissues which express the genes of interest must be used to provide the substrate mRNA, and these may be difficult to handle or contain only small amounts of message. While cDNA may be used without cloning, it is desirable to insert, or recombine, the cDNA into a plasmid or bacteriophage vector, transform the vector into a bacterial host, and produce clones which will yield large amounts of purified DNA.

The choice of cloning strategies takes into account such information as the size, amino acid sequence, and antigenic properties of the protein of interest. The synthesis of DNA from the mRNA template involves the use of reverse transcriptase isolated from RNA tumor viruses. The enzyme reverses the common direction of flow of information from DNA to protein, using the mRNA as a template for DNA synthesis. In addition to the mRNA template, reverse transcriptase requires a free 3'-hydroxyl group from a "primer" deoxynucleotide which is used for extension of the DNA strand by the enzyme. Most commonly, the primer used is a short stretch (12–20 nucleotides) of oligo(dT), which is annealed to the poly(A) tail of the mRNA. However, any oligomer that will hybridize to mRNA can be used a primer. Thus, if nucleotide sequence coding for the specific protein is known or can be deduced from the amino acid sequence, a primer can be chemically synthesized. The use of such an oligonucleotide primer allows cDNA transcription to begin at regions of the mRNA other than the 3' end, as when oligo(dT) is used as primer. After the synthesis of the DNA:RNA hybrid molecule, the RNA is removed by treatment with alkali and the single-stranded cDNA is used as a template for double-stranded cDNA synthesis with DNA polymerase or reverse transcriptase. The primer in this case is provided by the hairpin loop formed by the single-stranded cDNA. The double-stranded cDNA is combined with a plasmid vector, usually pBR322, by the annealing of complementary nucleotide homopolymer tracts ("sticky" ends) which have been added to the double-stranded cDNA and the plasmid. The recombined plasmid is inserted into a bacterial host, *Escherichia coli*, by transformation. Due to properties of this host–vector system, each bacterial colony contains only one species of plasmid. The population of transformed bacteria containing the new DNA is considered a "library" of cloned cDNAs. Theoretically, the composition of the library will reflect the mRNA population initially transcribed. A specific clone in the library can be identified by hybridization to radiolabeled nucleic acid probes such as purified mRNA, a related DNA, or complementary oligonucleotides. Methods for positive identification of a cDNA clone include matching DNA sequence with known amino acid sequence, and hybridization of the cloned DNA to mRNA, isolation of the mRNA, *in vitro* translation, and identification of the translated protein. The later technique is often called hybrid-selected translation.

In an alternative host–vector system recently developed by Young and Davis (1983), double-stranded cDNA is inserted within the β-galactosidase structural gene, *lacZ*, of the bacteriophage expression vector, λgt11. When expression of the *lacZ* gene is induced, the eukaryotic DNA will be transcribed and translated, producing a hybrid molecule which will contain the amino acid sequence of both proteins. The hybrid protein produced by a bacterial colony can be identified by reaction with antibodies to the protein of interest.

A summary of cloned cDNAs encoding proteoglycan core and link proteins is presented in Table II.

TABLE II
cDNA Clones for Proteoglycan Core and Link Proteins

Protein	Region encoded	Reference
Large core protein		
Chick cartilage core	C terminus	Sai *et al.* (1986)
Rat chondrosarcoma	C terminus	Doege *et al.* (1986)
Small core protein		
Rat L2 yolk sac tumor	Complete	Bourdon *et al.* (1985)
Human fibroblast	Complete	Rouslahti *et al.* (1986)
Bovine bone	C terminus	Day *et al.* (1986)
Link protein		
Chick	Complete	Deak *et al.* (1986)
Rat chondrosarcoma	C terminus	Doege *et al.* (1986)

A. Proteoglycan Core Proteins and Related Proteins

Cloning of the large cartilage aggregating proteoglycan core protein presented some obstacles rarely encountered in the cloning of other proteins: (1) the large size of the core protein mRNA made it difficult to synthesize a full-length cDNA, (2) the codon redundancy for amino acids in the known sequence precluded the use of mixed oligonucleotide primers or probes, and (3) the presence of a protein epitope at the COOH-terminal region of the molecule had not been determined. From the size of the cell-free translation product it was known that the size of the predominant cartilage core protein would be synthesized by a mRNA of approximately 9 to 11 kb. The core protein isolated from the matrix contains at least three domains: the hyaluronic acid-binding region (~65 kDa in size), a region containing keratin sulfate chains (~30 kDa in size), and a region that contains predominantly chondroitin sulfate chains (>100 kDa in size). As mentioned, the nascent core protein is larger and may contain additional domains. The amino terminus is located at the hyaluronic acid-binding region (Bonnet et al., 1983, Stevens et al., 1984), which contains the best-characterized antigenic regions (Caterson et al., this volume). Regardless of these difficulties, cDNAs for regions of the core protein from chicken embryo sternal cartilage (Sai et al., 1986) and rat chondrosarcoma (Doege et al., 1986b) have been cloned. Both of these clones were isolated from libraries made using oligo(dT)-primed mRNA inserted into protein expression vectors. In both clones, the polypeptide sequence expressed by the cloned cDNA reacted with polyclonal antisera to the proteoglycans. The chicken cDNA is 1230 nucleotides long and encodes 379 amino acids of the COOH terminus of the core protein and about 100 bp of the 3'-untranslated region. This cloned DNA hybridized to an mRNA of 8.1 kb on Northern blot analysis. Identification of the cDNA was confirmed by comparing the amino acid sequence deduced from nucleotide sequence to 20 amino acids of a bovine peptide (Perrin et al., 1984) (Fig. 7). When published, the location of the amino acid sequence in the protein was not known, although it was thought to be from a region free of chondroitin sulfate chains. This amino acid sequence can now be localized at the COOH terminus of the protein because the cDNA was synthesized from the 3' end of the mRNA using an oligo(dT) primer. Consequently, the sequence of this clone shows that an amino acid sequence free of chondroitin sulfate chains lies COOH-terminal to the chondroitin sulfate attachment region and contains an antigenic site.

The cDNA clone encoding a similar region of the rat chondrosarcoma core protein was isolated by screening a library capable of protein expression with a polyclonal antiserum raised to rat chondrosarcoma aggregating proteoglycan (Doege *et al.*, 1986b). This DNA is 872 nucleotides in length, encodes 269 amino acids of the COOH terminus, and was identified by comparing the deduced amino acid sequence of the bovine core protein peptide shown in Fig. 7. Both the rat and chicken cDNAs encode the bovine peptide with only one mismatch between the rat and bovine, although, throughout the remaining sequence the cDNAs diverge to some extent in both nucleotide sequence and deduced amino acid sequence. Both cDNAs contain approximately 10 cysteine residues within the COOH-terminal 300 amino acids, which may form intramolecular disulfide bonds. Toward the NH_2 terminus, the deduced amino acid sequence extends into a region containing serine and threonine residues, which may provide sites for attachment of chondroitin sulfate chains. As these cDNAs encode only a portion of the core protein, they will be used to construct and isolate other cDNAs containing more extensive coding regions. Interestingly, the amino acid sequences of the cloned portions of the rat and chicken core proteins demonstrates a high degree of homology with the amino acid sequence of a receptor present on the surface of liver cells. The receptor binds asialoglycoproteins (Ashwell and Harford, 1982), and the degree of homology between the two proteins may indicate that a functional activity, such as glycoconjugate binding, resides in the COOH-terminal domain of this core protein.

A cDNA clone coding for a small chondroitin sulfate proteoglycan core protein from the rat L2 yolk sac tumor was reported recently by Bourdon *et al.* (1985). The yolk sac tumor proteoglycan has a small core protein (~20 kDa) with about 14 chondroitin sulfate side chains

CHICKEN					CAA	C			C			C	T	C		T		G
RAT	ATG	ATC	TGG	CAT	GAG	AGG	GGT	GAA	TGG	AAC	GAT	GTC	CCC	TGC	AAT	TAC	CAG	CTG	CCC	TTC	ACA
BOVINE	[MET]	ILE	TRP	HIS	GLU	LYS	GLY	GLU	TRP	ASN	ASP	VAL	PRO	CYS	ASN	TYR	GLN	LEU	PRO	PHE	THR
CHICKEN						GLN											HIS				
RAT						ARG															

FIG. 7. Comparison of the amino acid sequence of a peptide isolated from bovine nasal cartilage proteoglycan (Perin *et al.*, 1984) with a portion of the deduced amino acids sequence from the chicken cDNA clone and the rat chondrosarcoma cDNA clone. The bracketed MET in the bovine sequence was inferred, as the peptide was obtained from a cyanogen bromide digest. Positions in the rat and chicken nucleotide and amino sequences which differ from the bovine sequence are indicated. The chicken DNA sequence is from Sai *et al.* (1986), and the rat chondrosarcoma DNA sequence is from Doege *et al.* (1986a).

(Oldberg *et al.*, 1981). A library was prepared from the yolk sac tumor mRNA population and screened by hybridization with oligonucleotide probes encoding the amino acid sequence from the NH_2-terminal region of the core protein. The DNA sequence of this clone gives us many insights into the nature of the core protein. The chondroitin sulfate chain attachment site appears to be composed of alternating serine and glycine residues spanning 49 amino acids. This alternating sequence is similar to the heparin attachment site in the mast cell heparin proteoglycan (Robinson *et al.*, 1978). Reported chondroitin sulfate attachment regions in bovine articular cartilage core protein contain the sequence Leu-Pro-Ser-Gly-Glu; in the porcine tracheal core protein they contain the sequence Leu-Pro-Ser-Gly-Glu-Gly-Pro-Glu (Isemura *et al.*, 1981); and in the bovine skin fibroblast core protein the region contains the sequence NH_2-Asp-Glu-Ala-Ser-Gly-Ile-Gly (Pearson *et al.*, 1983). Consequently, many sequences containing a Ser-Gly may serve as substrate for chondroitin sulfate chain initiation. This cDNA clone encodes a core protein of 104 amino acids containing a NH_2-terminal domain of 14 amino acids and a COOH-terminal domain of 41 amino acids flanking the repetitive Ser-Gly region. The first amino acid, arginine, is identical to the first amino acid in the mature core protein. Recent studies with additional cDNA clones coding for the L2 yolk sac tumor proteoglycan core protein have revealed that there is more coding region 5' from the arginine codon and that this core protein is synthesized in a precursor form with a 25-amino acid hydrophobic signal sequence and a 49-amino acid propeptide sequence which is thought to be removed to yield the mature core protein (Rouslahti *et al.*, 1986).

In order to determine the number of genes which encode the yolk sac core protein, the cDNA was hybridized to a nitrocellulose filter containing total genomic DNA cut into fragments by the restriction enzyme *Eco*RI (a Southern blot). When the entire cDNA was used as a hybridization probe, multiple genomic fragments were identified. When cDNA was cut into fragments with and without the (Ser-Gly)-encoding region, it was found that the DNA encoding the (Ser-Gly) region was responsible for hybridization to multiple bands, indicating that there are other similar DNA sequences in the genome. The non-(Ser-Gly) probe hybridized to a single band, indicating that there is a unique gene that contains elements repeated in other genes (Rouslahti *et al.*, 1986). The results are compatible with the concept that there is a single gene for the yolk sac tumor proteoglycan core protein.

DNA sequences strikingly similar to those which encode the repeating (Ser-Gly) of rat yolk sac core protein have been found in the gene responsible for regulation of circadian rhythms in Drosophila (Shin *et al.*, 1985). By analysis of mutant flies exhibiting aberrant sleep–wake cycles and courtship songs, the period (*per*) locus was defined in a specific region of the X chromosome (Konopka and Benzer, 1971). DNA from the *per* locus has been cloned and found to contain a functional copy of a gene which produces a mRNA of 4.5 kb (Bargiello and Young, 1984). Some mutations at this locus disrupt or eliminate the flies' circadian rhythms; the rhythms can be restored by insertion of the gene into germ-line cells of mutant flies (Zehring *et al.*, 1984). The mRNA transcribed from the *per* locus contains 24 copies of a tandemly repeated sequence ACNGGN (encoding Thr-Gly) and a total of 28 copies interspersed throughout the gene. This sequence is very close to the codons used to encode the Ser-Gly dipeptide (TCAGGC) which theoretically could provide a substrate for initiation of chondroitin sulfate chains. Sequences encoding repeated Ser-Gly were also found, although usually interrupted by other codons. In order to determine whether these sequences are present in higher eukaryotes, the Drosophila DNA was used to isolate homologous sequences from the mouse genome. A total of 192 copies of the Thr-Gly repeat was found as well as 11 uninterrupted copies of the Ser-Gly (Shin *et al.*, 1985). The organization of some of these sequences is . . . $(Thr\text{-}Gly)_{12}(Ser\text{-}Gly)_7(Thr\text{-}Gly)_3$ . . . $(Thr\text{-}Gly)_{13}(Ser\text{-}Gly)_{11}(Thr\text{-}Gly)_3$ (Shin *et al.*, 1985). Although it is not known whether these sequences are substituted with glycoaminoglycan chains in Drosophila and mouse or which regions of the putative protein are functional, the similarity to the core protein sequences suggest that proteins related to proteoglycans may participate in the regulation of biological rhythms.

A clone encoding a small core protein distinct from the rat yolk sac tumor proteoglycan core protein has been isolated from a bone cell cDNA library. This proteoglycan is a member of the group of small proteoglycans discussed in Section 11. cDNAs were synthesized from subperiosteal bone cell mRNA and inserted into the λ gt11 expression vector system (A. Day, M. Young, and J. Termine, personal communication). The clone was identified by hybrid-selected cell-free translation and immunoprecipitation of the product with an antiserum raised against the bone proteoglycan, PGII. As discussed previously, two bands were observed in the cell-free translate. In hybrid-selected translation experiments used to confirm the identity of the clone, cDNA selected mRNA encoding both core proteins. Northern blot analysis showed that the clone hybridized to two RNA species at 1.8 and 1.4 kb from bone, tendon, mature articular cartilage chondrocytes, and skin cells. These may be the two mRNAs which encode the two proteins immunoprecipitated by antisera. Preliminary evidence indicates that the protein made by the isolated recombinant clone reacts with antisera raised against small dermatan sulfate proteoglycans from tendon and from cartilage. These small proteoglycan core proteins, whether they contain chondroitin sulfate as in bone, or dermatan sulfate as in tendon and cartilage, may be identical. Cloning and sequencing the cDNAs for the other core proteins in this group will demonstrate the extent of homology between these proteins and help to determine whether the two mRNAs observed in Northern blot analysis are transcribed from the same gene or two different genes.

A related proteoglycan core protein has been cloned from a fibroblast cDNA library (Rouslahti et al., 1986; Krusius and Rouslahti, 1986). The approach used was similar to that described for isolation of the small bone core protein clone. The initial screening was performed using antibodies prepared against the fetal membrane form of the proteoglycan which were known to cross-react with the fibroblast core protein. Positive clones were further screened with a rabbit antiserum prepared within a 14-amino acid synthetic peptide from the NH$_2$-terminal sequence of the fibroblast core protein (Brennan et al., 1984). Identity of the clone was confirmed by showing that the cDNA encoded the NH$_2$ sequence of the fibroblast core protein. The core protein is synthesized in a precursor form with a signal peptide of 16 amino acids and a propeptide sequence of 14 amino acids which, in fetal membrane tissue, is processed into a 329-amino acid mature polypeptide. The cDNA clone encodes three Ser-Gly dipeptides, but no repeats of the kind present in the yolk sac tumor proteoglycan. The sequence also contains three potential sites for asparagine-linked glycosylation. Hybridization of this clone to genomic DNA restriction enzyme fragments suggests that there is one gene for the core protein and possibly one related gene (Krusius and Rouslahti, 1986).

Although these small proteoglycans discussed above and in Section II from cartilage, tendon, bone, skin fibroblasts, fetal membrane, and smooth muscle contain regions of antigenic cross-reactivity, have amino acid sequences in common, and show DNA cross-hybridization, it is not known whether they share identical core proteins. If it is found that the core proteins are encoded by one gene, then the diversity among tissues found in this proteoglycan group will be shown to depend on the extent and type of glycosylation, possibly on the enzyme system involved in the epimerization of glucuronic acid (characteristic of chondroitin sulfate) to iduronic acid (characteristic of dermatan sulfate).

One type of collagen, type IX, appears to behave as a proteoglycan core protein, providing additional evidence for the diversity of core protein genes. Noro *et al.* (1983) first observed that one of the nonaggregating proteoglycans isolated from chick embryo epiphyseal cartilage contained two glycosaminoglycan chains covalently bound to a collagenase-sensitive polypeptide. The same proteoglycan was subsequently isolated from chicken sternal cartilage and shown to be immunologically and structurally identical to type IX collagen (Vaughan *et al.*, 1985; Bruckner *et al.*, 1985). The structure of two of the three type IX collagen α-chains has been elucidated by nucleotide sequence analysis of recombinant cDNA clones derived from chicken embryo cartilage mRNA (Ninomiya and Olsen, 1984; van der Rest *et al.*, 1985) and by biosynthetic studies in chick embryo sternal organ cultures (Bruckner *et al.*, 1983; von der Mark *et al.*, 1984). Preliminary results indicate that the glycosaminoglycan chains are probably located in a pepsin-sensitive area of the amino-terminal region of the α_2-chain, where the Gly-X-Y repeat has not been maintained (P. Bruckner, personal communication). The "interruption" in the Gly-X-Y triplet repeat could be caused by a deletion, addition, or substitution of nucleotides. In such a case, it is possible that the disrupted collagen helix would have exposed a cryptic glycosylation site.

B. Link Proteins

Cloning of link protein cDNAs has begun to answer some of the questions concerning the biosynthesis of link proteins and, along with the cell-free translation studies, has exposed some unusual features of the system. The λgt11 expression system was used to isolate a link protein clone from a rat chondrosarcoma cDNA library (Doege *et al.*, 1986). The clone was identified using polyclonal and monoclonal antibodies raised against link protein. DNA isolated from this clone hybridized to four mRNAs on Northern blots which ranged in size from 1.5 to 5 kb (Fig. 8A) (Doege *et al.*, 1986a). Sequence analysis of this clone shows further indication that the two regions of the protein containing the epitope were derived by duplication of the DNA. The cloned cDNA was used as a hybridization probe in analysis of genomic DNA from rat chondrosaroma and rat liver. The simplicity of the restriction fragments and the size of the fragments suggests that there is only a single gene for link protein.

Multiple mRNAs ranging in size from 2.5 to 6 kb (Fig. 8B) have also been detected in chicken sternal mRNA by hybridization with a cloned cDNA coding for a link protein (Deak *et al.*, 1986). The cDNA was identified by hybridization to a synthetic oligonucleotide coding for

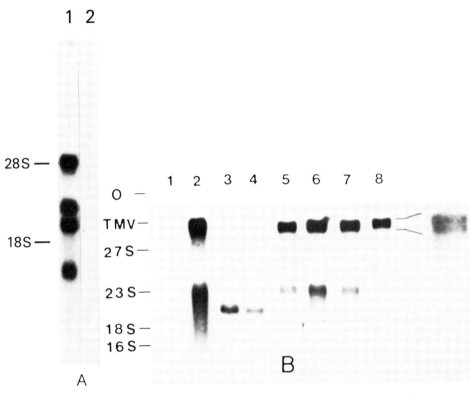

FIG. 8. (A) RNA transfer blot analysis of link protein mRNA. Poly(A)⁺ RNA from rat
chondrosarcoma (lane 1) and rat embryo fibroblast (lane 2) were electrophoresed on a
0.7% denaturing agarose gel, blotted to nitrocellulose, and probed with fragments of link
cDNA. Rat chondrosarcoma 18S and 28S rRNAs were used as size markers. The cloned
cDNA hybridized to mRNAs of approximately 1.5, 2.1, 2.5, and 5.5 kb. (B) RNA blot
hybridization analysis of poly(A)⁺ RNA isolated from chick embryo calvaria and sterna.
One microgram of RNA was separated on 0.8% agarose–2.2 *M* formaldehyde gels, blotted, and hybridized to radioactive probes. Lanes 1 and 3, RNA from chicken embryo
calvaria; lanes 2 and 4, RNA from chicken embryo sterna. Lanes 1 and 2 were hybridized
with the link protein cDNA, and lanes 3 and 4 were hybridized to a clone coding for
chicken actin. Lanes 5–8 were hybridized with various link protein DNA fragments (see
Deak *et al.*, 1986, for details). The extension of lane 8 shows that the upper band can be
resolved into two bands if electrophoresis is prolonged. The cloned cDNA hybridized to
mRNAs of approximately 3.0, 5.8, and 6.0 kb. (A) Reproduced from Doege *et al.* (1986b);
(B) reproduced from Deak *et al.* (1986).

seven amino acids present in both bovine and rat chondrosarcoma link proteins. This cDNA clone encodes 2.5 kb of mRNA and includes the entire link protein sequence. In the cloned cDNA, the sequences encoding link protein are located in the 5' portion and cover about 1000 nucleotides. The remaining sequences do not encode link protein and presumably are not translated. The sequence of the mRNAs larger than 2.5 kb is not known at this time, and the role played by the larger messages is unclear. One possibility is that the larger mRNAs synthesize a precursor protein; however, no detectably larger precursor to link has been observed by *in vitro* translation of mRNA. Alternatively, the larger mRNAs could include (1) nontranslated nucleotides encoded by the gene, (2) mRNA encoding an additional protein immunologically unrelated to link, or (3) mRNA encoding a larger link-related protein which is translated in tissues other than cartilage. The second possibility implies that the mRNA is polycistronic; however, such messages are very rare and all of viral origin (Kozak, 1983). The third possibility would require tissue-specific processing of the mRNA such that a larger protein could be translated. We have shown that, in cartilage, the larger mRNAs synthesize a 42 kDa link protein. Interestingly, canine synovial fibroblasts synthesize a 70-kDa protein antigenically related to the cartilage link protein (Fife *et al.*, 1985). Studies on the cell-free translation of this canine fibroblast mRNA will determine whether a larger translation product exists. At the present time, the most likely case is that the larger mRNAs include nontranslated sequence as has been observed in other protein systems. In fact, a group of myelin basic proteins are encoded by mRNAs which are much larger than predicted, and the extra sequence is located at the 3' end while the coding sequence is in the 5' region (deFerra *et al.*, 1985; Takahashi *et al.*, 1985). Further isolation of cDNA clones for the various mRNAs and the gene(s) encoding link protein will be necessary in order to determine why there appears to be a wide size difference in the mRNAs coding for link proteins, how the mRNAs are generated, and what the relationships are between multiple mRNAs and different species of link proteins.

Comparison of the link protein sequence provided by nucleotide and amino acid sequencing shows that the link proteins are highly conserved. The secondary structure predicted for the rat chondrosarcoma link protein is shown in Fig. 9. From analysis of the rat chondrosarcoma and avian sequences, it can be seen that even though most of the protein is conserved, divergence at the NH_2 terminus of the mature protein is considerable. Very close to the NH_2 terminus, the rat chondrosarcoma link protein contains one site for asparagine-linked carbo-

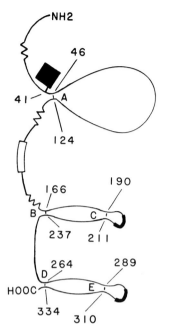

Fig. 9. Positions of the disulfide bonds and likely secondary structure within the sequence of rat chondrosarcoma link protein. Numbers refer to the positions of the cysteine residues. The solid square refers to the position of the oligosaccharide. Open squares are sections likely to be β-sheet, and zig-zag sections are sections likely to be α-helix. The regions containing the epitopes for the monoclonal antibody 8-A-4 are denoted with a heavy line. Reproduced from Neame *et al.* (1986).

hydrate addition (Neame *et al.*, 1986), and the avian contains two sites (Deak *et al.*, 1986). The two link proteins present in chicken cartilage could be generated by glycosylation at one or both of the Asn-X-(Thr, Ser) sites. At this time the possibility of multiple mRNAs containing a different number of potential glycosylation sites cannot be eliminated.

IV. PROSPECTIVES

The following concepts represent a few of the areas of research that can be addressed by molecular biological studies of proteoglycans and link proteins. Only a few years ago one was limited to the hypotheses that related proteins were encoded either by evolutionarily related genes or by posttranslational processing of a single polypeptide. However, now that genes can be studied directly, alternative mechanisms for the generation of diversity in the structure of similar proteins are quite numerous. The known possibilities include evolution of separate genes forming a gene family, as seen in the collagen genes; rearrangements of the genomic DNA, as has been demonstrated for immunoglobulins and T-cell receptor genes; transcription of different RNAs from a single gene either initiated at different promoters, as in myosin light chains (Periasamy *et al.*, 1984) or terminated at alternative 3' ends, as in the neuronal peptides (Rosenfeld *et al.*, 1984); alternative processing of the RNA transcript, as in myosin heavy chain (Rosek and Davidson, 1983) and fibronectins (Schwarzbauer *et al.*, 1983); or a combination of these mechanisms as in the troponin T isoforms (Breitbart *et al.*, 1985). The discovery of the intricate patterns of exon splicing generating the troponin T isoforms indicates that one gene can produce a great deal of protein diversity. In this case, at least 10 antigenically distinct proteins can be produced which are developmentally regulated and tissue specific. The mechanism used by proteoglycan genes to generate protein diversity is not known at this time. However, cDNA probes will be used to isolate the corresponding genes, and these questions will soon begin to be answered.

Elucidation of the structure of core and link protein genes will yield insights into the evolution of the gene. The gene encoding a particular protein will be larger than the mRNA and often will have more noncoding DNA than coding DNA. It has been hypothesized that the coding regions (exons) contain functional domains that were present in some primordial gene and that the modern protein is constructed by combining exons which code for specific protein domains important to the structure and function of the new protein (Gilbert, 1977, 1985). The gene structure of the elastin and fibronectin genes appear to reflect this hypothesis (Cecilia *et al.*, 1985; Odermatt *et al.*, 1985), as they are composed of exons containing distinct functional units. A new gene evolves from an ancestral gene by duplication of the gene and then, while selective pressure may still remain on the original gene, variations are allowed to occur in the duplicated gene by mutation and DNA rearrangements. New DNA may even be introduced by exon transfer between genes. An example of exon exchange is seen in the tissue plasminogen activator gene, which appears to have exons derived from fibronectin and epidermal growth hormone genes (Rogers, 1985). As discussed previously in this chapter, there is already evidence of common coding regions (exons) between a proteoglycan core protein and link protein and between a core protein and a receptor protein, indicating that these proteins likely evolved by integrating and maintaining exons of a specific function.

There is a strong indication that antigenically or functionally related proteoglycan core protein genes are encoded by gene families in which separate members are related by duplication and variation as described. The individual functional components of the gene, such as the various carbohydrate addition sites, the hyaluronic acid-binding or collagen-binding sites, or membrane insertion sequences, may have evolved from a common ancestral gene or exon. Further diversity in protein structure could be generated by differential RNA transcription or splicing and posttranslational modifications. Isolation of the genes encoding the proteoglycan core protein and link proteins will allow the determination of the DNA sequence and, consequently, protein sequence of the entire nascent protein, and thus provide a basis for the study of the evolution of core and link proteins.

It is of considerable interest to determine the number of copies of specific core and link protein genes for the analysis of genetic mechanisms of disease. If there is more than one copy of a gene, as in the case of α-globin genes, a change in the DNA in one gene may not result in a severe effects in the total population of molecules synthesized. However, if, as is generally the case for a structural gene, only one copy of the gene is present per haploid genome, a single base change, insertion, or deletion of DNA may have a significant effect on the mRNA and proteins synthesized. If more than one protein is generated from a single gene by variation in exon splicing, one change in the DNA could effect a set of proteins. This effect will be most apparent if it occurs in an important functional region of the protein. In the osteogenesis imperfecta syndromes, a variety of collagen gene alterations have been observed, including a single base change converting glycine to cysteine (Cohn et al., 1986) and a deletion of DNA removing a coding region (Chu et al., 1985; Barsch et al., 1985), resulting in production of abnormal collagen chains. These types of changes will likely be discovered for proteoglycan core and link protein genes. The determination of the number of link and core protein genes will help to determine which proteoglycans and link proteins contain unique core proteins and which are generated by various posttranslational modifications of the same protein.

The localization of the proteoglycan and link protein genes on human chromosomes may give some insight into how these genes are regulated. Most of the collagen genes are located on different chromosomes, presumably to reduce unequal crossing over of DNA during meiosis; however, a group of connective tissue genes coding for elastin (Cecelia et al., 1984), fibronectin (Kornblihtt et al., 1983), and type V collagen (Emannuel et al., 1985) are on chromosome 2, which may facilitate the expression of these genes as a set. The human type II collagen gene is located on chromosome 12 (Solomon *et al.*, 1985), and it is attractive to speculate that other genes which compose the chondrocyte phenotype are also located in close proximity. Through an analysis of rodent–human hybrid cell clones, a proteoglycan core protein gene which is related to a malignant phenotype of melanoma cells has been assigned to chromosome 15 (Rettig *et al.*, 1986). Proteoglycans and link proteins are differentially expressed during development and maturation of cartilage (Rosenberg *et al.*, 1985, Inerot and Heinegard, 1984; Thonar and Kuettner, this volume; Vasan and Lash, 1977). For example, during osteogenesis, a series of proteoglycans are synthesized which are characteristic of the developing tissue: mesenchyme, cartilage, hypertrophic cartilage, and bone (Carrino *et al.*, 1985). Therefore, sets of genes may be regulated by tissue-specific factors to generate the proper proteoglycan phenotype. Changes in types of proteoglycans synthesized have been observed in disorders such as osteoarthrosis (Thonar and Kuettner, this volume), atherosclerosis (Wight *et al.*, this volume), and cancer (Harper and Reisfeld, this volume), indicating inappropriate gene expression. Abnormal synthesis of proteoglycan core protein has been suggested in two chondrodysplasias, the cartilage matrix-deficient (cmd/cmd) mouse (Kimata *et al.*, 1981) and the nanomelic chick (McKeown-Longo *et al.*, 1982) and in various human osteochondrodysplasias (Stanescu *et al.*, 1984). In the animal models, the synthesis of core protein is greatly reduced if not absent, indicating a defective gene or regulatory molecule (Argraves *et al.*, 1981; Kimata *et al.*, 1981). In one form of human pseudoachondroplasia, one of the cartilage core proteins accumulates in the endoplasmic reticulum perhaps as a result of a defect in the protein (Stanescu *et al.*, 1984). In the Coffin–Lowry syndrome (Beck *et al.*, 1983), an X-linked inherited disorder, an abnormal core protein of the dermatan sulfate proteoglycan has been found in fibroblasts from these patients. An interesting abnormality has been reported in an inherited form of chondrodysplasia in Alaskian malamute dogs (Bingel *et al.*, 1985), in which the growth plate chondrocytes synthesize proteoglycans characteristic of immature cartilage rather than mature cartilage. This disease could be the result of a failure to express the mature proteoglycan phenotype due to a genetic defect in a regulatory molecule or the core protein. Other genetically related abnormalities of proteoglycan synthesis and degradation exist that are not strictly related to core and link protein biosynthesis. Defects in the enzymes involved in "intermediary metabolism" such as sulfation (Sugahara and Schwartz, 1979; Bingel *et al.*, 1986) and glycosylation

(Dannenberg *et al.*, 1983) will produce a significant impact on the biosynthesis of extracellular matrix.

cDNA or genomic clones for core and link protein will provide sensitive probes for the determination of gene expression. The techniques for detection of mRNA are extremely sensitive and accurate, and nucleic acids are relatively easy to isolate. Northern blot analysis requires only nanogram quantities of mRNA, "dot" blots require cytoplasmic RNA from as few as 2.5×10^4 cells, and *in situ* hybridization to tissue sections resolves individual molecules within cells. Thus, the developmental stage at which a cell initiates synthesis of core and link proteins or changes a proteoglycan type can be more precisely determined with molecular biological techniques than has been possible with protein biochemical methodologies.

Elucidation of the mechanism of proteoglycan synthesis will indicate the level at which regulation occurs, but the more significant questions relating to the mechanism of regulation will necessitate studies on chromatin, gene interaction, and specific regulatory proteins. In the future, the use of gene transfer techniques will greatly enhance our ability to define the mechanisms that act at the level of transcription and posttranslational processing. For example, deletion mapping followed by gene transfer into cells could determine which specific nucleotide sequences in or around a core protein gene are involved in transcriptional control. Gene transfer techniques may also play a major role in characterizing various aspects of the processing reactions of the nascent core proteins. This approach could be extended to analyze the effects of site-specific *in vitro* mutagenesis (e.g., altering carbohydrate addition sites) followed by gene transfer. These types of experiments will allow us to determine the specificity of enzymes involved in posttranslational modification. For example, could core proteins be properly processed in cells where they are not naturally found? These types of studies will eventually yield insights into the function of proteoglycans in the extracellular matrix, in membranes, and within cells.

ACKNOWLEDGMENTS

The author wishes to thank Ms. Madelyn Sieraski and Dr. Tom Hering for invaluable help in preparation of the manuscript, and Ed Dudek for expert technical assistance in the original research reported in this article. For communicating data prior to publication, the author is very grateful to Drs. Paul Goetinck, Marian Young, John Termine, Errki Ruoslahti, Kurt Doege, Yoshihiko Yamada, Marvin Tanzer, and Peter Bruckner. Critical reading of the manuscript and the valuable comments of Drs. David Schwartz,

Vince Hascall, Peter Byers, and Tom Wight were sincerely appreciated. Comments on an early form of the manuscript by Drs. Chris Handley and Charles Boyd were also appreciated. This research was supported by a grant from the National Institutes of Health (AM 34142) and a Basil O'Connor Starter Grant from the March of Dimes Birth Defects Foundation.

REFERENCES

Adams, S., Boettiger, D., Focht, R. J., Holtzer, H., and Pacifici, M. (1982). *Cell* **30,** 373–384.
Argraves, W. S., McKeown-Longo, P. J., and Goetinck, P. F. (1981). *FEBS Lett.* **131,** 265–268.
Ashwell, G., and Harford, J. (1982). *Annu. Rev. Biochem.* **51,** 531–549.
Baker, J. R., and Caterson, B. (1979). *J. Biol. Chem.* **254,** 2387–2393.
Baker, J. R., Caterson, B., and Christner, J. E. (1983). *Prog. Clin. Biol. Res.* **110B,** 17–24.
Bargiello, T. A., and Young, M. W. (1986). *Proc. Natl. Acad. Sci. U.S.A.* **81,** 2142–2146.
Barsch, G. S., Roush, C. L., Bonadia, J., Byers, P. H., and Gelinas, R. E. (1985). *Proc. Natl. Acad. Sci. U.S.A.* **82,** 70–74.
Beck, M., Glossl, J., Ruter, R., and Kresse, H. (1983). *Pediatr. Res.* **17,** 926–929.
Bingel, S. A., Sande, R. D., and Wight, T. N. (1985). *Lab. Invest.* **53,** 479–485.
Bingel, S. A., Sande, R. D., and Wight, T. N. (1986). *Connect. Tissue Res.,* **15,** 283–302.
Bonnet, F., Le Gledic, S., Perin, J., Jolles, J., and Jolles, P. (1983). *Biochim. Biophys. Acta* **743,** 82–90.
Bourdon, M. A., Oldberg, A., Pierschbacher, M., and Rouslahti, E. (1985). *Proc. Natl. Acad. Sci. U.S.A.* **82,** 1321–1325.
Breitbart, R. E., Nguyen, H. T., Medford, R. M., Destree, A. T., Mahdavi, V., and Nadal-Ginard, B. (1985). *Cell* **41,** 67–82.
Brennan, M. J., Oldberg, A., Pierschbacher, M. D., and Rouslahti, E. (1984). *J. Biol. Chem.* **259,** 13742–13750.
Bruckner, P., Mayne, R., and Tuderman, L. (1983). *Eur. J. Biochem* **136,** 333–339.
Bruckner, P., Vaughan, L., and Winterhalter, K. H. (1985). *Proc. Natl. Acad. Sci. U.S.A.* **82,** 2608–2612.
Carrino, D. A., Weitzhandler, M., and Caplan, A. I. (1985). *In* "The Chemistry and Biology of Mineralized Tissues" (W. Butler, ed.). Ebsro Media, Birmingham, Alabama.
Caterson, B., Baker, J. R., Christner, J. E., Lee, Y., and Lentz, M. (1985). *J. Biol. Chem.* **260,** 11348–11356.
Cecila, G., Yoon, K., Ornstein-Goldstein, N., Indik, Z. K., Boyd, C., May, M., Cannizzaro, L. A., Emanuel, B. S., and Rosenbloom, J. (1985). *In* "Extracellular Matrix Structure and Function," pp. 333–350. Liss, New York.
Chang, Y., Yanagishita, M., Hascall, V. C., and Wight, T. N. (1983). *J. Biol. Chem.* **258,** 5679–5688.
Chu, M.-L., Gargiulo, V., Williams, C. J., and Ramirez, F. (1985). *J. Biol. Chem.* **260,** 691–694.
Cohn, D. H., Byers, P. H., Steinmann, B., and Gelinas, R. E. (1986). *Proc. Natl. Acad. Sci. U.S.A.,* **83,** 6045–6047.
Day, A., Young, M., and Termine, J., submitted.
Dannenberg, A., Buss, E. G., and Goetinck P. F. (1985). *Prog. Clin. Biol. Res.* **110B,** 85–95.
Deak, F., Kiss, I., Sparks, K. J., Argraves, W. S., Hampikian, G., and Goetinck, P. F. (1986). *Proc. Natl. Acad. Sci. U.S.A.* **83,** 3766–3770.

deFerra, F., Engh, H., Hudson, L., Kamholz, J., Puckett, C., Molineaux, S., and Lazzarini, R. A. (1985). *Cell* **43**, 721–727.

Doege, K. J., Hassell, J. R., Caterson, B., and Yamada, Y. (1986a). *Proc. Natl. Acad. Sci. U.S.A.* **83**, 3761–3765.

Doege, K. J., Fernandez, P., Hassell, J. R., Sasaki, M., and Yamada, Y. (1986b). *J. Biol. Chem.* **261**, 8108–8111.

Emannuel, B. S., Cannizzaro, L. A., Seyer, J. M., and Myers, J. C. (1985). *Proc. Natl. Acad. Sci. U.S.A.* **82**, 3385–3389.

Fife, R. S., Caterson, B., and Myers, S. L. (1985). *J. Cell Biol.* **100**, 1050–1055.

Fisher, L. W., Termine, J. D., Dejter, S. W., Whitson, W. S., Yanagishita, M., Kimura, J. H., Hascall, V. C., Kleinman, H. K., Hassell, J. R., and Nilsson, B. (1983). *J. Biol. Chem.* **258**, 6588–6594.

Frichauf, A. M., Lehrach, H., Rosner, C., and Boedtker, H. (1978). *Biochemistry* **17**, 3243–3249.

Geetha-Habib, M., Campbell, S. C., and Schwartz, N. B. (1984). *J. Biol. Chem.* **259**, 7300–7310.

Gilbert, W. (1978). *Nature (London)* **271**, 501.

Gilbert, W. (1985). *Science* **228**, 823–824.

Glossl, J., Beck, M., and Kresse, H. (1984). *J. Biol. Chem.* **259**, 14144–14150.

Hascall, V. C., and Sajdera, S. W. (1970). *J. Biol. Chem.* **245**, 4529–4538.

Hassel, J. R., Kimura, J. H., and Hascall, V. C. (1986). *Annu. Rev. Biochem.*, in press.

Heinegard, D., and Paulsson, M. (1984). "Extracellular Matrix Biochemistry" (K. Piez and A. Reddi, eds.), pp. 227–328. Elsevier, New York.

Hering, T. M., and Sandell, L. J. (1986a). *Fed. Proc. Fed. Am. Soc. Exp. Biol.* **45**, 1907.

Hering, T. M., and Sandell, L. J. (1986b). *J. Cell Biol.* **103**, 387a.

Hering, T. M., and Sandell, L. J. (1987). *Trans. Orth. Res. Soc.* **12**, 86.

Isemura, M., Hanyu, T., Kosaka, H., Ono, T., and Ikenaka, T. (1981). *J. Biochem.* **89**, 1113–1119.

Kimata, K., Barrach, H.-J., Brown, K. S., and Pennypacker, J. P. (1981). *J. Biol. Chem.* **256**, 6961–6968.

Kimura, J. H., Hardingham, T. E., and Hascall, V. C. (1980). *J. Biol. Chem.* **255**, 7134–7143.

Kimura, J. H., Thonar, E. J. M., Hascall, V. C., Reiner, A., and Poole, A. R. (1981). *J. Biol. Chem.* **256**, 7890–7897.

Konopka, R., and Benzer, S. (1971). *Proc. Natl. Acad. Sci. U.S.A.* **68**, 2112–2116.

Kornblihtt, A. R., Vibe-Pedersen, K., and Baralle, F. E. (1983). *Proc. Natl. Acad. Sci. U.S.A.* **80**, 3218–3222.

Kozak, M. (1983). *Microbiol. Rev.* **47**, 1–45.

Krusius, T., and Rouslahti, E. (1986). *Proc. Natl. Acad. Sci.* **83**, 7683–7687.

Le Gledic, S., Perin, J.-P., Bonnet, F., and Jolles, P. (1983). *J. Biol. Chem.* **258**, 14759–14761.

Levitt, D., and Dorfman, A. (1974). *Cur. Top. Dev. Biol.* **8**, 103–107.

Lohmander, L. S., Fellini, S. A., Kimura, J. H., Stevens, R. L., and Hascall, V. C. (1983). *J. Biol. Chem.* **258**, 12280–12286.

McKeown-Longo, P. J., Sparks, K. J., and Goetinck, P. F. (1982). *Coll. Rel. Res.* **2**, 231–244.

McKeown-Longo, P. J., and Goetinck, P. F. (1982). *Biochem. J.* **201**, 387–394.

Milstein, S., Brownless, G. G., Harrison, T. M., and Mathews, M. G. (1972). *Nature (London)* **239**, 117–120.

Neame, P. J., Perin, J. P., Bonnet, F., Christner, J. E., Jolles, P., and Baker, J. R. (1985). *J. Biol. Chem.* **260**, 12402–12404.

Neame, P. J., Christner, J. E., and Baker, J. R. (1986). *J. Biol. Chem.* **261**, 3519–3535.

Ninomiya, Y., and Olsen, B. R. (1984). *Proc. Natl. Acad. Sci. U.S.A.* **81**, 3014–3018.

Noro, A., Kimata, K., Shinomura, T., Maeda, N., Yano, S., Takahashi, N., and Suzuki, S. (1983). *J. Biol. Chem.* **258**, 9323–9331.

Odermatt, E., Tamkun, J. W., Hynes, R. O. (1985). *Proc. Natl. Acad. Sci. U.S.A.* **82**, 6571–6575.

Oldberg, A., Hayman, E. G., and Rouslahti, E. (1981). *J. Biol. Chem.* **256**, 10847–10852.

Pearson, C. H., Winterbottom, N., Fackre, D. S., Scott, P. G., and Carpenter, M. R. (1983). *J. Biol. Chem.* **258**, 15101–15104.

Periasamy, M., Strehler, E. E., Garfinkel, L. I., Gubits, R. M., Ruiz-Opazo, N., and Nadal-Ginard, B. (1984). *J. Biol. Chem.* **259**, 13595–13604.

Perin, J., Bonnet, F., Pizon, V., Jolles, J., and Jolles, P. (1984). *FEBS Lett.* **176**, 37–42.

Poole, A. R. (1986). *Biochem. J.* **236**, 1–14.

Rettig, W. J., Real, F. X., Spengler, B. A., Biedler, J. L., and Old, L. J. (1986). *Science* **231**, 1281–1284.

Robinson, H. C., Horner, A. A., Höök, M., Ogren, S., and Lindahl, U. (1978). *J. Biol. Chem.* **253**, 6687–6693.

Rogers, J. (1985). *Nature (London)* **315**, 458–459.

Rosenfeld, M. G., Amara, S. G., and Evans, R. M. (1984). *Science* **225**, 1315–1320.

Rosenberg, L. C., Choi, H. U., Tang, L., Johnson, T. L., Paul, S., Webber, C., Reiner, A., and Poole, A. R. (1985). *J. Biol. Chem.* **260**, 6304–6313.

Rozek, C. E., and Davidson, N. (1983). *Cell* **32**, 23–34.

Roughley, P. J., Poole, A. R., and Mort, J. A. (1982). *J. Biol. Chem.* **257**, 11908–11914.

Rouslahti, E., Bourdon, M., and Krusius, T. (1986). *Ciba Found. Symp.* **124**, 266–272.

Sai, S., Tanaka, T., Kosher, R. A., and Tanzer, M. L. (1986). *Proc. Natl. Acad. Sci. U.S.A.* **83**, 5081–5085.

Sandell, L. J., Sawhney, R. S., Yeo, T.-K., and Wight, T. N. (1986), submitted.

Sandmeyer, S., Gallis, B., and Bornstein, P. (1981). *J. Biol. Chem.* **256**, 5022–5028.

Sawhney, R. S., and Sandell, L. J. (1986). *J. Cell. Biol.,* **103**, 382a.

Sawhney, R. S., and Sandell, L. J. (1987). *Trans. Orth. Res. Soc.* **12**, 83.

Schwarzbauer, J. E., Tamkun, J. W., Lemischka, I. R., and Hynes, R. O. (1983). *Cell* **35**, 421–431.

Shin, H.-S., Bargiello, T. A., Clark, B. T., Jackson, F. R., and Young, M. W. (1985). *Nature (London)* **317**, 445–448.

Solomon, E., Hiorns, L. R., Spurr, N., Kurkinen, M., Barlow, D., Hogan, B. L. M., and Dalgleish, R. (1985). *Proc. Natl. Acad. Sci. U.S.A.* **82**, 3330–3334.

Stanescu, V., Stanescu, R., and Maroteaus, P. (1984). *J. Bone Jt. Surg.* **66A**, 817–836.

Stevens, J. W., Oike, Y., Handley, C., Hascall, V. C., Hampton, A., and Caterson, B. (1984). *J. Cell. Biochem.* **26**, 247–259.

Sugahara, K., and Schwartz, N. (1979). *Proc. Natl. Acad. Sci. U.S.A.* **76**, 6615–6618.

Takahashi, N., Roach, A., Teplow, D. B., Prusiner, S. B., and Hood, L. (1985). *Cell* **42**, 139–148.

Tang, L., Rosenberg, L., Reiner, A., and Poole, A. (1979). *J. Biol. Chem.* **254**, 10523–10531.

Treadwell, B. V., Mankin, D. P., Ho, P. K., and Mankin, H. J. (1980). *Biochemistry* **19**, 2269–2275.

Upholt, W. B., Vertel, B. M., and Dorfman, A. (1979). *Proc. Natl. Acad. Sci. U.S.A.* **76**, 4847–4851.

Upholt, W. B., Vertel, B. M., and Dorfman, A. (1981). *Ala. J. Med. Sci.* **18**, 35–40.

van der Rest, M., Mayne, R., Ninomiya, Y., Seidah, N. G., Chretin, M., and Olsen, B. R. (1985). *J. Biol. Chem.* **260**, 220–225.

Vasan, N. S., and Lash, J. W. (1977). *Biochem. J.* **164,** 179–183.

Vaughan, L., Winterhalter, K. H., and Bruckner, P. (1985). *J. Biol. Chem.* **260,** 4758–4763.

Vertel, B. M., Upholt, W. B., and Dorfman, A. (1984). *Biochem. J.* **217,** 259–263.

Vogel, K. G., and Fisher, L. W. (1986). *J. Biol. Chem.* **261,** 11334–11340.

Vogel, K. G., and Heinegard, D. (1985). *J. Biol. Chem.* **260,** 9298–9306.

von der Mark, K., van Menxel, M., and Wiedemann, H. (1984). *Eur. J. Biochem.* **138,** 629–633.

Vuorio, E., Sandell, L. J., Kravis, D., Sheffield,, V. C., Vuorio, T., Dorfman, A., and Upholt, W. (1982). *Nucleic Acids Res.* **10,** 1174–1192.

Young, R. A. and Davis, R. W. (1983). *Proc. Natl. Acad. Sci. U.S.A.* **80,** 1194–1198.

Zehring, W. A., Wheeler, D. A., Reddy, P., Konopka, R. J., Kyriacou, C. P., Rosbash, M., and Hall, J. C. (1984). *Cell* **39,** 369–176.

Biosynthesis of Heparin and Heparan Sulfate

Ulf Lindahl and Lena Kjellén

Department of Veterinary Medical Chemistry, Swedish University of Agricultural Sciences, The Biomedical Center, S-751 23 Uppsala, Sweden

I. Introduction

Heparin and heparan sulfate comprise a group of polysaccharides within the glycosaminoglycan family, distinguished by extensive structural heterogeneity and complexity. Widely distributed throughout the animal kingdom, as well as within the mammalian organism (Kennedy, 1979; Gomes and Dietrich, 1982), these compounds have been attributed a multitude of biological functions, most of which involve interactions with other biological macromolecules. The best known example is the blood anticoagulant activity of heparin, which depends on specific binding of the polysaccharide to the protease inhibitor, antithrombin. Studies of such systems have led to a growing realization that ligand binding may be critically dependent on the fine structure of the polysaccharide molecule. Consequently, these studies have stimulated the interest in polysaccharide biosynthesis and its regulation.

In general terms, our knowledge of glycosaminoglycan biosynthesis has reached a level that allows analysis of regulatory mechanisms at the molecular level (Lindahl *et al.*, 1977, 1986; Rodén, 1980; Feingold *et al.*, 1981; Gallagher *et al.*, 1986). Basic features of the reactions involved have been defined. Further, information regarding the subcellular organization of the biosynthetic apparatus is beginning to accumulate. Yet the biosynthesis of glycosaminoglycans appears largely unexplored compared to protein biosynthesis, for which the mechanisms, organization, and control have been elucidated in depth. Contrary to the latter process, the formation of a polysaccharide chain is not under direct control of a template-dependent mechanism (which instead determines the properties of the biosynthetic enzymes). As a result the product typically displays structural heterogeneity, which in

59

turn will complicate attempts to characterize the biosynthetic process in detail.

The present review will deal with the biosynthesis of heparin and heparan sulfate, the structurally most complex of the glycosaminoglycans. We will emphasize aspects of organization and regulation of the biosynthetic process, and will in particular focus on mechanisms that determine the fine structure of the polysaccharide chains. For a more detailed account of older and related work in this area, the reader is referred to a comprehensive review by Rodén (1980) and to a monograph by Comper (1981). Before discussing the biosynthetic mechanisms, we will review the structural properties and some functional aspects of heparin and heparan sulfate proteoglycans.

II. The Proteoglycans—Structures and Interactions

A presentation of the structural properties of heparin and heparan sulfate will immediately run into a problem of definition: how can we discriminate between the two polysaccharides? In most previous literature differentiation is based on a combined assessment of structural properties (see Gallagher and Walker, 1985) and anticoagulant activity. This approach no longer holds, as shown by the occurrence of "heparin," by structural criteria, that essentially lacks anticoagulant activity, and of "heparan sulfate" with significant anticoagulant activity. In this article we will use a provisional nomenclature, aimed primarily at defining the biosynthetic systems to be discussed in the following sections. The term "heparin" will thus be restricted to a sulfated, glucosamine-containing polysaccharide present in connective tissue-type mast cells, that is synthesized as a proteoglycan in which the polysaccharide-substituted portion of the core protein contains an extended sequence of alternating serine and glycine residues. All other structurally related polysaccharides will be denoted "heparan sulfate." We are aware of the limitations of this definition, when applied to polysaccharide chains that have been separated from their respective core proteins.

A. Macromolecular Properties and Core Protein Structure

1. Heparin Proteoglycans

Heparin proteoglycans (M_r ~750,000 to 1,000,000) have been isolated from rat skin (Horner, 1971; Robinson et al., 1978) and from rat peritoneal mast cells (Horner, 1977; Yurt *et al.*, 1977; Metcalfe *et al.*, 1980). Each molecule contains 10–15 polysaccharide chains (M_r 60,000–100,000) that are attached to a peptide core. The polysaccharide-substituted portion of the core structure is resistant to degradation by proteolytic enzymes. Amino acid analysis revealed that this region is composed of serine and glycine residues, about 20–25 residues of each, where at least 2 out of 3 serine units carry polysaccharide chains. This region would thus have a molecular weight of only about 3500. The linkage of polysaccharide to serine is mediated by a galactosyl-galactosyl-xylosyl trisaccharide sequence (Fig. 1) that is common not only to proteoglycans of heparin and heparan sulfate but occurs also in chondroitin sulfate and dermatan sulfate proteoglycans (Rodén, 1980). It has been proposed that the serine and glycine residues are arranged in alternating sequence, so as to minimize crowding of the chains (Robinson *et al.*, 1978). More conclusive proof for such an arrangement was obtained in studies of a chondroitin sulfate proteoglycan that was isolated from a rat yolk sac tumor (Bourdon *et al.*, 1985). Cloning of the core protein cDNA revealed a nucleotide sequence corresponding to exclusively alternating serine and glycine residues, in all 49 amino acids. A cDNA from a mouse mastocytoma cDNA library shows a high sequence homology with the rat yolk sac core protein cDNA, and thus presumably corresponds to the coding sequence of a core protein gene (L. Kjellén, L. Hellman, I. Pettersson, P. Lillhager, E. Ruoslahti, and U. Pettersson, unpublished observation). Hybrid proteoglycans containing both heparin–heparan sulfate and chondroitin sulfate have been identified in some cell types including a rat basophilic granulocyte line (David and Van den Berghe, 1985; Rapraeger *et al.*, 1985; Seldin *et al.*, 1985; Skinner and Fritz, 1985), demonstrating that the same core protein in certain cells can be utilized as primers for synthesis of both chondroitin sulfate and heparin–heparan sulfate. As mast cells are closely related to basophils, it is possible that the (Ser-Gly)$_n$-containing protein, corresponding to the mastocytoma cDNA clone, serves as a core protein in heparin as well as in chondroitin sulfate proteoglycans.

In certain tissues (pig intestinal mucosa, bovine lung and liver capsule, mouse mastocytoma) heparin does not occur primarily as a proteoglycan but rather as single polysaccharide fragments (commercially available "heparin") with an M_r of 5000–30,000 (Lindahl *et al.*, 1965; Ögren and Lindahl, 1971; Jansson *et al.*, 1975). Various tissues contain endoglycosidases capable of depolymerizing the extended chains of the heparin proteoglycan to such fragments (Ögren and Lindahl, 1975; Robinson *et al.*, 1978; Young and Horner, 1979) or to smaller oligosaccharides (see Thunberg *et al.*, 1982a, for references). One of

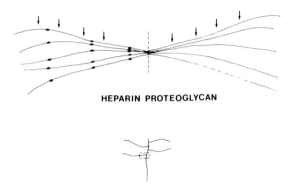

HEPARIN PROTEOGLYCAN

HEPARAN SULFATE PROTEOGLYCAN

FIG. 1. Schematic representation of a heparin proteoglycan and of a heparan sulfate proteoglycan (small-size type, see the text). The scale is chosen to indicate roughly the relative molecular dimensions of the polysaccharide and core protein components. Since the size of the heparin proteoglycan core protein is unknown, the peptide which extends beyond the polysaccharide-substituted region has been indicated by a dashed line. The antithrombin-binding regions (thicker segments) are accumulated in a nonrandom fashion in about half of the chains (arbitrarily selected; however, see Fig. 7); a fraction of the heparin proteoglycan molecules lacks this region altogether. The arrows indicate the cleavage sites for the mast cell endo-β-D-glucuronidase, which generates polysaccharide fragments ("heparin") with high or low affinity for antithrombin. Cleavage is specific and does not involve the β-glucuronidic linkage of the antithrombin-binding region. The polysaccharide–protein linkage region (indicated for one of the chains in the heparan sulfate proteoglycan only, but similar in the heparin proteoglycan) is shown in detail in the lower part of the figure. A neutral galactosyl-galactosyl-xylosyl trisaccharide sequence links a terminal glucuronic acid residue in the polysaccharide chain to a serine residued in the peptide core. For further information see the text.

these enzymes, an endo-β-D-glucuronidase (Ögren and Lindahl, 1975), degrades the newly synthesized proteoglycan in the mast cell (Fig. 1) to products that are stored in granules in the cell (Ögren and Lindahl, 1976). It may be noted that polysaccharides indistinguishable from heparin with regard to structure and biological activity occur also in invertebrates such as certain marine clams (Dietrich *et al.*, 1985; Jordan and Marcum, 1986). The corresponding proteoglycans have not yet been characterized.

2. HEPARAN SULFATE PROTEOGLYCANS

Heparan sulfate proteoglycans occur as cell surface components on a variety of cells, but are also found in the extracellular matrix and in specialized structures such as basement membranes. Certain common features distinguish these structurally heterogeneous macromolecules from the heparin proteoglycans of mast cells (Fig. 1). The heparan sulfate proteoglycans typically display higher protein contents, and fewer and shorter polysaccharide chains; generally they can be degraded to single polysaccharide chains by treatment with proteolytic enzymes. The heparan sulfate proteoglycans isolated range in size from M_r ~55,000 to >400,000. Examples of the smaller-sized species include proteoglycans isolated from liver (Oldberg et al., 1979), hepatoma (Mutoh et al., 1980; Robinson et al., 1984), brain microsomal membranes (Miller and Waechter, 1984; Klinger et al., 1985), glomerular basement membranes (Kanwar et al., 1984; Parthasarathy and Spiro, 1984), aortic endothelial cells (Oohira et al., 1983), and the Engelbreth–Holm–Swarm mouse sarcoma (Fujiwara et al., 1984). Large-sized proteoglycans have been obtained from fibroblasts (Vogel and Peterson, 1981; Carlstedt et al., 1983), glial cells (Norling et al., 1981), ovarian granulosa cells (Yanagishita and Hascall, 1983), colon carcinoma cells (Iozzo, 1984), the Engelbreth–Holm–Swarm sarcoma (Hassell et al., 1980; Fujiwara et al., 1984), and aortic endothelial cells (Oohira et al., 1983). Proteoglycans produced by mammary epithelial cells (Rapraeger and Bernfield, 1983), parietal yolk sac cells (Oohira et al., 1982; Tyree et al., 1984), and Morris rat hepatoma (Oldberg et al., 1982) appear to be of intermediate size. Certain tissues apparently produce heparan sulfate proteoglycans of more than one size class.

The variation in molecular size of the proteoglycans may be partly due to differences in the number as well as the length of the polysaccharide chains attached to the protein core. By and large, however, the molecular size of the intact macromolecule would seem to correlate with that of the corresponding core protein. Estimates of the molecular weights of core proteins range from 5000 to 240,000 (Mutoh et al., 1980; Kjellén and Höök, 1983; Kanwar et al., 1984; Fujiwara et al., 1984; Iozzo, 1984). One of the high molecular weight heparan sulfate proteoglycans, derived from fibroblasts, was found to occur as a disulfide-linked dimer, each subunit containing a core protein with an M_r of 80,000–100,000 (Cöster et al., 1983).

Heparan sulfate proteoglycans have a complex amino acid composition, generally dominated by serine, glycine, aspartic acid, and glutamic acid (Mutoh et al., 1978; Oldberg et al., 1979; Hassell et al., 1980; Oldberg et al., 1982; Fujiwara et al., 1984; Kanwar et al., 1984; Iozzo,

1984; Parthasarathy and Spiro, 1984). While proteoglycans from different sources show some variation in amino acid composition, there is no consistent correlation with molecular weight.

Antibodies have been raised against a variety of purified heparan sulfate proteoglycans (Hassell *et al.*, 1980; Oldberg *et al.*, 1982; Buonassisi and Colburn, 1983; Fujiwara *et al.*, 1984; Stow and Farquhar, 1984; Stow *et al.*, 1985). Studies on immunological cross-reactivity suggested that the proteoglycans from basement membrane are distinct from those of liver plasma membrane (Stow and Farquhar, 1984; Dziadek *et al.*, 1985). These results, together with the variation in molecular weight, point to the occurrence of more than one core protein gene.

Proteoglycans of heparan sulfate, similar to those of the galactosaminoglycans, contain N- and O-linked oligosaccharide substituents on their core proteins (Yanagishita and Hascall, 1983; Parthasarathy and Spiro, 1984; Klinger *et al.*, 1985). Such oligosaccharides have been found in large- as well as in small-sized proteoglycans, and may account for a considerable portion of the total carbohydrate. Thus Parthasarathy and Spiro (1984) calculated that up to one-fifth of the total sugar units in a heparan sulfate proteoglycan from glomerular basement membranes occurred in oligosaccharides.

B. *Polysaccharide Structure*

The polysaccharide chains in heparin and heparan sulfate proteoglycans are composed of hexuronic (D-glucuronic or L-iduronic) acid and D-glucosamine residues that are arranged in alternating sequence. The glucosamine residues are either N-acetylated or N-sulfated, very rarely N-unsubstituted. O-Sulfate substituents may be located at C-6 of the glucosamine and at C-2 of the iduronic acid units (see Comper, 1981, for references). More recently, O-sulfate groups have been found in small quantity at C-3 (Lindahl *et al.*, 1980) and C-2 (Bienkowski and Conrad, 1985), of glucosamine and glucuronic acid residues, respectively. In all, four different uronic acid and six variously substituted glucosamine monosaccharide units have been identified to date (Table I).

In heparin or heparan sulfate chains the various uronic acid and glucosamine units are combined to form a multitude of different saccharide sequences. However, mechanistic constraints imposed during biosynthesis favor certain combinations of neighboring monosaccharide units, thus limiting the sequence heterogeneity. Our current conception of structural variability is summarized by the scheme of nearest-neighbor compatibility shown in Table I. The structures so far identified involve 17 different →UA→GlcN→ and 10 →GlcN→UA→ disaccharide sequences (see Table I for abbreviations). The structural studies behind these results will not be discussed here; in general terms, heparin or heparan sulfate were selectively degraded by enzymic or chemical methods (see Silverberg *et al.*, 1985), followed by fractionation and characterization of the resulting fragments. Selected, comprehensive recent references to such studies are given in Table I; for reference to older work, see Lindahl (1976), Rodén (1980), Comper (1981), and Casu (1985).

The difference in polysaccharide structure between heparin and heparan sulfate is essentially quantitative with few, if any, qualitative distinctions. Recent findings place 3-*O*-sulfated glucosamine residues not only in heparin but also, by the definition used in this review, in heparan sulfate (Marcum *et al.*, 1986; Pejler *et al.*, 1987). Further, 2-*O*-sulfated glucuronic acid units have been demonstrated in heparin (Bienkowski and Conrad, 1985) as well as in heparan sulfate (Fedarko and Conrad, 1986) preparations. In general, heparin contains more *N*- and *O*-sulfate and more iduronic acid, but less *N*-acetyl and glucuronic acid than does heparan sulfate (Taylor *et al.*, 1973; Höök *et al.*, 1974b; Gallagher and Walker, 1985). The predominant disaccharide unit in heparin is →IdUA(2-OSO_3)→GlcNSO_3(6-OSO_3)→ which can occur in extended block structures, whereas the nonsulfated →GlcUA→ GlcNAc→ sequence is commonly found in heparan sulfate. N-Acetylated disaccharide residues generally occur as isolated units (i.e., surrounded on both sides by N-sulfated disaccharide units) in heparin (Cifonelli and King, 1972, 1973, 1977), whereas they tend to form more extended block structures in heparan sulfate (Cifonelli, 1968; see also, for example, Sjöberg and Fransson, 1980; Gallagher and Walker, 1985). The first one or two glucosamine units immediately adjacent to the polysaccharide–protein linkage region (Fig. 1) are regularly N-acetylated (Lindahl, 1966; Knecht *et al.*, 1967; Cifonelli and King, 1972). While iduronic acid residues and *O*-sulfate groups are generally accumulated in N-sulfated regions of the polysaccharide chains (Höök *et al.*, 1974b; Linker and Hovingh, 1975; Cifonelli and King, 1977), *N*-acetylglucosamine residues located next to an N-sulfated disaccharide unit often carry a 6-*O*-sulfate group (Rosenberg and Lam, 1979; Lindahl *et al.*, 1983; Sanderson *et al.*, 1983; Bienkowski and Conrad, 1985). Heparan sulfates of low *O*-sulfate contents generally show a preponderance of iduronosyl 2-*O*-sulfate over glucosaminyl 6-*O*-sulfate groups (e.g., see Taylor *et al.*, 1973; Cifonelli and King, 1977).

TABLE I

DISACCHARIDE UNITS IN HEPARIN AND HEPARAN SULFATE[a]

A \ B	CH_2OH / $HNCOCH_3$		$CH_2OSO_3^-$ / $HNCOCH_3$		CH_2OH / $HNSO_3^-$		$CH_2OSO_3^-$ / $HNSO_3^-$		CH_2OH / OSO_3^- / $HNSO_3^-$		$CH_2OSO_3^-$ / OSO_3^- / $HNSO_3^-$	
	A→B	B→A	A→B	B→A	A→B	B→A	A→B	B→A	A→B	B→A	A→B	B→A
COO^- / OH / OH	+	+	+	+	+	+	+	+	+	−	+	−
COO^- / OH / OSO_3^-	−	−	−	−	+	(+)	+	(+)	−	−	−	−
COO^- / OH / OH	+	−	+	−	+	+	+	+	−	−	−	+
COO^- / OH / OSO_3^-	+	−	+	−	+	+	+	+	−	+	−	+

[a] The B monosaccharide units (horizontal display) are all D-glucosamine residues; the A units (vertical display) are D-glucuronic acid (upper two) and L-iduronic acid (lower two) residues. The various →UA→GlcN→ (A→B) and →GlcN→UA→ (B→A) disaccharide units that have been described in heparin or heparan sulfate are indicated by + (plus), whereas − (minus) denotes combinations that have not been found. The postulation of a particular disaccharide sequence is based either on actual isolation of the corresponding disaccharide derivative (following deamination or eliminase digestion of the polysaccharide), on sequence data pertaining to larger fragments, or on more circumstantial evidence such as the structures of biosynthetic polymeric intermediates. The two (A→B) units containing sulfated glucuronic acid have been reported to occur in heparin (Bienkowski and Conrad, 1985); at least one of the two reverse sequences, indicated by (+), is likely to exist but has not yet been identified. The →IdUA→ GlcNSO₃(3,6-di-OSO₃)→ sequence was recently identified in mollusc "heparin" (see the text). The table summarizes information from the following reports: Jacobsson et al. (1979b), Linker (1979), Jacobsson and Lindahl (1980), Sjöberg and Fransson (1980), Fransson et al. (1980), Delaney et al. (1980), Lindahl (1980), Casu et al. (1981), Hopwood and Elliott (1981), Weissmann and Chao (1981), Kosakai and Yosizawa (1982), Ototani et al. (1982), Atha et al. (1984), Lindahl et al. (1984), Linker and Hovingh (1984), Bienkowski and Conrad (1985). Abbreviations: UA, unspecified hexuronic acid; GlcUA, D-glucuronic acid; IdUA, L-iduronic acid; GlcN, 2-deoxy-2-amino-D-glucose (glucosamine); GlcNH₃, GlcNSO₃, GlcNAc—glucosamine residue with unsubstituted, sulfated and acetylated amino group, respectively. The location of O-sulfate groups is indicated in parentheses.

C. Interactions with Other Macromolecules

The biological effects and putative functions attributed to heparin-related polysaccharides generally involve binding of other macromolecules, mostly proteins. Such interactions range from precise lock-and-key fit, which depends not only on the density of the charged groups but also on their specific location and on the fine structure of the carbohydrate backbone, to cooperative electrostatic binding of fairly low specificity. The latter is generally promoted by a large molecular size of the polysaccharide and by high contents of sulfate and iduronic acid (see Lindahl and Höök, 1978); typically, the various glycosaminoglycans can be arranged in a series according to increasing affinity, heparin being the high-affinity extreme. Since the major topic of this chapter is polysaccharide biosynthesis, the functional aspects will be covered only briefly; instead, we will emphasize the role of polysaccharide structure, to be discussed later in relation to regulation of biosynthesis. Hence, interactions dependent on specific saccharide structure will be considered in some detail, less specific (or poorly characterized) effects only in brief. For more comprehensive reviews of older literature see Lindahl and Höök (1978), Comper (1981), and Höök et al. (1984).

The best-known example of lock-and-key binding involving a heparinlike compound is the interaction with antithrombin, a protease inhibitor which mediates the blood anticoagulant activity of the polysaccharide (see Björk and Lindahl, 1982). The antithrombin-binding region in heparin has been identified as a pentasaccharide sequence (Fig. 2) composed of two uronic acid and three glucosamine residues (Lindahl et al., 1980; Casu et al., 1981; Ototani and Yosizawa, 1981; Thunberg et al., 1982b; Atha et al., 1984, 1985; Lindahl et al., 1984). Of the two uronic acid units, one (B) appears to be invariably glucuronic acid, the other (D) iduronic acid. The nonreducing-terminal glucosamine residue (A) is mostly (but not always) N-acetylated, whereas the two remaining glucosamine units are N-sulfated. The iduronic acid residue D and the glucosamine units are O-sulfated at C-2 and C-6, respectively; again, 6-O-sulfation of the internal amino sugar unit C is not a constant finding. The 3-O-sulfate group on the latter unit is not present in other regions of the heparin molecule, and is thus a distinguishing structural feature of the antithrombin-binding pentasaccharide sequence. Proposed structure–function relationships imply that at least four of the sulfate groups, including the unique 3-O-sulfate, are essential to the heparin–antithrombin interaction and hence to the anticoagulant activity (Fig. 2). These proposals have been

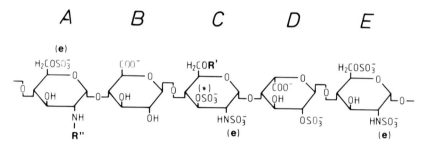

FIG. 2. Structure of the antithrombin-binding region in heparin. Structural variants are indicated by **R'** (-H or -SO₃⁻ in unit C) and **R"** (-SO₃⁻ or -COCH₃ in unit A). The 3-O-sulfate group marked by an asterisk (unit C) is unique to the antithrombin-binding region of the heparin molecule. In addition, each of the three sulfate groups indicated by (e) has been proved essential to the high-affinity binding of antithrombin. The functional roles of the two O-sulfate groups on units D and E are unknown.

confirmed by chemical synthesis of the pentasaccharide (Sinaÿ *et al.*, 1984; Petitou, 1984). For further details regarding the anticoagulant action, the reader is referred to reviews by Rosenberg (1977) and by Björk and Lindahl (1982). The 3-O-sulfated glucosamine residues found in heparan sulfate from cultured endothelial cells (Marcum *et al.*, 1986) and from basement membrane (Pejler *et al.*, 1987) likewise correlate with affinity for antithrombin. A recent characterization of the heparinlike polysaccharide from the mollusc *Anomalocardia brasiliana* showed an increase in affinity for (human!) antithrombin with increasing contents of the →GlcUA→GlcNSO₃(3,6-di-OSO₃)→ disaccharide sequence (G. Pejler, Å. Danielsson, I. Björk, U. Lindahl, and C. P. Dietrich, unpublished data; see also Jordan and Marcum, 1986), which is of critical importance to the interaction between mammalian heparin and antithrombin. Interestingly, the clam polysaccharide also contained significant amounts (~3% of total mass) of the sequence →IdUA→GlcNSO₃(3,6-di-OSO₃)→, previously not identified; this component did not correlate with affinity for antithrombin.

Characterization of heparin proteoglycans isolated from rat skin showed that the antithrombin-binding regions were distributed in a nonrandom fashion (Horner and Young, 1982; Jacobsson *et al.*, 1986; see Fig. 1). Only about one-third of the proteoglycan preparation had high affinity for antithrombin; in these molecules about half of the polysaccharide chains were of high-affinity type, whereas the remaining chains had low affinity. Structural characterization showed, on the average, three 3-O-sulfated glucosamine residues for each high-affinity chain, whereas all low-affinity chains, as expected, lacked any detectable 3-O-sulfate groups. The endo-β-D-glucuronidase that cleaves these chains, in the mast cell, to polysaccharide fragments ("heparin"; see Section II,A,1) requires both N- and O-sulfate groups, apparently in specific (but as yet undefined) configuration, for substrate recognition, and thus does not attack the β-glucuronidic linkage in the antithrombin-binding region (Thunberg et al., 1982a). The distribution of the latter regions and of the recognition sequences for the endoglucuronidase in the heparin proteoglycan will determine the proportion of fragments with high and with low affinity for antithrombin, obtained after completed degradation (Fig. 1).

Another interaction of potential functional importance is implied by the finding of Fedarko and Conrad (1986) that nuclei of hepatocytes accumulate a heparan sulfate with a high content of the unusual →GlcUA(2-OSO$_3$)→GlcNSO$_3$(6-OSO$_3$)→ disaccharide unit. No ligand interacting with this polysaccharide has yet been identified. It may be noted in this context that ligands other than proteins may be involved in specific interactions with polysaccharides. Fransson (1981) proposed that the so-called self-association of heparan sulfate chains involves "contact zones" which contain glucuronic acid and iduronic acid in approximately equal proportions. When these mixed sequences are replaced by iduronic acid-containing block structures, the interaction is abolished.

The common feature of the interactions discussed so far is the apparent dependence on a specific (albeit not always fully characterized) polysaccharide structure. Additional proteins with ability to bind heparinlike polysaccharides include various factors concerned with the hemostatic mechanism [coagulation enzymes (see Björk and Lindahl, 1982), the heparin-neutralizing proteins platelet factor 4 (Handin and Cohen, 1976; Jordan et al., 1982; Lane et al., 1984, 1986) and histidine-rich glycoprotein (Lijnen et al., 1983; Lane et al., 1986), heparin cofactor II (Tollefsen and Pestka, 1985)], the lipolytic enzyme lipoprotein lipase (Bengtsson et al., 1980; Cheng et al., 1981; Shimada et al., 1981), complement factors (Kazatchkine et al., 1981), connective-tissue matrix components [collagen (e.g., see Koda and Bernfield, 1984; Fujiwara et al., 1984), fibronectin (Yamada, 1983), laminin (Sakashita et al., 1980), vitronectin (Hayman et al., 1983)], and the plasma membrane components which interact with the polysaccharide chains of cell surface heparan sulfate proteoglycans (see Höök et al., 1984, for references). While it is entirely possible that some of these interactions may be as specific as that between heparin and antithrombin, no polysaccharide sequences have yet been implicated; in most cases binding is probably relatively nonspecific. Some of these interaction systems are likely to be *in vitro* artifacts without biological relevance. On the other hand, a low specificity per se has nothing to say regarding the functional importance of a polysaccharide–protein interaction. Consider, for instance, the binding of lipoprotein lipase to heparan sulfate, apparently an interaction of fairly low specificity (Bengtsson *et al.*, 1980). Nonetheless, the heparan sulfate proteoglycan has been attributed a vital function in anchoring lipoprotein lipase at the surface of vascular endothelial cells (Olivecrona *et al.*, 1977; Cheng *et al.*, 1981; Shimada *et al.*, 1981). The importance of the biosynthetic control mechanism that ensures adequate sulfation of the endothelial heparan sulfate is readily appreciated.

Also, the binding of heparan sulfate to cell surfaces is promoted by a high sulfate content of the polysaccharide (Kjellén *et al.*, 1980). Neoplastic transformation leads to a decrease in sulfation of the heparan sulfate produced by cultured cells (Keller *et al.*, 1980; Winterbourne and Mora, 1981; see also Höök *et al.*, 1984). Accordingly, heparan sulfate proteoglycans synthesized by rat hepatoma cells were unable to interact with these cells, contrary to the proteoglycans obtained from normal rat liver (Robinson *et al.*, 1984). The biological significance of the interaction between cell surface components and heparinlike polysaccharides is unclear. Such binding may possibly mediate some of the effects observed on addition of heparinlike polysaccharides to cultured cells, for instance, inhibition of growth of smooth muscle cells (Castellot *et al.*, 1984; Fritze *et al.*, 1985).

Finally, we may note some interactions of potential functional importance involving the core protein of heparan sulfate proteoglycans. The core may anchor proteoglycans to the cell surface, presumably by a hydrophobic portion intercalated in the lipid bilayer of the plasma membrane (Kjellén *et al.*, 1981; Norling *et al.*, 1981; Rapraeger and Bernfield, 1983). This hydrophobic region is probably missing from matrix proteoglycans as well as from those cell surface proteoglycans which interact with plasma membrane components via their polysaccharide chains (Höök *et al.*, 1984). A different and intriguing kind of interaction was described by Fransson *et al.* (1984), who found that the core protein of a heparan sulfate proteoglycan associated with the cell surface of fibroblasts was identical or closely similar to the transferrin receptor. The proteoglycan was able to bind transferrin *in vitro* and was also precipitable with some of the monoclonal antibodies available against the transferrin receptor.

III. OUTLINE OF THE BIOSYNTHETIC PROCESS

Most of the information available regarding the biosynthesis of heparin and heparan sulfate relates to experiments with cell-free systems. In particular, microsomal fractions from murine, heparin-producing mast cell tumors have been employed, and the results may therefore have less direct bearing on the formation of heparan sulfate. On the other hand, the biosynthesis of heparin and heparan sulfate clearly follow the same general pathway, and the main features of the mastocytoma system probably apply to both processes. Throughout this presentation the distinguishing features of either process will be emphasized as required.

An outline of heparin (or heparan sulfate) biosynthesis can be appropriately subdivided into three sections, dealing with (A) formation of the protein core and of the polysaccharide–protein linkage region, (B) polymerization of monosaccharide residues into polysaccharide chains, and (C) modification of the polysaccharide chains. This disposition is in accord with the chronological order of the different steps, and also with the experimental segregation of these steps that is obtained in cell-free systems. The concerted action and regulation of the various reactions, and the overall organization of the biosynthetic apparatus will be discussed largely in the subsequent sections.

A. Initial Glycosylation of the Protein Core

The polysaccharide–protein linkage region, a galactosyl-galactosyl-xylosyl trisaccharide sequence, in heparin and heparan sulfate proteoglycans is identical to that of the chondroitin sulfate proteoglycan (Fig. 1). The enzymes responsible for the biosynthesis of this sequence include a xylosyl transferase and two different galactosyl transferases, which have been studied in heparin-synthesizing (Grebner *et al.,*. 1966; Helting, 1971) as well as in chondroitin sulfate-synthesizing (Rodén, 1980) systems. Esko *et al.* (1985) described a Chinese hamster ovary cell mutant with a drastically decreased level of xylosyl transferase activity. This mutant was defective in synthesis of both chondroitin sulfate and heparan sulfate, providing evidence that the same xylosyl transferase is responsible for the initiation of chondroitin sulfate and heparan sulfate. An intriguing feature is the phosphorylation at C-2 of the xylose residue, which apparently occurs in the biosynthesis of both chondroitin sulfate (Oegema *et al.*, 1984) and heparan sulfate (Fransson *et al.*, 1985). It seems likely that also

the two galactose units of the linkage regions in different proteoglycans are synthesized by the same enzymes (Rodén, 1980). Formation of the linkage region is concluded by transfer of glucuronic acid to
the second galactose residue, catalyzed by a glucuronosyl transferase
which is different from that involved in the formation of the polysaccharide chain proper (Helting, 1972).

B. Polymer Formation

Experiments originally deigned by Silbert (1963) showed that polysaccharide chain formation occurred when a mastocytoma microsomal
fraction was incubated with UDP-N-acetylglucosamine and UDP-
glucuronic acid (see also Lindahl et al., 1973). Studies using well-defined oligosaccharides or carbohydrate–serine compounds (containing
the polysaccharide–protein linkage region) as sugar acceptors in the
glycosyl transferase reactions established a mechanism of polymerization involving stepwise transfer of D-glucuronic acid (Helting and Lindahl, 1971, 1972) and N-acetyl-D-glucosamine (Helting and Lindahl,
1972; Forsee and Rodén, 1981) units to the nonreducing terminus of
the growing polysaccharide chain (see also Rodén, 1980). The alternating sequence of the two different sugar residues in the product could
be attributed to the substrate specificities of the corresponding transferases, glucuronosyl transfer occuring only to sequences terminating
with N-acetylglucosamine, and vice versa. These findings do not support the involvement of intermediary lipid-linked disaccharides, such
as the glucuronosyl-N-acetylglucosaminyl-pyrophosphoryldolichol
found in lung fibroblasts (Turco and Heath, 1977). The mechanism of
chain elongation thus is similar to that operating in chondroitin sulfate biosynthesis (Rodén, 1980), but differs from the formation of
hyaluronic acid, which apparently occurs by the addition of monosaccharide units at the reducing end (Prehm, 1983).

The role of the protein core in the polymerization reaction is unclear.
β-D-Xylosides, which serve as efficient exogenous initiators of
chondroitin sulfate biosynthesis (Rodén, 1980), appeared relatively ineffective as primers for heparin (Robinson and Lindahl, 1981; Stevens
and Austen, 1982) or heparan sulfate (Hopwood and Dorfman, 1977;
Johnston and Keller, 1979; Sudhakaran et al., 1981). Addition of ben-
zyl β-D-xyloside to mouse mastocytoma cells, pretreated with cyclohex-
imide to inhibit biosynthesis of heparin and chondroitin sulfate proteoglycans, restored chondroitin sulfate synthesis (as single chains) to
levels well above those of control cultures, but had a marginal stimulatory effect on the formation of heparin polysaccharide (Robinson

and Lindahl, 1981). Instead, large amounts of heparinlike fragments (M_r 1500–5000) were formed, apparently due to a breakdown of the usual process of chain assembly. This result might suggest that the polymerization reaction in the biosynthesis of heparin is dependent on the presence of a core protein. On the other hand, an exogenous octasaccharide of the appropriate structure was capable of serving as a primer for polysaccharide formation (Forsee and Rodén, 1981). A covalent linkage between this primer sequence and the core protein therefore was not a prerequisite to polymerization. The role of the core protein becomes particularly intriguing in view of the findings, cited in Section II,A,2, of hybrid proteoglycans containing both heparan sulfate and chondroitin sulfate chains bound to the same core. Since the linkage region to protein is identical to both polysaccharides, the committing step toward the formation of either heparin–heparan sulfate or chondroitin sulfate would be the incorporation of the first hexosamine moiety, N-acetylglucosamine and N-acetylgalactosamine, respectively, in each type of chain. Interestingly, the N-acetylgalactosaminyl transferase which catalyzes this particular step in chondroitin sulfate biosynthesis appears to differ from that involved in the formation of the more peripheral portions of the polysaccharide chain (Rohrmann et al., 1985).

Solubilization of the microsomal membranes in 1% Triton X-100 abolished polymerization on both endogenous (J. Riesenfeld and U. Lindahl, unpublished observation) and exogenous (Forsee and Rodén, 1981) primers, and led to solubilization of active glucuronosyl and N-acetylglucosaminyl transferases. Polymerization—that is, the concerted action of the two enzymes—thus occurs only when the enzymes are in the membrane-bound state.

Little is known regarding the mode of action of the membrane-bound polymerase complex. Previous studies by Richmond et al. (1973) on the biosynthesis of chondroitin sulfate (in a microsomal system from chick cartilage) suggested that the formation of an individual polysaccharide chain may be completed within a few minutes. More recent studies demonstrated that the rate of formation of heparin precursor polysaccharide (rate of incorporation of ^{3}H from UDP-[^{3}H]glucuronic acid into mastocytoma microsomal polysaccharide) could be modulated by varying the concentration of UDP-N-acetylglucosamine (K. Lidholt, J. Riesenfeld, D. S. Feingold, and U. Lindahl, unpublished data). The size of the newly formed polysaccharide chains was directly related to the rate of polymerization. These results agree with the notion that the assembly of an individual polysaccharide chain occurs during a defined period of time.

C. Polymer Modification

The products of the polymerization reaction are proteoglycans with polysaccharide chains composed of $[\rightarrow GlcUA\beta \overset{1,4}{\rightarrow} GlcNAc\alpha \overset{1,4}{\rightarrow}]_n$ sequences (in the following denoted PS-NAc). Several modifications of these chains are required before they have assumed the structures typical of a completed heparin or heparan sulfate molecule. Attempts to characterize this process were first made by Silbert (1967a,b), who found that the nonsulfated polysaccharide produced by a mastocytoma microsomal fraction (by incubation with the appropriate UDP sugars) was converted into a sulfated product resembling heparin when 3'-phosphoadenylylsulfate (PAPS) was added to the incubations. Further work in our laboratory focused on the structural characterization of polysaccharide products formed during this process (Lindahl et al., 1973; Höök et al., 1974a, 1975; Jacobsson and Lindahl, 1980; see also reviews by Lindahl et al., 1977, 1986; Rodén, 1980; Feingold et al., 1981; Gallagher et al., 1986). Anion-exchange chromatography of mastocytoma microsomal polysaccharide, produced in the absence of PAPS, showed two peaks, corresponding to the initial product of the polymerization reaction (PS-NAc) and its partially N-deacetylated derivative (PS-NH$_3$; Fig. 3). Following the addition of PAPS to a microsomal fraction containing such preformed polysaccharide, the two nonsulfated components were progressively replaced by three sulfated products (PS-NSO$_3$, PS-N/O-SO$_3$-a, and PS/N/O-SO$_3$-b). The results of a structural analysis of these components could be interpreted in terms of a biosynthetic scheme, involving the following reactions (in chronological order of occurrence; Fig. 3): deacetylation of N-acetylglucosamine residues in PS-NAc (yielding the product PS-NH$_3$); sulfation of the resulting unsubstituted amino groups (PS-NSO$_3$); epimerization at C-5 of D-glucuronic acid residues to yield L-iduronic acid units, and 2-O-sulfation of the latter units (PS-N/O-SO$_3$-a); O-sulfation of glucosamine residues at C-6 (PS-N/O-SO$_3$-b). Additional O-sulfation at C-3 of glucosamine (Riesenfeld et al., 1983b) and at C-2 of glucuronic acid (M. Kusche, L.-G. Oscarsson, R. Reynertson, L. Rodén, and U. Lindahl, unpublished data) units has been demonstrated in the mastocytoma microsomal system using appropriate oligosaccharide substrates (see also Section IV,A,3).

Further studies on the kinetics and extent of polymer modification, using similar experimental systems, afforded the following major results and conclusions:

1. Throughout the modification process the microsomal polymeric intermediates remained at constant elution positions (Fig. 3); the rela-

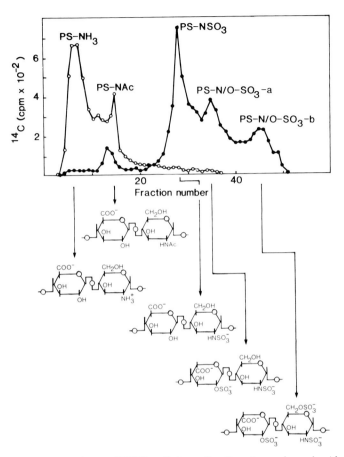

FIG. 3. Chromatography on DEAE–cellulose of radioactive polysaccharide formed upon incubation of a mastocytoma microsomal fraction with UDP-[14C]glucuronic acid and UDP-N-acetylglucosamine. Products were formed in the absence of PAPS (○), or in the presence of PAPS added to the microsomal fraction containing preformed, nonsulfated [14C]polysaccharide (●). The product of the initial polymerization reaction (PS-NAc) is composed of the disaccharide unit shown immediately below this fraction (at the top of the structural scheme). The structures indicated further down this scheme define in sequential order the changes introduced in the subsequent consecutive modification reactions; the elution positions of the corresponding intermediates are indicated by arrows. Modified from I. Jacobsson, Ph.D. thesis, The Swedish University of Agricultural Sciences, Uppsala (1979).

tive amounts of the various intermediates changed, but without any accumulation of material of intermediate charge density. The reactions thus proceed in an ordered fashion through several discrete stages. The stepwise nature of the process is reflected by the structures of the intermediates; for instance, PS-NSO$_3$ is N-sulfated to the same extent as the final product but is essentially devoid of O-sulfate groups, suggesting that the N-sulfation reaction is terminated before O-sulfation is initiated. Similar intermediates were found in studies on heparan sulfate biosynthesis using microsomal fractions from rat liver (Riesenfeld *et al.*, 1982b) or from a mouse sarcoma (Silbert and Baldwin, 1984), in accord with the notion that heparin and heparan sulfate are synthesized by closely related mechanisms.

2. The most retarded, fully sulfated product (PS-N/O-SO$_3$-b) could be detected after only 30 sec of incubation with PAPS. Polymer modification is thus a rapid process, the individual polysaccharide chain passing through the complete series of reactions within a period of seconds. Hence, the enzymatic apparatus will accommodate only a limited number of proteoglycan substrate molecules at a time, the process being continuously repeated with new precursor.

3. Structural analysis of the various intermediates revealed that polymer modification is incomplete, in the sense that most of the reactions involved will engage only a fraction of the potential target residues. This fundamental property of the biosynthetic mechanism provides the very basis for the elaboration of polysaccharide sequences of different structure, and hence is responsible not only for the structural difference between heparin and heparan sulfate but also for the heterogeneity observed within purified polysaccharide preparations.

The regulation of individual polymer modification reactions and of their concerted action will form a recurring theme throughout the following discussion. However, before considering these aspects some pertinent properties of the individual reactions will be reviewed.

1. DEACETYLATION OF *N*-ACETYLGLUCOSAMINE RESIDUES

During heparin biosynthesis in the cell-free mastocytoma system, as outlined above, the N-deacetylation reaction takes place in two distinct steps. The initial product of the polymerization reaction (PS-NAc) is converted into a partially N-deacetylated derivative (PS-NH$_3$) in which ~50% of the amino groups are unsubstituted (Höök *et al.*, 1975). Subsequently, in the course of sulfation half of the remaining *N*-acetyl groups are removed and replaced by *N*-sulfate groups, the latter substituents promoting further N-deacetylation (Riesenfeld et $al.$, 1982a). The less N-deacetylated intermediate, PS-NH$_3$, is formed also when sulfation is initiated at the same time as the polymerization reaction (by including PAPS along with the UDP sugars in the incubation mixtures; J. Riesenfeld and U. Lindahl, unpublished observation), suggesting that the stepwise N-deacetylation is not necessarily an artifact due to experimental segregation of the two processes. The difference between heparin and heparan sulfate with regard to the contents and distribution of N-acetylated disaccharide units (see Section II,B) must reflect differences in the extent and mode of biosynthetic N-deacetylation of the corresponding precursor proteoglycans. The regular occurrence in both polysaccharides of at least one N-acetylated glucosamine unit immediately adjacent to the polysaccharide–protein linkage region (Lindahl, 1966; Cifonelli, 1968; Cifonelli and King, 1972) suggests that this portion of the proteoglycan molecule is inaccessible to the N-deacetylase (Fig. 1).

A simple, single-vial assay for the N-deacetylase has been developed (Navia et $al.$, 1983), based on the release of [^{3}H]acetic acid from the chemically N-^{3}H-acetylated intermediate, PS-NH$_3$ (Riesenfeld et $al.$, 1980). Conveniently, a bacterial polysaccharide (isolated from $Esche$-$richia$ $coli$ K5) was found to have a structure identical to that of the herapin precursor, PS-NAc, thus providing a more easily available source of substrate for the assay. The labeled N-acetyl groups required for the assay were introduced into the K5 polysaccharide by reaction with [^{3}H]acetic anhydride, following partial N-deacetylation by hydrazinolysis (Navia et $al.$, 1983). The mastocytomal deacetylase requires manganese ions for full activity and has a pH optimum around 6.5. It is readily solubilized from the microsomal membranes using the nonionic detergent Triton X-100 (J. Riesenfeld and U. Lindahl, unpublished observation), but has not yet been purified.

2. N- AND O-SULFATION REACTIONS

Enzymatic N-sulfation of various N-desulfated, heparin-related polysaccharides has been demonstrated in a number of cell-free systems, and partial purification of N-sulfotransferases has been reported (see Rodén, 1980, for an account of previous studies, and Göhler et $al.$, 1984, for a more recent report). However, progress in this area has been slow, and has in particular been hampered by the lack of suitable assay procedures for the various sulfotransferases. Assays that distinguish between N- and O-sulfation, using N-desulfated heparin and heparan sulfate, respectively, as sulfate acceptors in incubations with [^{35}S]PAPS, were used in a preliminary study of the basic kinetic properties of enzymes solubilized from a mastocytoma microsomal fraction (Jansson *et al.*, 1975). Differences in metal ion requirements, susceptibility to heat inactivation, and inhibition by salt strongly suggested that the *N*- and *O*-sulfate transfer reactions should be ascribed to different enzymes.

Moreover, incorporation of *O*-[^{35}S]sulfate groups into the two major different positions, at C-2 of iduronic acid and C-6 of glucosamine residues, could be demonstrated by identification of disaccharides obtained after deaminative cleavage of labeled polysaccharide with nitrous acid (Jacobsson and Lindahl, 1980). In a similar fashion, 3-*O*-sulfation of glucosamine units was established using an *O*-desulfated decasaccharide (derived from the antithrombin-binding region of heparin) as sulfate acceptor (Riesenfeld *et al.*, 1983b). Unfortunately, the laborious degradation procedures involved in these experiments have precluded the development of specific assays suitable for routine analysis of the various *O*-sulfotransferases.

3. Uronosyl C-5-Epimerization

All the available evidence suggests that epimerization of D-glucuronic acid residues to L-iduronic acid units occurs after N-sulfation but before the O-sulfation reactions (for further information see Section IV,A,2). More specifically, most of the iduronic acid units become 2-O-sulfated in direct association with their formation from glucuronic acid residues, prior to the concluding 6-O-sulfation reaction (Jacobsson and Lindahl, 1980).

The mechanism of the actual epimerization reaction was elucidated by determining the fate of the hydrogen atom at C-5 of susceptible hexuronic acid residues (Lindahl *et al.*, 1976; Jacobsson *et al.*, 1979a; Prihar *et al.*, 1980; Jensen *et al.*, 1983). Incubation of polysaccharide substrates containing 5-^{3}H-labeled uronic acid residues with the microsomal epimerase resulted in release of ^{3}H into the medium. Conversely, incubation of unlabeled polysaccharides (of the appropriate structure; see Section IV,A,2) with ^{3}H$_2$O led to ^{3}H incorporation at C-5 of both glucuronic and iduronic acid residues. These results indicated a freely reversible reaction mechanism involving abstraction of a proton from C-5, followed by readdition of a proton from the medium to the resulting carbanion intermediate, in either the D-glucuronosyl or L-iduronosyl configuration. The equilibrium favored retention of the former configuration, as reflected by a glucuronosyl/iduronosyl ratio of approximately 4:1 (Prihar *et al.*, 1980; Jensen *et al.*, 1983).

These findings raised a question as to the role of 2-O-sulfation in relation to the formation of iduronic acid residues (Jacobsson et al., 1979a). O-Sulfation of iduronic acid units yields structures which are not recognized as substrate by the epimerase (see Section IV,A,2) and which therefore would be withdrawn from the D-glucuronosyl–L-iduronosyl equilibrium. A promoting effect of O-sulfation on iduronic acid formation was indeed verified in experiments using the polysaccharide intermediate PS-NSO$_3$ as an *exogenous* (soluble) substrate for microsomal enzymes (Jacobsson et al., 1979a). The addition of PAPS to such incubations resulted in appreciable O-sulfation of this substrate, and at the same time approximately doubled the yield of iduronic acid. However, while these effects were established over a period of hours, it is recalled that the processing of an individual *endogenous* microsomal (membrane-bound) polysaccharide molecule through the corresponding modification reactions occurs within less than 1 min. Indeed, analysis of the endogenous O-sulfated product, PS-N/O-SO$_3$, showed that whereas the iduronic acid residues, as expected, had lost the ^{3}H label at C-5 initially incorporated via UDP-[5-^{3}H]glucuronic acid, the glucuronic acid units retained the label quantitatively throughout the modification process (Jacobsson et al., 1979a). This result showed that with the endogenous polysaccharide substrate epimerization is never allowed to approach equilibrium. Instead, only a fraction of the potentially susceptible D-glucuronic acid residues is attacked by the epimerase, each attack leading to the formation of an L-iduronic acid unit which then remains stable. Likewise, once the spared glucuronic acid residues have escaped epimerization they remain stable. Nevertheless, when polysaccharide containing such seemingly unreactive units were extracted from the microsomal membranes and were then reintroduced to the microsomal enzyme system as an *exogenous* substrate, those glucuronic acid residues that satisfied the substrate specificity of the epimerase were readily attacked by the enzyme (as evidenced by the release of ^{3}H). In the intact biosynthetic system the association between the C-5 epimerase and its polysaccharide substrate thus appears to be transient, and the resulting reaction is, in effect, irreversible. The precise interrelation between the epimerization and 2-O-sulfation of uronic acid residues remains elusive and, at present, it can only be stated with certainty that the two processes are strongly interconnected.

The release of ^{3}H from 5-^{3}H-labeled uronic acid residues in polysaccharides of appropriate structure was utilized to develop a simple assay procedure for the epimerase (Jacobsson et al., 1979a; Jensen et al., 1983). The [3]H$_2$O formed may be quantified either after separation from incubation mixtures by distillation or, more conveniently, by a single-vial adaptation of the procedure involving partition of reaction products in a biphasic scintillation system (Campbell *et al.*, 1983). Enzyme activity was found to be strongly dependent on the ionic strength of the medium; however, no requirement for specific metal ions was observed. The pH optimum was close to 7.4. A substantial proportion of the total epimerase activity of the mouse mastocytoma was found in a high-speed supernatant fraction, and the enzyme has been purified approximately 8000-fold from this source (Malmström *et al.*, 1980).

IV. REGULATION OF POLYMER MODIFICATION

The structures of polysaccharide chains reflect the combined action of biosynthetic polymerization (glycosyl transfer) and polymer modification reactions. As already noted in previous sections, the latter reactions generally do not proceed to completion, and the final product therefore contains regions which remain at lower levels of modification. Such regions may be included in highly specific structures of functional importance, such as the antithrombin-binding sequence in heparin (see Section II,C). We thus impinge on a central problem of regulation: by which mechanisms are the target units selected in the various modification reactions? Due to the stepwise nature of polymer modification, the polysaccharide substrate for a given reaction will be the product of the preceding reaction. Hence, the substrate specificities of the various enzymes will determine how the corresponding reactions will be influenced by modifications introduced in previous steps. With these relationships in mind we will discuss (A) the substrate specificities of the various modification enzymes; (B) the potential regulatory role of the initial modification (i.e., the deacetylation of *N*-acetylglucosamine residues), and (C) the selection of target groups for modification.

A. Substrate Specificity in Polymer Modification Reactions

1. N-DEACETYLATION AND N-SULFATION OF GLUCOSAMINE RESIDUES

The *N*-deacetylase that initiates the series of polymer modification reactions is specific for the polysaccharide precursor of heparin–

heparan sulfate, and will thus recognize sequences with the structure $[\rightarrow GlcUA\beta\overset{1,4}{\rightarrow}GlcNAc\alpha\overset{1,4}{\rightarrow}]_n$. The related polysaccharide, hyaluronic acid $([\rightarrow GlcUA\beta\overset{1,3}{\rightarrow}GlcNAc\beta\overset{1,4}{\rightarrow}]_n)$, is composed of the same monosaccharide components but is resistant to the enzyme, obviously by virtue of its differently positioned glycosidic linkages (Riesenfeld et al., 1980). Since all (except the terminally located) N-acetylglucosamine residues in the heparin precursor PS-NAc are structurally equivalent, an exogenous, soluble substrate will be attacked in a random fashion (Riesenfeld et al., 1980). Such a reaction is self-limiting and comes to a stop when about one-third of the N-acetyl groups have been released. While the introduction of N-sulfate groups into the partially N-deacetylated product will promote further deacetylation (Riesenfeld et al., 1982a; see also Section III,C,1), the precise mechanism behind this effect is unknown. For instance, it has not been established whether a $\rightarrow GlcNSO_3\rightarrow GlcUA\rightarrow GlcNAc\rightarrow$ structure is preferred over the reverse sequence as a substrate for the deacetylase (Table II).

N-Deacetylation is, for obvious reasons, prerequisite to N-sulfation of glucosamine residues. Apart from a possible inhibitory effect of adjacent iduronic (instead of glucuronic) acid residues (Göhler et al., 1984; Table II), there is little direct information regarding the substrate recognition properties of N-sulfotransferases. The structural properties of, in particular, heparan sulfate indicate a highly versatile N-sulfation process, which provides N-sulfated disaccharide units both in extended sequences and as isolated elements surrounded by N-acetylated block structures (Table II). N-Unsubstituted glucosamine residues thus appear to become N-sulfated regardless of the N-substituent pattern of adjacent structures. The efficiency of this process is reflected by the very low proportion of glucosamine residues bearing free amino groups in the final biosynthetic product, heparin or heparan sulfate. In fact, N-sulfation is unique among the polymer modification reactions in that the involvement of available substrate groups is virtually quantitative.

2. Uronosyl C-5 Epimerization

Experiments aimed at elucidating the substrate specificity of the epimerase were designed to identify susceptible uronic acid residues by their ability to release C-5-^{3}H atoms (from prelabeled polysaccharide), or to incorporate such atoms from ^3H_2O, during incubation with the enzyme (Jacobsson et al., 1979a, 1984; Jensen et al., 1983). The results of such studies are summarized in Table II. The N-sulfated intermediate, PS-NSO$_3$, was found to be a substrate for the enzyme,

TABLE II

Substrate Specificity in Polymer Modification Reactions

Enzyme	Accepted as substrate[a]	Rejected as substrate	References and comments
N-Acetyl-D-glucosaminyl deacetylase	→GlcNAc→GlcUA→GlcNAc→ →GlcN→GlcUA→GlcNAc→ 　　$\mid$ 　　SO_3 and/or →GlcNAc→GlcUA→GlcN→ 　　　　　　　　$\mid$ 　　　　　　　　SO_3		Riesenfeld et al. (1980), (1982a). While the deacetylation of N-acetylglucosamine units is stimulated by adjacent N-sulfate groups, the exact relation between the promoting and the target units has not been established. The effects of iduronic acid or O-sulfate residues on substrate recognition by the N-deacetylase have not been studied.
D-Glucosaminyl N-sulfotransferase	→GlcN→GlcUA→GlcNH_3→GlcUA→GlcN→ 　$\mid$　　　　　　　　　　　$\mid$ 　R　　　　　　　　　　　R GlcUA→GlcNH_3→UA→aMan	IdUA→GlcNH_3→UA→aMan	The target glucosamine unit (in the middle) may become N-sulfated when the adjacent glucosamine residues (GlcNR) are either N-acetylated, N-unsubstituted, or N-sulfated (K. Lidholt, J. Riesenfeld, and U. Lindahl, unpublished data). Göhler et al. (1984). aMan = 2,5-anhydro-D-mannose.
Uronosyl C-5 epimerase	→GlcN→UA→GlcN→ 　$\mid$　　　　$\mid$ 　SO_3　　　SO_3	→GlcNAc→UA→GlcNAc→	Jacobsson et al. (1979a), (1984); Jensen et al. (1983). The target UA residues may be either

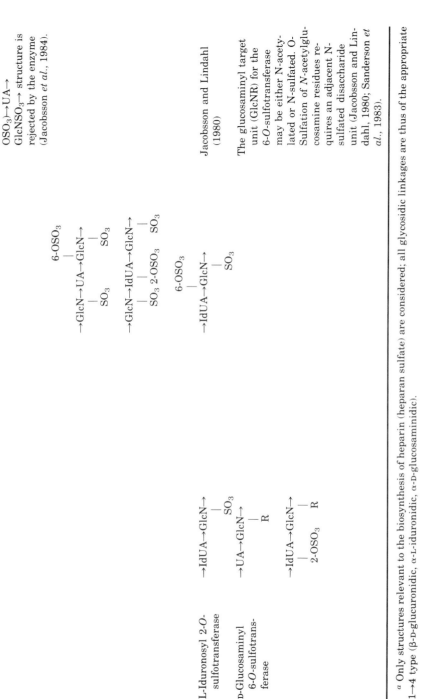

iduronic acid. Circumstantial evidence suggests that also a →GlcNSO$_3$(6-OSO$_3$)→UA→ GlcNSO$_3$→ structure is rejected by the enzyme (Jacobsson et al., 1984).

Jacobsson and Lindahl (1980)

The glucosaminyl target unit (GlcNR) for the 6-O-sulfotransferase may be either N-acetylated or N-sulfated. O-Sulfation of N-acetylglucosamine residues requires an adjacent N-sulfated disaccharide unit (Jacobsson and Lindahl, 1980; Sanderson et al., 1983).

L-Iduronosyl 2-O-sulfotransferase

D-Glucosaminyl 6-O-sulfotransferase

^{a} Only structures relevant to the biosynthesis of heparin (heparan sulfate) are considered; all glycosidic linkages are thus of the appropriate 1→4 type (β-D-glucuronidic, α-L-iduronidic, α-D-glucosaminidic).

whereas the N-acetylated precursor, PS-NAc, was not, demonstrating that N-sulfate groups are required for substrate recognition. This requirement was confirmed and qualified in experiments using polysaccharide substrates with artificial, mixed N-sulfate/N-acetyl patterns; a $\rightarrow$GlcNSO$_3$$\rightarrow$GlcUA$\rightarrow$GlcNAc$\rightarrow$ sequence was attacked by the enzyme, whereas the reverse structure, $\rightarrow$GlcNAc$\rightarrow$GlcUA$\rightarrow$GlcNSO$_3$$\rightarrow$, was resistant. This conclusion is in accord with established structural features of heparin and heparan sulfate. The sequence $\rightarrow$GlcNSO$_3$ $\rightarrow$IdUA$\rightarrow$GlcNAc$\rightarrow$ thus has been conclusively identified in both polysaccharides (Lindahl, 1966; Höök et al., 1974a; Linker, 1979; Rosenberg and Lam, 1979; Hopwood and Elliott, 1981; Bienkowski and Conrad, 1985), whereas $\rightarrow$GlcNAc$\rightarrow$IdUA$\rightarrow$GlcNSO$_3$$\rightarrow$ has not been found (however, see Linker, 1979; Hopwood and Elliott, 1981). Further, O-sulfate groups on adjacent glucosamine residues or on the potential target (iduronic acid) unit itself were inhibitory to epimerase action (Table II). These relationships are illustrated in Fig. 4, which shows the processing of a hypothetical heparin nonasaccharide sequence through the series of polymer modification reactions. The susceptibility of individual uronic acid residues to the epimerase is indicated at various levels of modification. The scheme illustrates the critical importance of N-sulfation for epimerase activity, the selective inhibitory effect of residual N-acetyl groups, and the blocking of target residues by subsequent O-sulfation of these or of adjacent units. It is tempting to speculate on a regulatory role of O-sulfate groups, in locking the C-5 configuration of uronic acid residues of intermediates that are past the epimerization stage. However, as pointed out previously (Section III,C,3), such a role appears unlikely in view of the transient association between the C-5 epimerase and its substrate.

3. O-SULFATION

Structural analysis of heparin and heparan sulfate has shown that the occurrence of O-sulfate groups is essentially restricted to the N-sulfated regions of the polysaccharides (Section II,B). The importance of previously incorporated N-sulfate groups for substrate recognition by the O-sulfotransferases was demonstrated more directly by incubating appropriate polysaccharides (heparan sulfate or partially O-desulfated heparin) with a mastocytoma microsomal fraction in the presence of [35S]PAPS (Jacobsson and Lindahl, 1980). Characterization of the resulting labeled polysaccharides showed the occurrence of O-[35S]sulfate groups in the vicinity of N-sulfated glucosamine residues, but not in the internal regions of N-acetylated block sequences.

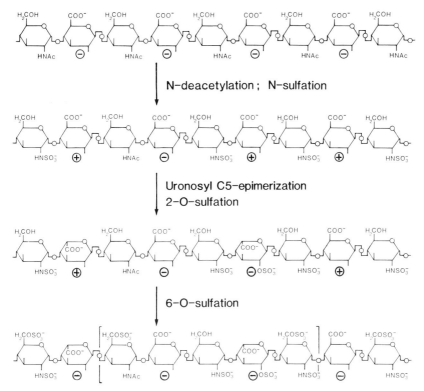

FIG. 4. Sequence of polymer modification reactions involved in the biosynthesis of a hypothetical nonasaccharide sequence in a heparin molecule. The substrate specificity of the uronosyl C-5 epimerase is illustrated by indicating uronic acid residues that are susceptible to attack by the enzyme (+) and those that are not (−). Note the glucuronosyl unit (next to the reducing-terminal glucosamine residue) that remains unmodified in spite of being a potential substrate for the enzyme. The pentasaccharide sequence within brackets corresponds to the antithrombin-binding region, except for a 3-O-sulfate group on the internal glucosamine residue which has not been included.

This constraint is readily explained as regards the iduronosyl 2-O-sulfate groups, since the substrate specificity of the uronsyl C-5 epimerase requires that at least one of the glucosamine units adjacent to the target glucuronic acid carry an N-sulfate group (Section IV,A,2). While N-acetylated glucosamine units may apparently become 6-O-sulfated (see, e.g., Rosenberg and Lam, 1979; Linker, 1979; Lindahl *et al.*, 1983; Bienkowski and Conrad, 1985), this occurs only in juxtaposition to an N-sulfated disaccharide unit (Jacobsson and Lindahl, 1980; Sanderson *et al.*, 1983). Moreover, 2-O-sulfation of the iduronic acid residue in a →IdUA→GlcNSO$_3$→ structure is precluded when the glucosamine unit carries a 6-O-sulfate group, whereas, in contrast, 6-O-sulfation of the latter unit may take place regardless of whether the iduronic acid is sulfated or not (Jacobsson and Lindahl, 1980). The substrate specificity of the enzyme that catalyzes the critical 3-O-sulfation of one of the glucosamine residues in the antithrombin-binding region (see Section II,C) has not yet been defined in detail. However, in recent experiments using the pentasaccharide, $GlcNSO_3$ (6-OSO_3)→GlcUA→$GlcNSO_3$(6-OSO_3)→IdUA(2-OSO_3)→$GlcNSO_3$(6-OSO_3), as exogenous sulfate acceptor with mastocytoma microsomal enzymes, 3-O-sulfation of the internal glucosamine unit was found to conclude the formation of a functional antithrombin-binding sequence (M. Kusche, U. Lindahl, M. Petitou, and J. Choay, unpublished data). Since the 3-O-sulfate group appears to be strictly confined to the anti-thrombin-binding region of mammalian heparin, this result suggests that other components of the pentasaccharide sequence, also essential to the interaction with antithrombin, may be required for substrate recognition by the 3-O-sulfotransferase. Finally, O-sulfation of glucuronic acid units (presumably at C-2; see Bienkowski and Conrad, 1985) has been demonstrated with microsomal polysaccharide (M. Kusche and U. Lindahl, unpublished data) as well as with exogenous oligosaccharides as sulfate acceptor. Double-isotope labeling of the microsomal polysaccharide (see Jacobsson *et al.*, 1979a) confirmed that the sulfated glucuronic acid units were indeed formed through a distinct sulfotransferase reaction and not by "back-epimerization" of sulfated iduronic acid residues.

   The accumulated information regarding the substrate specificities of the enzymes involved in the biosynthetic polymer modification reactions has been compiled in Table II. The modification of a polysaccharide precursor introduced in each of these reactions may be either prerequisite to, compatible with, or inhibitory to other reactions. The functional significance of these relationships is determined by the sequential order of the reactions, as illustrated in Figs. 5 and 6. N-Sulfation is prerequisite to all subsequent modification reactions, either directly, through the requirement for N-sulfate groups in the glucuronosyl C-5 epimerization and glucosaminyl 6-O-sulfation reactions, or indirectly, through the requirement for iduronic acid acceptor units in the 2-O-sulfotransferase reaction [so far, also 2-O-sulfated glucuronic acid residues have been found only adjacent to N-sulfated glucosamine units (Bienkowski and Conrad, 1985)]. The extent of N-sulfation will therefore set the upper limit to the potential overall polymer modification, and will also influence the location in the polysaccharide chain of iduronic acid residues and O-sulfate groups. This

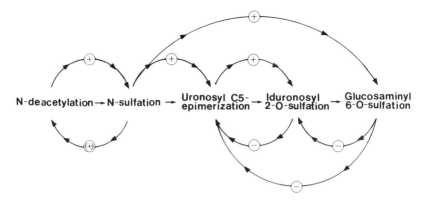

Fig. 5. Scheme of potential functional interrelationships between polymer modification reactions. The curved arrows indicate whether the structural modification introduced in a particular reaction is prerequisite (+), just stimulatory [(+)], or inhibitory (−) to other reactions. For further information see the text.

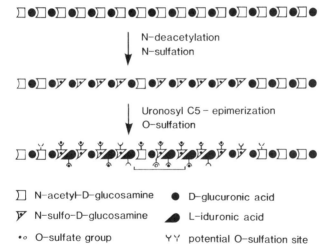

☐ N–acetyl–D–glucosamine   ● D–glucuronic acid

▼ N–sulfo–D–glucosamine   ◢ L–iduronic acid

•○ O–sulfate group   Y Y̌ potential O–sulfation site

Fig. 6. Scheme of polymer modification reactions illustrating the directory effect of N-sulfate groups on all subsequent reactions. The upper sequence shows the product of the initial polymerization reaction. The symbols are chosen to illustrate the substrate specificity of the uronosyl C-5 epimerase in relation to the N-substituents of glucosamine residues (the reducing terminus of the sequence is to the right). O-Sulfation occurs only within N-sulfated regions or (for 6-O-sulfation of N-acetylglucosamine units) within one disaccharide from the nearest N-sulfate group (probably only one of the two N-acetylglucosamine residues shown with potential but unoccupied 6-O-sulfate sites may actually become sulfated). Polymer modification is incomplete; in each reaction (with the possible exception of N-sulfation), a fraction of the potential target residues escape modification. The pentasaccharide sequence indicated by the horizontal bar corresponds to the antithrombin-binding region; the unique 3-O-sulfate group has been indicated (small open circle).

is in contrast to the inhibitory effects of 2- and 6-O-sulfate groups, which are restricted to reactions that precede the incorporation of each of these groups, and thus would not seem to exert any regulatory function in the intact biosynthetic system (see the discussion on the relationship between uronosyl C-5 epimerization and 2-O-sulfation in Sections III,C,3 and IV,A,2).

B. Regulation of the
N-Acetylglucosaminyl Deacetylase

The importance of N-sulfate groups in polymer modification points to a key regulatory role for the N-deacetylase which creates the target sites for N-sulfation. Without deacetylation, the initial polysaccharide product will be unable to continue through the series of reactions (N-sulfation, uronosyl C-5 epimerization, and O-sulfation) which leads to the fully completed heparin or heparan sulfate molecule. Moreover, the regulation of deacetylase activity may be the prime factor in determining the final distribution of extensively modified, highly sulfated, as opposed to unmodified regions in such molecules (Fig. 6). The mode of regulation of N-deacetylation is essentially unknown. Since N-sulfation promotes further release of N-acetyl groups, the concerted action of the two processes (Höök et al., 1975; Riesenfeld et al., 1982a,b) may in itself involve an element of regulation.

Direct evidence for the modulation of N-deacetylase activity in heparin biosynthesis was obtained in recent studies on the effect of n-butyrate on cultured murine mastocytoma cells (Jacobsson et al., 1985). Heparin synthesized in the presence of 2.5 mM butyrate displayed an N-acetyl:N-sulfate ratio of ~1:3, as compared to ~1:9 for the corresponding control material produced in the absence of butyrate. Furthermore, this highly N-acetylated (and less N-sulfated) preparation showed a 3-fold larger proportion (54% vs 17% for the control) of components with high affinity for antithrombin. A similar stimulatory effect of butyrate on the biosynthesis of antithrombin-binding heparan sulfate was recently observed by Marcum et al. (1986) in experiments with cloned aortic endothelial cells. The presence of butyrate thus leads to an inhibition of the N-deacetylation/N-sulfation process in heparin biosynthesis, along with an augmented formation of the specific antithrombin-binding region. In accordance with these results, Heifetz and Prager (1981) previously reported that butyrate decreased the sulfation of heparin-related polysaccharide in human kidney tumor cells. The rationale for performing these experiments was the well-established inhibitory action of butyrate on histone deacetylation, expressed (presumably via hyperacetylation of histones) by a multitude of effects in various cellular systems (ranging from modulation of the general growth characteristics to specific changes of the phenotypic expression of the cells; see Jacobsson *et al.*, 1985, for references). However, the effect of butyrate on heparin biosynthesis could not simply be ascribed to direct inhibition of the N-deacetylase, since assays for mastocytoma microsomal deacetylase activity were only marginally affected even at butyrate concentrations much higher than those found to modulate heparin biosynthesis in intact mastocytoma cells (Jacobsson *et al.*, 1985). Moreover, polysaccharide formed on incubating the microsomal fraction with nucleotide sugars and PAPS was extensively deacetylated, irrespective of whether butyrate was present or not. Finally, the effect on heparin N-deacetylation/N-sulfation in the cultured cells was observed only after preincubation with butyrate for 10–20 hr. Taken together, these findings suggest that the expression of modulated N-deacetylation depends somehow on the metabolism of the intact cell. It is of interest to note in this context that heparan sulfate produced *in vivo* or *in vitro* by hepatocytes from diabetic rats showed a higher *N*-acetyl and hence a lower *N*-sulfate content than the control polysaccharide derived from normal cells (Kjellén *et al.*, 1983; E. Unger and L. Kjellén, unpublished observation). It is tempting to correlate this finding with the increased triglyceride degradation which is prevalent in diabetes, and which leads to increased formation of low molecular weight fatty acids in the liver.

These findings may relate further to the control of N-deacetylation in heparan sulfate biosynthesis. Cultured rat hepatocytes produce a typical heparan sulfate, in which N-acetylated glucosamine residues, amounting to approximately half of the total glucosamine, occur both as isolated units and in more extended block sequences (Oldberg *et al.*, 1977). Yet the polysaccharide produced on incubating a microsomal fraction from the same cells with the appropriate UDP sugars and PAPS was extensively N-deacetylated, retaining only ~25% of the *N*-acetyl groups initially present (Riesenfeld *et al.*, 1982b). Furthermore, the residual *N*-acetyl groups occurred largely as isolated units and showed essentially no accumulation in block structures; in fact, the N-substitution pattern of this polysaccharide was indistinguishable from that of a microsomal intermediate in cell-free heparin biosynthesis. While heparin produced by intact mastocytoma cells is thus as extensively N-deacetylated as the corresponding microsomal polysaccharide, the heparan sulfate synthesized by hepatocytes is less deacetylated than the microsomal homolog produced under cell-free conditions. The modulation of N-deacetylation in the latter system apparently depends on the integrity of the cell. Interestingly, the inhibition by butyrate of the N-deacetylation step in heparin biosynthesis also depended on the metabolism of the intact cell (see above). Could it be that the same mechanism of inhibition which is normally suppressed in the mast cell (but may be activated by butyrate) is a constitutive property of the hepatocyte?

C. Selection of Target Units

As outlined previously (Section II,C), heparin and heparan sulfate are capable of interacting with a variety of other macromolecules. While our knowledge in this area is still limited, it is obvious that some of these interactions are specific in their requirement for a particular polysaccharide structure. Examples of such structures are the antithrombin-binding region in the heparin molecule (Fig. 2), the apparently specific (but as yet poorly defined) regions in heparan sulfate chains which mediate self-association, and the cleavage sites for the endoglucuronidase involved in processing the newly synthesized heparin proteoglycan in the mast cell (Section II,C). Each of these regions appear to contain glucuronic acid residues, the presence of which implies a restraint in biosynthetic polymer modification, as functionally important components. Conversely, formation of the antithrombin-binding region requires that modification also be specifically extended to accommodate a unique component, the glucosamine 3-O-sulfate group. These observations raise an obvious question: which are the mechanisms that control the selection of individual target units in the various polymer modification reactions?

The regulatory role of N-deacetylation in conjunction with the substrate specificities of the enzymes involved in the subsequent reactions has already been discussed; while we do not understand the mechanism by which certain N-acetyl groups are selected for deacetylation (to be followed by N-sulfation), whereas others are retained, we can predict in some detail how the resulting pattern of N-acetyl and N-sulfate groups may influence the C-5 epimerization of glucuronic acid residues and the incorporation of O-sulfate groups (Fig. 6). However, each of these latter reactions will again involve only a fraction of the potential substrate units available. The proportion of N-sulfated disaccharide units which thus may escape O-sulfation varies greatly, from almost nil in certain heparin preparations to ~75% in a heparan sulfate produced by the Engelbreth–Holm–Swarm mouse sarcoma (Pejler *et al.*, 1987). Restraints in individual reactions will show up as isolated glucuronic acid units surrounded by N-sulfated glucosamine residues, or as iduronic acid or N-sulfated glucosamine units lacking *O*-sulfate groups (Fig. 6). The mode of selection is unknown and may conceivably involve an element of randomness.

The biosynthesis of the antithrombin-binding region in the heparin proteoglycan poses some intriguing problems of selectivity in polymer modification. The N-substituent (acetyl or sulfate group) of the non-reducing-terminal glucosamine unit in this region (*A* in Fig. 2) is variable (Lindahl *et al.*, 1984), in accord with the notion that this group does not significantly influence the interaction with antithrombin (Thunberg *et al.*, 1982b). Yet the extent of N-acetylation of this particular unit is much greater than is that of the average glucosamine residue in the heparin polymer. This apparent contradiction can be rationalized by assuming an absolute functional requirement for a D-glucuronic (rather than L-iduronic) acid unit at the adjacent position *B*. Such a requirement is implied by the finding that unit *B* is invariably D-glucuronic acid even when unit *A* is N-sulfated (Lindahl *et al.*, 1984; note the →IdUA→GlcNSO$_3$(3-OSO$_3$)→ sequence in clam heparin (Section II,C), apparently without relation to antithrombin binding). A D-glucuronic acid residue predestined to become unit *B* of the antithrombin-binding sequence thus has to escape C-5 epimerization even when inserted in a →GlcNSO$_3$→GlcUA→GlcNSO$_3$→ sequence, the preferred substrate for the epimerase (Section IV,A,2). Since a →GlcNAc→GlcUA →GlcNSO$_3$→ structure is not recognized as a substrate by the enzyme, an *N*-acetyl group at position *A* will secure unit *B* in D-gluco configuration (see Figs. 4 and 6). Indeed, such a mechanism might explain why the increased retention of *N*-acetyl groups in heparin, induced by butyrate treatment of mast cells, is accompanied by the increase in the proportion of molecules having high affinity for antithrombin (Section IV,B).

On the other hand, we do not know at which stage of polymer modification a particular region of the heparin precursor polysaccharide becomes committed to conversion into an antithrombin-binding site. The only structural feature which clearly distinguishes this site from other regions in the heparin molecule is the 3-O-sulfated glucosamine residue (Fig. 6); all other components, including the sulfate groups, that are essential to antithrombin binding occur also elsewhere in the polysaccharide chain, as well as in heparin molecules with low affinity for antithrombin. It is conceivable that these latter components be introduced before incorporation of the 3-*O*-sulfate group, and indeed, that they may be required for substrate recognition by the 3-*O*-sulfotransferase. The assembly of such a hypothetical recognition site (Fig. 4), composed of ubiquitous monosaccharide and sulfate residues, might occur by random polymer modification (Laurent *et al.*, 1978), within the restrictions imposed by the substrate specificities of the enzymes involved in the process. On the other hand, the completed antithrombin-binding regions are not randomly distributed in the heparin proteoglycan (see Section II,C) but rather accumulated in a limited fraction of the polysaccharide chains (Fig. 1). Whatever mechanism that commits polymer modification to the formation of these regions apparently operates in a nonrandom fashion. Interestingly, the antithrombin-binding regions in the rat skin heparin proteoglycan were largely N-sulfated rather than N-acetylated at position A ($R'' = SO_3^-$ in Fig. 2; Jacobsson *et al.*, 1986). The nonrandom distribution of the binding sites therefore would not seem to depend on the occurrence of residual N-acetyl groups located at specific positions.

V. ORGANIZATION OF THE BIOSYNTHETIC APPARATUS

The biosynthesis of a proteoglycan requires the coordinated action of a large number of different enzymes. For example, the formation of a fully substituted heparin proteoglycan will involve at least 12 separate reactions, not counting the assembly of the polypeptide core. Little is known regarding the organization of these enzymes, in the intact cell, into a functional biosynthetic machinery; in fact, all that can be stated with some confidence is that most of the enzymes are membrane bound. Moreover, it appears probable that the initial reactions (formation of the peptide core) and the concluding events (sulfation of the polysaccharide chains) take place in the rough endoplasmic reticulum and the Golgi apparatus, respectively. In the preceding sections we have described the properties of individual reactions as well as their concerted action, and have discussed possible mechanisms of regulation in relation to the formation of specific saccharide structures. In particular, the kinetics of polysaccharide chain formation and modification point to a biosynthetic system operating under strictly ordered conditions. Most of these observations pertain to cell-free systems; however, it seems reasonable to assume that disruption of the cell should, if anything, create disorder rather than the opposite, and hence, that the level of organization revealed by the cell-free experiments should prevail also in the intact cell. In the remaining part of this review we will consider some aspects of the overall organization of the biosynthetic process.

A. Compartmentalization of the Biosynthetic Process

Most of the information currently available regarding the subcellular organization of proteoglycan biosynthesis derives from studies on the biosynthesis of chondroitin sulfate. Labeling experiments with chondrocytes indicated an intracellular half-life of 1.5–2 hr for the core protein of the chondroitin sulfate proteoglycan, and further suggested that formation and sulfation of the polysaccharide chains occur (presumably in the Golgi complex) during the last few minutes before export of the completed macromolecule from the cell (Thonar *et al.*, 1983; Schwartz and Habib, 1983; however, see Glössl and Kresse, 1984, for apparently much faster intracellular processing of a dermatan sulfate proteoglycan). More recent data based on labeling kinetics for individual sugar residues imply that formation of the entire polysaccharide–protein linkage region, including xylosylation of the core protein, occurs in close conjunction with elongation of the polysaccharide chains proper, hence late in the overall biosynthetic process (Lohmander *et al.*, 1985). This conclusion seems at variance with the finding of Hoffman *et al.* (1984) that the xylosyl transferase in embryonic chick cartilage cells is located within the cisternae of the rough endoplasmic reticulum. The biosynthesis of heparin/heparan sulfate is essentially unexplored as regards these particular aspects. Moreover, the various products of this process, destined for export from the cell, intercalation into the plasma membrane, or intracellular storage, presumably utilize different routes of transportation in the cell, and it seems likely that the localization of precursor molecules may vary already at the stage of polysaccharide chain initiation.

Is the heparin–protein linkage region synthesized at the same time and at the same subcellular locus as is the polysaccharide chain proper? As outlined previously (Section III,B), a mastocytoma microsomal fraction will catalyze the formation of full-sized polysaccharide chains when supplied with UDP-glucuronic acid and UDP-*N*-acetylglucosamine. The endogenous sugar acceptor must therefore include a preformed oligosaccharide "primer," consisting of at least the galactosyl-galactosyl-xylosyl trisaccharide linkage region bound to the core protein. These findings are compatible with a compartmentalized glycosyl transfer process, in which the polysaccharide–protein linkage region is assembled separately from the bulk of the polysaccharide chain.

Compartmentalization does not necessarily imply extensive physical separation of processes, for example between different subcellular organelles. Since the formation and modification of polysaccharide chains in heparin biosynthesis occur within such a short time interval

(Section III), any major transportation of precursor proteoglycans from one subcellular locus to another appears unlikely. Yet the two processes seem to be at least partly separated, since extended, nonsulfated polysaccharide chains are formed in cell-free biosynthesis, even when PAPS is present from the outset of the experiment (Section III,C,1). Also the individual polymer modification reactions are functionally segregated as reflected by the stepwise progression of the concerted process (Section III,C). This is strikingly apparent in the case of the uronosyl C-5 epimerase reaction, which involves a transient, precisely timed interaction between the enzyme and its proteoglycan substrate; proteoglycan molecules at the pre- or postepimerization stage are not touched by the enzyme, even if they contain susceptible uronic acid residues (Section III,C,3). Riesenfeld *et al.* (1983a) showed that the membrane-bound state of the system is prerequisite to the highly ordered kinetics of polymer modification in heparin biosynthesis. In the presence of the nonionic detergent Triton X-100 (at 1% concentration), sulfation of preformed, microsomal polysaccharide occurred by simultaneous slow, progressive modification of all available precursor molecules, in marked contrast to the stepwise kinetics displayed by the unperturbed system. Similar disordered reactions could be observed in a high-speed supernatant prepared from the detergent-treated microsomal fraction, indicating extensive solubilization of both the particulate enzymes and their proteoglycan substrate. While the products synthesized by the intact or the solubilized microsomal systems were about equally sulfated, they showed different N-substituent patterns. The polysaccharide obtained with intact microsomes contained extended sequences of consecutive N-sulfated as well as of N-acetylated disaccharide units, indicative of a nonrandom *N*-acetyl–*N*-sulfate exchange process. In contrast, the product sulfated by the solubilized system displayed a seemingly random distribution of *N*-acetyl and *N*-sulfate groups, with a large proportion of alternating N-acetylated and N-sulfated disaccharide units. These findings suggest that the kinetics and regulation of the polymer modification process depend heavily on the organization of the enzymes in the intracellular membranes of the cell.

B. Model of Polymer Modification

The previous discussion has focused in particular on the reactions involved in polymer modification and on their concerted action. Major aspects of this important process remain obscure, for instance the mode of interaction between enzymes and macromolecular substrates, the means of transportation of proteoglycan precursor molecules, and mechanisms of regulation. Nevertheless, we will attempt to rationalize the scattered pieces of information available into a model that may at least illustrate some conspicuous features of the process (Fig. 7). According to this proposal, the newly formed proteoglycan precursor traverses a series of multienzyme complexes, such that the enzyme molecules, rather than migrating along the polysaccharide chains, are being translocated from one polysaccharide chain to the next. The composition of each enzyme complex reflects the course of polymer modification: N-deacetylation apparently occurs in conjunction with N-sulfation, uronosyl C-5 epimerization is largely associated with 2-O-sulfation of iduronic acid residues, and so on. The spacing of the various enzyme complexes in the membrane scaffold is such that a proteoglycan substrate will interact with only one complex at a time. The arrangement of enzyme molecules within these complexes (outlined in arbitrary array in the figure) could in part account for some of the characteristic structural features of the completed polysaccharide chains, such as the blockwise distribution of N-acetyl and N-sulfate groups. Indeed, by a similar line of argument we may explain the nonrandom distribution of the antithrombin-binding regions in the heparin proteoglycan; could it be, as suggested in Fig. 7, that the selective 3-O-sulfation of glucosamine residues, in a restricted fraction of the polysaccharide chains, reflects the distribution of 3-O-sulfotransferase molecules in the membrane?

The model applies primarily to the heparin proteoglycan but may, at least in part, be relevant also to the biosynthesis of heparan sulfate or other proteoglycans. While clearly speculative, it explains in a simple way some characteristic properties of polymer modification that are difficult to reconcile with more random modes of enzyme–substrate interaction. The model thus helps visualize how a proteoglycan precursor molecule may pass through the various steps of polymer modification within a few seconds. It is easily appreciated that the machinery will accommodate only a limited number of substrate molecules at a time. Moreover, the stepwise progression of polymer modification and the generation of structurally distinct intermediates is readily explained by the topological segregation of the various enzyme complexes in the endoplasmic membrane. In fact, this segregation of enzymes and their substrates would account for the strictly compartmentalized interactions observed, for instance, in the uronosyl C-5 epimerization reaction. The epimerase will gain access to the susceptible D-glucuronic acid residues only during the swift transfer of proteoglycan substrate molecules from the locus of the precursor intermediate, PS-

NSO_3, to that of the product, $PS-N/O-SO_3$-a. Finally, the model may provide a useful frame of reference for evaluating alternative schemes of proteoglycan biosynthesis. Suggestions mainly in older literature that heparan sulfate (low-sulfated) might serve as a precursor of heparinlike polysaccharides (highly sulfated) would thus imply a recycling of proteoglycans through the process of polymer modification. Metcalfe *et al.* (1980) has proposed that new heparin chains may be assembled on preformed, fully sulfated, heparin proteoglycans over a period of several hours. Such a pathway would involve even more extensive recycling of the proteoglycan, involving polymerization as well as polymer modification. Further experimental evidence is required to support this proposal.

VI. PERSPECTIVES FOR THE FUTURE

Many of the basic unsolved problems in heparin/heparan sulfate biosynthesis are common to biosynthetic processes involving other glycosaminoglycans. We will conclude this presentation by outlining the major trends of investigation in the glycosaminoglycan field, which may be expected to resolve some of the current enigmas of biosynthesis. In essence, these areas of research can be subdivided into four main categories.

First, information is now beginning to accumulate regarding the structures of proteoglycan core proteins. The notoriously slow progress in this area is mainly attributed to difficulties in sequencing polypeptides containing large proportions of carbohydrate-substituted amino acids. Recent reports have shown that these problems can be overcome by use of recombinant DNA technology (Bourdon *et al.*, 1985; Sai *et al.*, 1986; Doege *et al.*, 1986). A detailed knowledge of core protein structure may help to clarify the potential roles of this component in anchoring the proteoglycan precursor to the endoplasmic membrane, directing the precursor to the appropriate membrane-bound enzymes, and initiating polysaccharide chain formation. Specific antibodies against core proteins will be used to locate proteoglycan precursors, at various levels of modification, and in various subcellular compartments (see, e.g., Schwartz and Habib, 1983; Ratcliffe *et al.*, 1984). Classification of proteoglycans with regard to core protein structure may reveal previously unknown biosynthetic relationships (consider, for instance, the repetitive [Ser-Gly]$_n$ sequence in the heparin proteoglycan and in the chondroitin sulfate proteoglycan described by Bourdon *et al.*, 1985).

Another important area of investigation that has so far progressed at a fairly modest pace is the isolation and characterization of the biosynthetic enzymes. To our knowledge the only enzymes that to date have been obtained in a state anywhere near homogeneity are xylosyl transferase from cartilage (Schwartz and Rodén, 1974; Rodén, 1980) and the uronosyl C-5 epimerase from mouse mastocytoma (Malmström et al., 1980). The isolation procedures are generally complicated by the fact that most of the enzymes are firmly bound to intracellular membranes. Progress will probably be facilitated by the continued development of simple and specific assay procedures and by the use of monoclonal antibodies in the isolation and functional characterization of the enzymes. The potential of immunochemical approaches has been demonstrated, for instance, in studies of the interactions (Schwartz, 1975), metabolism (Schwartz, 1976), and intracellular location (Hoffman et al., 1984) of glycosyltransferases involved in the biosynthesis of chondroitin sulfate.

The temporal and topological analysis of proteoglycan assembly in cultured intact cells constitute a third field of experimentation that is likely to expand rapidly in the near future. Whereas such studies have the obvious advantage of approaching in vivo conditions, they generally do not permit the direct probing of selected individual reactions or reaction sequences. This limitation can, however, be partly overcome by selectively perturbing various steps of the overall biosynthetic process by use of inhibitors affecting, for instance, core protein synthesis (Rodén, 1980; Kimura et al., 1981), intracellular transport processes, or sulfation (Tajiri et al., 1980). The formation of polysaccharide chains may be experimentally segregated from core protein biosynthesis by means of exogenous initiators (xylosides or galactosides) of the polymerization process (see, e.g., Hardingham, 1982). Moreover, the temporal aspects of proteoglycan assembly, regarding, for instance, the various glycosylation reactions, may be elucidated by analysis of labeling kinetics (with radioactively labeled amino acids, sugars, or sulfate; for instructive examples, see Thonar et al., 1983, or Lohmander et al., 1985, and references therein).

Finally, we may anticipate the development of further refined cellfree systems, specifically designed for studying the concerted action of enzymes involved in polymerization and polymer modification reactions. Most of the cell-free experiments reported so far have employed fairly crude subcellular fractions (a notable exception is reported by Freilich et al., 1975), and it is essential that the fractionation procedures be improved to furnish better defined, purified biosynthetic preparations. Hopefully, these systems will permit a more detailed analysis of major unresolved problems, such as the arrangement of the membrane-bound biosynthetic enzymes, the mode of transportation of proteoglycan precursors between these enzymes, and the regulatory mechanisms behind the generation of specific polysaccharide structures. Among a number of yet distant goals to these endeavors, we may visualize the reconstruction of biosynthetic assembly lines from purified enzymes in artificial membrane systems.

REFERENCES

Atha, D. H., Stephens, A. W., Rimon, A., and Rosenberg, R. D. (1984). *Biochemistry* **23**, 5801–5812.
Atha, D. H., Lormeau, J.-C., Petitou, M., Rosenberg, R. D., and Choay, J. (1985). *Biochemistry* **24**, 6723–6729.
Bengtsson, G., Olivecrona, T., Höök, M., Riesenfeld, J., and Lindahl, U. (1980). *Biochem. J.* **189**, 625–633.
Bienkowski, M. J., and Conrad, H. E. (1985). *J. Biol. Chem.* **260**, 356–365.
Björk, I., and Lindahl, U. (1982). *Mol. Cell. Biochem.* **48**, 161–182.
Bourdon, M. A., Oldberg, Å., Pierschbacher, M., and Ruoslahti, E. (1985). *Proc. Natl. Acad. Sci. U.S.A.* **82**, 1321–1325.
Buonassisi, V., and Colburn, P. (1983). *Biochim. Biophys. Acta* **760**, 1–12.
Campbell, P., Feingold, D. S., Jensen, J. W., Malmström, A., and Rodén, L. (1983). *Anal. Biochem.* **131**, 146–152.
Carlstedt, I., Cöster, L., Malmström, A., and Fransson, L.-Å. (1983). *J. Biol. Chem.* **258**, 11629–11636.
Castellot, J. J., Beeler, D. L., Rosenberg, R. D., and Karnovsky, M. J. (1984). *J. Cell. Physiol.* **120**, 315–320.
Casu, B. (1985). *Adv. Carbohydr. Chem. Biochem.* **43**, 51–134.
Casu, B., Oreste, P., Torri, G., Zopetti, G., Choay, J., Lormeau, J.-C., and Petitou, M. (1981). *Biochem. J.* **197**, 599–609.
Cheng, C.-F., Oosta, G. M., Bensadoun, A., and Rosenberg, R. D. (1981). *J. Biol. Chem.* **256**, 12893–12898.
Cifonelli, J. A. (1968). *Carbohydr. Res.* **8**, 233–242.
Cifonelli, J. A., and King, J. (1972). *Carbohydr. Res.* **21**, 173–186.
Cifonelli, J. A., and King, J. (1973). *Biochim. Biophys. Acta* **320**, 331–340.
Cifonelli, J. A., and King, J. (1977). *Biochemistry* **16**, 2137–2141.
Comper, W. D. (1981). "Heparin (and Related Polysaccharides). Structural and Functional Properties." Gordon and Breach Science Publishers, New York.
Cöster, L., Malmström, A., Carlstedt, I., and Fransson, L.-Å. (1983). *Biochem. J.* **215**, 417–419.
David, G., and Van den Berghe, H. (1985). *J. Biol. Chem.* **260**, 11067–11074.
Delaney, S. R., Leger, M., and Conrad, H. E. (1980). *Anal. Biochem.* **106**, 25–34.
Dietrich, C. P., de Paiva, J. F., Moraes, C. T., Takahashi, H. K., Porcionatto, M. A., and Nader, H. B. (1985). *Biochim. Biophys. Acta* **843**, 1–7.
Doege, K., Fernandez, P., Hassell, J. R., Sasaki, M., and Yamada, Y. (1986). *J. Biol. Chem.* **261**, 8108–8111.
Dziadek, M., Fujiwara, S., Paulsson, M., and Timpl, R. (1985). *EMBO J.* **4**, 905–912.
Esko, J. D., Stewart, T. E., and Taylor, W. H. (1985). *Proc. Natl. Acad. Sci. U.S.A.* **82**, 3197–3201.

Fedarko, N. S., and Conrad, H. E. (1986). *J. Cell Biol.* **102**, 587–599.

Feingold, D. S., Rodén, L., Forsee, T., Jacobsson, I., Jensen, J. W., Lindahl, U., Malmström, A., and Prihar, H. (1981). In "Chemistry and Biology of Heparin" (R. L. Lundblad, W. V. Brown, K. G. Mann, and H. R. Roberts, eds.), pp. 157–171. Elsevier North Holland, Amsterdam.

Forsee, W. T., and Rodén, L. (1981). *J. Biol. Chem.* **256**, 7240–7247.

Fransson, L.-Å. (1981). *Eur. J. Biochem.* **120**, 251–255.

Fransson, L.-Å., Sjöberg, I., and Havsmark, B. (1980). *Eur. J. Biochem.* **106**, 59–69.

Fransson, L.-Å., Carlstedt, I., Cöster, L., and Malmström, A. (1984). *Proc. Natl. Acad. Sci. U.S.A.* **81**, 5657–5661.

Fransson, L.-Å., Silverberg, I., and Carlstedt, I. (1985). *J. Biol. Chem.* **260**, 14722–14726.

Freilich, L. S., Lewis, R. G., Reppucci, A. C., and Silbert, J. E. (1975). *Biochem. Biophys. Res. Commun.* **63**, 663–668.

Fritze, L. M. S., Reilly, C. F., and Rosenberg, R. D. (1985). *J. Cell Biol.* **100**, 1041–1049.

Fujiwara, S., Wiedemann, H., Timpl, R., Lustig, A., and Engel, J. (1984). *Eur. J. Biochem.* **143**, 145–157.

Gallagher, J. T., and Walker, A. (1985). *Biochem. J.* **230**, 665–674.

Gallagher, J. T., Lyon, M., and Steward, W. P. (1986). *Biochem. J.* **236**, 313–325.

Glössl, J., and Kresse, H. (1984). *Hoppe-Seyler's Z. Physiol. Chem.* **365**, 991.

Göhler, D., Niemann, R., and Buddecke, E. (1984). *Eur. J. Biochem.* **138**, 301–308.

Gomes, P. B., and Dietrich, C. P. (1982). *Biochem. Physiol.* **73B**, 857–863.

Grebner, E. E., Hall, C. W., and Neufeld, E. F. (1966). *Arch. Biochem. Biophys.* **116**, 391–398.

Handin, R. I., and Cohen, H. J. (1976). *J. Biol. Chem.* **251**, 4273–4282.

Hardingham, T. E. (1982). In "Cell Function and Differentiation," Part A, pp. 423–434. Liss, New York.

Hassell, J. R., Gehron Robey, P., Barrach, H.-J., Wilczek, J., Rennard, S. I., and Martin, G. R. (1980). *Proc. Natl. Acad. Sci. U.S.A.* **77**, 4494–4498.

Hayman, E. G., Pierschbacher, M. D., Öhgren, Y., and Ruoslahti, E. (1983). *Proc. Natl. Acad. Sci. U.S.A.* **80**, 4003–4007.

Heifetz, A., and Prager, M. D. (1981). *J. Biol. Chem.* **256**, 6529–6532.

Helting, T. (1971). *J. Biol. Chem.* **246**, 815–822.

Helting, T. (1972). *J. Biol. Chem.* **247**, 4327–4332.

Helting, T., and Lindahl, U. (1971). *J. Biol. Chem.* **246**, 5442–5447.

Helting, T., and Lindahl, U. (1972). *Acta Chem. Scand.* **26**, 3515–3523.

Hoffmann, H.-P., Schwartz, N. B., Rodén, L., and Prockop, D. J. (1984). *Connect. Tissue Res.* **12**, 151–164.

Höök, M., Lindahl, U., Bäckström, G., Malmström, A., and Fransson, L.-Å. (1974a). *J. Biol. Chem.* **249**, 3908–3915.

Höök, M., Lindahl, U., and Iverius, P.-H. (1974b). *Biochem. J.* **137**, 33–43.

Höök, M., Lindahl, U., Hallén, A., and Bäckström, G. (1975). *J. Biol. Chem.* **250**, 6065–6071.

Höök, M., Kjellén, L., Johansson, S., and Robinson, J. (1984). *Annu. Rev. Biochem.* **53**, 847–869.

Hopwood, J. J., and Dorfman, A. (1977). *J. Biol. Chem.* **252**, 4777–4785.

Hopwood, J. J., and Elliott, H. (1981). *Carbohydr. Res.* **91**, 165–190.

Horner, A. A. (1971). *J. Biol. Chem.* **246**, 231–239.

Horner, A. A. (1977). *Fed. Proc. Fed. Am. Soc. Exp. Biol.* **36**, 35–39.

Horner, A. A., and Young, E. (1982). *J. Biol. Chem.* **257**, 8749–8754.

Iozzo, R. V. (1984). *J. Cell. Biol.* **99**, 403–417.

Jacobsson, I., and Lindahl, U. (1980). *J. Biol. Chem.* **255**, 5094–5100.

Jacobsson, I., Bäckström, G., Höök, M., Lindahl, U., Feingold, D. S., Malmström, A., and Rodén, L. (1979a). *J. Biol. Chem.* **254**, 2975–2982.

Jacobsson, I., Höök, M., Pettersson, I., Lindahl, U., Larm, O., Wirén, E., and von Figura, K. (1979b). *Biochem. J.* **179**, 77–87.

Jacobsson, I., Lindahl, U., Jensen, J. W., Rodén, L., Prihar, H., and Feingold, D. S. (1984). *J. Biol. Chem.* **259**, 1056–1063.

Jacobsson, K.-G., Riesenfeld, J., and Lindahl, U. (1985). *J. Biol. Chem.*, **260**, 12154–12159.

Jacobsson, K.-G., Horner, A. A., and Lindahl, U. (1986). *Biochem. J.*, **240**, 625–632.

Jansson, L., Höök, M., Wasteson, Å., and Lindahl, U. (1975). *Biochem. J.* **149**, 49–55.

Jensen, J. W., Rodén, L., Jacobsson, I., Lindahl, U., Prihar, H., and Feingold, D. S. (1983). *Carbohydr. Res.* **117**, 241–253.

Johnston, L. S., and Keller, J. M. (1979). *J. Biol. Chem.* **254**, 2575–2578.

Jordan, R. E., and Marcum, J. A. (1986). *Arch. Biochem. Biophys.* **248**, 690–695.

Jordan, R. E., Favreau, L. V., Braswell, E. H., and Rosenberg, R. D. (1982). *J. Biol. Chem.* **257**, 400–406.

Kanwar, Y. S., Veis, A., Kimura, J. H., and Jakubowski, M. L. (1984). *Proc. Natl. Acad. Sci. U.S.A.* **81**, 762–766.

Kazatchkine, M. D., Feacon, D. T., Metcalfe, D. D., Rosenberg, R. D., and Austen, K. F. (1981). *J. Clin. Invest.* **67**, 223–228.

Keller, K. L., Keller, J. M., and Moy, J. N. (1980). *Biochemistry* **19**, 2529–2536.

Kennedy, J. F. (1979). Proteoglycans—biological and chemical aspects of human life. *In* "Studies in Organic Chemistry 2," pp. 45–54. Elsevier, Amsterdam.

Kjellén, L., and Höök, M. (1983). *In* "Glycoconjugates" (M. A. Chester, D. Heinegård, A. Lundblad, and S. Svensson, eds.), pp. 94–95. Proc. *7th Int. Symp. Glycoconjugates, Lund.*

Kjellén, L., Oldberg, Å., and Höök, M. (1980). *J. Biol. Chem.* **255**, 10407–10413.

Kjellén, L., Pettersson, I., and Höök, M. (1981). *Proc. Natl. Acad. Sci. U.S.A.* **78**, 5371–5375.

Kjellén, L., Bielefeld, D., and Höök, M. (1983). *Diabetes* **32**, 337–342.

Kimura, J. H., Caputo, C. B., and Hascall, V. C. (1981). *J. Biol. Chem.* **256**, 4368–4376.

Klinger, M. M., Margolis, R. U., and Margolis, R. K. (1985). *J. Biol. Chem.* **260**, 4082–4090.

Knecht, J., Cifonelli, J. A., and Dorfman, A. (1967). *J. Biol. Chem.* **242**, 4652–4661.

Koda, J. E., and Bernfield, M. (1984). *J. Biol. Chem.* **259**, 11763–11770.

Kosakai, M., and Yosizawa, Z. (1982). *J. Biochem.* **92**, 295–303.

Lane, D. A., Denton, J., Flynn, A. M., Thunberg, L., and Lindahl, U. (1984). *Biochem. J.* **218**, 725–732.

Lane, D. A., Pejler, G., Flynn, A. M., Thompson, E. A., and Lindahl, U. (1986). *J. Biol. Chem.* **261**, 3980–3986.

Laurent, T. C., Tengblad, A., Thunberg, L., Höök, M., and Lindahl, U. (1978). *Biochem. J.* **175**, 691–701.

Lijnen, H., Hoylaerts, M., and Collen, D. (1983). *J. Biol. Chem.* **258**, 3803–3808.

Lindahl, U. (1966). *Biochim. Biophys. Acta* **130**, 368–382.

Lindahl, U. (1976). *MTP Int. Rev. Sci. Org. Chem. Ser. Two, Carbohydr. Chem.* **7**, pp. 283–312.

Lindahl, U., and Höök, M. (1978). *Annu. Rev. Biochem.* **47**, 385–417.

Lindahl, U., Citonelli, J. A., Lindahl, B., and Rodén, L. (1965). *J. Biol. Chem.* **240**, 2817–2820.

Lindahl, U., Bäckström, G., Jansson, L., and Hallén, A. (1973) J. Biol. Chem. 248, 7234–7241.

Lindahl, U., Jacobsson, I., Höök, M., Bäckström, G., and Feingold, D. S. (1976). Biochem. Biophys. Res. Commun. 70, 492–499.

Lindahl, U., Höök, M., Bäckström, G., Jacobsson, I., Riesenfeld, J., Malmström, A., Rodén, L., and Feingold, D. S. (1977). Fed. Proc. Fed. Am. Soc. Exp. Biol. 36, 19–24.

Lindahl, U., Bäckström, G., Thunberg, L., and Leder, I. G. (1980). Proc. Natl. Acad. Sci. U.S.A. 77, 6551–6555.

Lindahl, U., Bäckström, G., and Thunberg, L. (1983). J. Biol. Chem. 258, 9826–9830.

Lindahl, U., Thunberg, L., Bäckström, G., Riesenfeld, J., Nordling, K., and Björk, I. (1984). J. Biol. Chem. 259, 12368–12376.

Lindahl, U., Feingold, D. S., and Rodén, L. (1986). TIBS 11, 221–225.

Linker, A. (1979). Biochem. J. 183, 711–720.

Linker, A., and Hovingh, P. (1975). Biochim. Biophys. Acta 385, 324–333.

Linker, A., and Hovingh, P. (1984). Carbohydr. Res. 127, 75–94.

Lohmander, L. S., Hascall, V. C., Yanagishita, M., Kuettner, K. E. and Kimura, J. H. (1986). Arch. Biochem-Biophys. 250, 211–227.

Malmström, A., Rodén, L., Feingold, D. S., Jacobsson, I., Bäckström, G., and Lindahl, U. (1980). J. Biol. Chem. 255, 3878–3883.

Marcum, J. A., and Rosenberg, R. D. (1985). Biochem. Biophys. Res. Commun. 126, 365–372.

Marcum, J. A., Atha, D. H., Fritze, L. M. S., Nawroth, P., Stern, D., and Rosenberg, R. D. (1986). J. Biol. Chem. 261, 7507–7517.

Metcalfe, D. D., Smith, J. A., Austen, K. F., and Silbert, J. E. (1980). J. Biol. Chem. 255, 11753–11758.

Miller, R. R., and Waechter, C. J. (1984). Arch. Biochem. Biophys. 228, 247–257.

Mutoh, S., Funakoshi, I., and Yamashina, I. (1978). J. Biochem. 84, 483–489.

Mutoh, S., Funakoshi, I., Ui, N., and Yamashina, I. (1980). Arch. Biochem. Biophys. 202, 137–143.

Navia, J. L., Riesenfeld, J., Vann, W. F., Lindahl, U., and Rodén, L. (1983). Anal. Biochem. 135, 134–140.

Norling, B., Glimelius, B., and Wasteson, Å. (1981). Biochem. Biophys. Res. Commun. 103, 1265–1272.

Oegema, T. R., Jr., Kraft, E. L., Jourdian, G. W., and Van Valen, T. R. (1984). J. Biol. Chem. 259, 1720–1726.

Ögren, S., and Lindahl, U. (1971). Biochem. J. 125, 1119–1129.

Ögren, S., and Lindahl, U. (1975). J. Biol. Chem. 250, 2690–2697.

Ögren, S., and Lindahl, U. (1976). Biochem. J. 154, 605–611.

Oldberg, Å., Höök, M., Öbrink, B., Pertoft, H., and Rubin, K. (1977). Biochem. J. 164, 75–81.

Oldberg, Å., Kjellén, L., and Höök, M. (1979). J. Biol. Chem. 254, 8505–8510.

Oldberg, Å., Schwartz, C., and Ruoslahti, E. (1982). Arch. Biochem. Biophys. 216, 400–406.

Olivecrona, T., Bengtsson, G., Marklund, S.-E., Lindahl, U., and Höök, M. (1977). Fed. Proc. Fed. Am. Soc. Exp. Biol. 36, 60–65.

Oohira, A., Wight, T. N., McPherson, J., and Bornstein, P. (1982). J. Cell Biol. 92, 357–367.

Oohira, A. Wight, T. N., and Bornstein, P. (1983). J. Biol. Chem. 258, 2014–2021.

Ototani, N., and Yosizawa, Z. (1981). J. Biochem. 90, 1553–1556.

Ototani, N., Kikuchi, M., and Yosizawa, Z. (1982). Biochem. J. 205, 23–30.

Parthasarathy, N., and Spiro, R. G. (1984). *J. Biol. Chem.* **259**, 12749–12755.
Pejler, G., Bäckström, G., Lindahl, U., Paulsson, M., Dziadek, M., Fujiwara, S., and Timpl, R. (1987). *J. Biol. Chem.*, in press.
Petitou, M. (1984). *Nouv. Rev. Fr. Hematol.* **26**, 221–226.
Prehm, P. (1983). *Biochem. J.* **211**, 191–198.
Prihar, H. S., Campbell, P., Feingold, D. S., Jacobsson, I., Jensen, J. W., Lindahl, U., and Rodén, L. (1980). *Biochemistry* **19**, 495–500.
Rapraeger, A. C., and Bernfield, M. (1983). *J. Biol. Chem.* **258**, 3632–3636.
Rapraeger, A., Jalkanen, M., Endo, E., Koda, J., and Bernfield, M. (1985). *J. Biol. Chem.* **260**, 11046–11052.
Ratcliffe, A., Fryer, P. R., and Hardingham, T. E. (1984). *J. Histochem. Cytochem.* **32**, 193–201.
Richmond, M. E., DeLuca, S., and Silbert, J. E. (1973). *Biochemistry* **12**, 3904–3910.
Riesenfeld, J., Höök, M., and Lindahl, U. (1980). *J. Biol. Chem.* **255**, 922–928.
Riesenfeld, J., Höök, M., and Lindahl, U. (1982a). *J. Biol. Chem.* **257**, 421–425.
Riesenfeld, J., Höök, M., and Lindahl, U. (1982b). *J. Biol. Chem.* **257**, 7050–7055.
Riesenfeld, J., Pettersson, I., Vann, W. F., Rodén, L., and Lindahl, U. (1983a). In "Glycoconjugates" (M. A. Chester, D. Heinegård, A. Lundblad, and S. Svensson, eds.), pp. 388–389. *Proc. 7th Int. Symp. Glycoconjugates, Lund.*
Riesenfeld, J., Thunberg, L., and Lindahl, U. (1983b). In "Glycoconjugates" (M. A. Chester, D. Heinegård, A. Lundblad, and S. Svensson, eds.), pp. 390–391. *Proc. 7th Int. Symp. Glycoconjugates, Lund.*
Robinson, H. C., and Lindahl, U. (1981). *Biochem. J.* **194**, 575–586.
Robinson, H. C., Horner, A. A., Höök, M., Ögren, S., and Lindahl, U. (1978). *J. Biol. Chem.* **253**, 6687–6693.
Robinson, J., Viti, M., and Höök, M. (1984). *J. Cell Biol.* **98**, 946–953.
Rodén, L. (1980). In "The Biochemistry of Glycoproteins and Proteoglycans" (W. J. Lennarz, ed.), pp. 267–371. Plenum, New York.
Rohrmann, K., Niemann, R., and Buddecke, E. (1985). *Eur. J. Biochem.* **148**, 463–469.
Rosenberg, R. D. (1977). *Fed. Proc. Fed. Am. Soc. Exp. Biol.* **36**, 10–18.
Rosenberg, R. D., and Lam, L. (1979). *Proc. Natl. Acad. Sci. U.S.A.* **76**, 1218–1222.
Sai, S., Tanaka, T., Kosher, R. A., and Tanzer, M. L. (1986). *Proc. Natl. Acad. Sci. U.S.A.* **83**, 5081–5085.
Sanderson, P. N., Nieduszynski, I. A., and Huckerby, T. N. (1983). *Biochem. J.* **211**, 677–682.
Sakashita, S., Engvall, E., and Ruoslahti, E. (1980). *FEBS Lett.* **116**, 243–246.
Schwartz, N. B. (1975). *FEBS Lett.* **49**, 342–345.
Schwartz, N. B. (1976). *J. Biol. Chem.* **251**, 3346–3351.
Schwartz, N. B., and Habib, G. (1983). In "Glycoconjugates" (M. A. Chester, D. Heinegård, A. Lundblad, and S. Svensson, eds.), p. 395. *Proc. 7th Int. Symp. Glycoconjugates, Lund.*
Schwartz, N. B., and Rodén, L. (1974). *Carbohydr. Res.* **37**, 167–180.
Seldin, D. C., Austen, K. F., and Stevens, R. L. (1985). *J. Biol. Chem.* **260**, 11131–11139.
Shimada, K., Gill, P. J., Silbert, J. E., Douglas, W. H. J., and Fanbury, B. L. (1981). *J. Clin. Invest.* **68**, 995–1002.
Silbert, J. E. (1963). *J. Biol. Chem.* **238**, 3542–3546.
Silbert, J. E. (1967a). *J. Biol. Chem.* **242**, 5146–5152.
Silbert, J. E. (1967b). *J. Biol. Chem.* **242**, 5153–5157.
Silbert, J. E. and Baldwin, C. T. (1984). *Glycoconjugate J.* **1**, 63–71.
Silverberg, I., Havsmark, B., and Fransson, L.-Å. (1985). *Carbohydr. Res.* **137**, 227–238.

104   ULF LINDAHL AND LENA KJELLÉN

Sinaÿ, P., Jacquinet, J.-C., Petitou, M., Duchaussoy, P., Lederman, I., Choay, J., and Torri, G. (1984). *Carbohydr. Res.* **132**, C5-C9.

Sjöberg, I., and Fransson, L.-Å. (1980). *Biochem. J.* **191**, 103–110.

Skinner, M. K., and Fritz, I. B. (1985). *J. Biol. Chem.* **260**, 11874–11883.

Stevens, R. L., and Austen, K. F. (1982). *J. Biol. Chem.* **257**, 253–259.

Stow, J. L., and Farquhar, M. G. (1984). *J. Cell Biol.* **99**, (4, Pt 2), 78a (Abstr.).

Stow, J. L., Kjellén, L., Unger, E., Höök, M., and Farquhar, M. G. (1985). *J. Cell Biol.* **100**, 975–980.

Sudhakaran, R., Sinn, W., and von Figura, K. (1981). *Hoppe-Seyler's Z. Physiol. Chem.* **362**, 39–46.

Tajiri, K., Uchida, N., and Tanzer, M. L. (1980). *J. Biol. Chem.* **255**, 6036–6039.

Taylor, R. L., Shively, J. E., Conrad, H. E., and Cifonelli, J. A. (1973). *Biochemistry* **12**, 3633–3636.

Thonar, E. J.-M. A., Lohmander, L. S., Kimura, J. H., Fellini, S. A., Yanagishita, M., and Hascall, V. C. (1983). *J. Biol. Chem.* **258**, 11564–11570.

Thunberg, L., Bäckström, G., Wasteson, Å., Robinson, H. C., Ögren, S., and Lindahl, U. (1982a). *J. Biol. Chem.* **257**, 10278–10282.

Thunberg, L., Bäckström, G., and Lindahl, U. (1982b). *Carbohydr. Res.* **100**, 393–410.

Tollefsen, D. M., and Pestka, C. A. (1985). *J. Clin. Invest.* **75**, 496–501.

Turco, S. J., and Heath, E. C. (1977). *J. Biol. Chem.* **252**, 2918–2928.

Tyree, B., Horigan, E. A., Klippenstein, D. L., and Hassell, J. R. (1984). *Arch. Biochem. Biophys.* **231**, 328–335.

Vogel, K. G., and Peterson, D. W. (1981). *J. Biol. Chem.* **256**, 13235–13242.

Weissmann, B., and Chao, H. (1981). *Carbohydr. Res.* **92**, 255–268.

Winterbourne, D. J., and Mora, P. T. (1981). *J. Biol. Chem.* **256**, 4310–4320.

Yamada, K. M. (1983). *Annu. Rev. Biochem.* **52**, 761–800.

Yanagishita, M., and Hascall, V. C. (1983). *J. Biol. Chem.* **258**, 12857–12864.

Young, E., and Horner, A. A. (1979). *Biochem. J.* **180**, 587–596.

Yurt, R. W., Leid, R. W., Jr., Austen, K. F., and Silbert, J. E. (1977). *J. Biol. Chem.* **252**, 518–521.

Proteoglycan Metabolism by Rat Ovarian Granulosa Cells *in Vitro*

Masaki Yanagishita and Vincent C. Hascall

Bone Research Branch, National Institute of Dental Research, National Institutes of Health, Bethesda, Maryland 20892

I. INTRODUCTION

Ovarian follicles, basic functional units of the ovary, consist of granulosa cells, thecal cells, and oocytes. Under the influence of cyclic pituitary gonadotropin secretion, the granulosa cells and the thecal cells interact in a highly concerted manner to produce sex steroid hormones (i.e., estrogens and progestagens) and to produce a fertilizable ovum. Although pituitary gonadotropins [follicle-stimulating hormone (FSH), luteinizing hormone (LH), and prolactin)] are the major regulators of follicular development (1), other intraovarian control mechanisms are known to modulate follicular development (2).

When immature female rats are injected with pituitary gonadotropins to induce maturational development of ovarian follicles, proteoglycan synthesis in ovarian tissues *in vivo* as monitored by incorporation of $^{35}SO_4$ into glycosaminoglycans is significantly stimulated (3). Based on this observation, which suggested that proteoglycans might have a role in follicle maturation, we initiated a study of the chemistry of proteoglycans by isolated granulosa cells in cultures. During the course of this investigation, several distinct types of proteoglycans, both secretory and cell membrane associated, were identified and characterized (4–9). This chapter summarizes the results of these studies and speculates about their general significance for understanding aspects of proteoglycan metabolism in other cell systems.

BIOLOGY OF PROTEOGLYCANS

II. Characterization of Granulosa Cell Proteoglycans

A. Isolation of Cells

When immature female rats (day 24) are injected with 5 units of pregnant mare's serum gonadotropin, ovarian granulosa cells are stimulated to divide, to produce follicular fluid, and to undergo maturation with the development of LH receptors and the production of sex steroid hormones (1, 2). After 48 hr, many follicles in the ovary are enlarged and can readily be punctured under a dissecting microscope to permit the granulosa cells inside to be extruded into a collecting medium. Because the interior of the follicle is separated from surrounding connective tissue by a well-defined basement membrane, essentially a pure population of granulosa cells can be recovered in the medium without contaminating fibroblastic connective tissue cells. Once in the medium the loosely adhering cells can be mechanically separated to yield a dispersed-cell suspension without protease treatment. The cells can then be plated out into tissue culture dishes (2–3 × 10^6 cells per 35-mm dish); 10 animals yield about 60 × 10^6 cells. The cells readily attach to the dish within a few hours when plated in medium 199 with 10% fetal calf serum. They do not appear to make a typical connective tissue matrix in culture nor do they undergo a significant amount of cell division. Thus, the cultures represent a uniform, stable population of purified granulosa cells.

B. Kinetics of Labeling

When cultures are labeled for different times with $^{35}SO_4$ and [³H]glucosamine, precursors for proteoglycan synthesis, the total incorporation of these labels into macromolecules becomes linear after a lag period of a few hours (Fig. 1), and linearity continues for at least 72 hr (5). The amount of labeled macromolecules associated with the cell layer increases with time but at a decreasing rate until it reaches a plateau at 16–20 hr, whereas labeled macromolecules in the medium continue to accumulate at a constant rate. It takes 16–20 hr for the labeled species in the cell layer to reach steady-state conditions where synthesis of proteoglycans is balanced by the combination of secretion into the medium and intracellular degradation. Therefore, the relative distribution of label in the different cell-associated species after 16–20 hr labeling reflects their actual concentrations in this compartment.

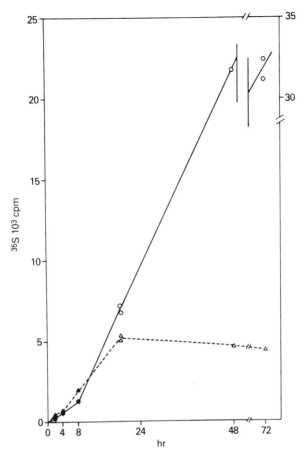

FIG. 1. Accumulation of ³⁵S-labeled proteoglycans in the culture medium (○) and cell layer (△) after the initiation of cell culture. Modified from Ref. 5.

C.   Identification of Proteoglycans and Degradation Products in the Cell Layer

Complete solubilization and dissociation of proteoglycans in the cell layer is most effectively achieved by extracting the cell layer with a combination of a detergent (typically 2% Triton X-100 or 2% of the zwitterionic detergent, CHAPS) with the effective denaturant, 4 M guanidine HCl (9). Small Sephadex G-50 columns equilibrated with 8 M urea in a solvent with 0.15 M NaCl are used to remove the unincorporated isotope and to replace the salt, guanidine HCl, with the uncharged denaturant, 8 M urea. This solvent effectively keeps the macromolecules dissociated and permits the use of an ion-exchange step to separate the proteoglycans from other macromolecules.

When extracts from cultures labeled 20 hr with $^{35}SO_4$ and either [^{3}H]glucosamine or [^{3}H]serine were applied to DEAE–Sephacel in the 8 M urea solvent and eluted with a continuous NaCl gradient, the results shown in Fig. 2 were obtained. Almost all of the incorporated ^{35}S activity binds to the ion-exchange resin, eluting as two proteoglycan peaks during the gradient. This indicates that the granulosa cells synthesize little or no sulfated glycoproteins, because these either would not bind to the column or would elute early in the gradient. Thus, the radioactive sulfate is a specific probe for proteoglycans. For [^{3}H]glucosamine as a precursor, ~35% of the incorporated label was recovered in the proteoglycan peak. Most of the remainder did not bind and represents glycoproteins synthesized by the cells. Little if any label was incorporated into hyaluronic acid, which would normally bind to the column but elute at a NaCl concentration of ~0.2 M, well before the proteoglycan peaks. With [^{3}H]serine as a precursor, <1% of the incorporated label was present in the proteoglycan peaks, with the vast majority eluting in the unbound fraction. This indicates that only

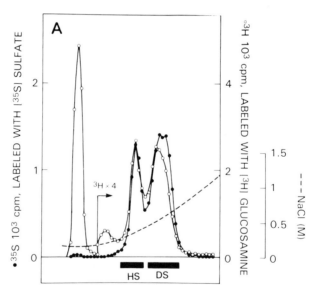

FIG. 2. DEAE–Sephacel chromatography of cell layer fraction. Cultures were labeled with (A) $^{35}SO_4$ (●) and [^{3}H]glucosamine (○), and (B) $^{35}SO_4$ (●) and [^{3}H]serine (○). NaCl concentration (- - - -). Inset in panel B shows the profile of ^{3}H activity with an expanded scale across the ^{35}S-labeled peaks. Modified from Ref. 9.

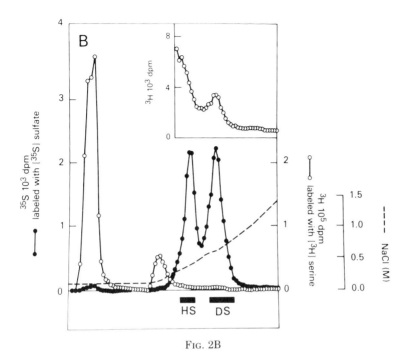

FIG. 2B

a small proportion of the total cell protein synthesis is directed toward proteoglycan core proteins and that the DEAE–chromatography step is very effective at purifying the proteoglycans from most of the other proteinaceous macromolecules in the cell extract. As the expanded scale in Fig. 2B indicates, however, it is still likely that some of the ^{3}H-labeled material in the proteoglycan peaks represents nonproteoglycan material as shown by the tailing of the ^{3}H peak through the proteoglycan region (9). This indicates the difficulty in achieving complete purification of cell-associated proteoglycans.

When the two proteoglycan peaks were recovered and analyzed, the first contained heparan sulfate (HS) species while the second contained dermatan sulfate (DS) species. Subsequent analyses indicated that the heparan sulfate chains contained ~0.8 sulfate ester groups per disaccharide whereas the dermatan sulfate contained 1.2–1.3 sulfate ester groups per disaccharide, providing the charge density difference leading to the resolution of the two proteoglycan classes on the ion-exchange resin (7, 8).

Molecular-sieve chromatography on Sepharose CL-4B (Fig. 3) revealed two distinct size classes for the heparan sulfate species and three for the dermatan sulfate species. Subsequent analyses led to the

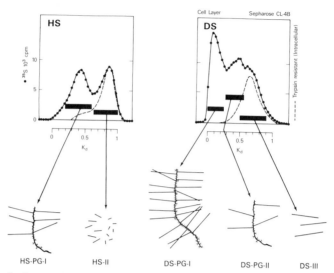

FIG. 3. Sepharose CL-4B chromatography of heparan sulfate (HS) proteoglycans and dermatan sulfate (DS) proteoglycans from the cell layer. Profiles in broken lines show each proteoglycan species isolated from the cell layer after trypsin treatment, indicating intracellular species. Models of proteoglycans eluting in the different fractions as indicated by bars are presented. Modified from Ref. 9.

model structures shown in Fig. 3 for each of these species. The first, partially excluded dermatan sulfate proteoglycan (DS-PG-I) is very large with hydrodynamic characteristics on Sepharose CL-2B similar to the $2-3 \times 10^6$ M_r cartilage monomer proteoglycan (5). After chondroitinase digestion, a core protein band of ~400,000 is observed on SDS–PAGE (7). The core protein is substituted with an average of 20 dermatan sulfate chains of 33,000 average M_r, $2-3 \times 10^2$ O-linked oligosaccharides, and 20–30 N-linked oligosaccharides. The molecule lacks the hyaluronic acid-binding region typical of the large cartilage proteoglycan, as indicated by its inability to bind to hyaluronate (4), and the absence of the epitope for the 1-C-6 monoclonal antibody (10), which is highly conserved for the hyaluronic acid-binding region (unpublished observations). This proteoglycan is a secretory product with essentially all of it being secreted into the medium in an apparently unaltered form (11). It is a major component of the follicular fluid and is thought to be involved in aspects of follicle development and ovulation (4, 5). It is probably a specialized proteoglycan characteristic of the ovarian granulosa cell.

The second dermatan sulfate proteoglycan (DS-PG-II) eluting with a K_d of 0.45 on Sepharose CL-4B, is smaller, with a core protein of 230,000 M_r after chondroitinase digestion (7). It contains only 4–5 dermatan sulfate chains, 20–30 O-linked oligosaccharides, and 4–5 N-linked oligosaccharides. Most of this molecule is associated with the cell layer and appears to be intercalated into the cell membrane, presumably with a hydrophobic peptide portion characteristic of integral membrane components and analogous to other cell surface proteoglycans observed in other systems (12–14). In analogy with the metabolic fate of the heparan sulfate proteoglycan, described in detail below, this dermatan sulfate proteoglycan is lost from the cell layer (with a $t_{1/2}$ of 4–5 hr) by two processes. These lead either to release into the medium as a form of slightly smaller size, presumably via a processing step which cleaves the ectodomain from its intercalated portion, or to a series of intracellular degradation steps which ultimately culminate by complete depolymerization of the glycosaminoglycan chains in a lysosomal compartment (11). About 30% of the proteoglycans are eventually released into the medium while the rest are eventually totally degraded. Very little if any of the proteoglycan species released into the medium are subsequently taken up by the cells.

The smallest dermatan sulfate species (DS-III), with a K_d of 0.65 on Sepharose CL-4B, consists of single dermatan sulfate chains with small amounts of covalently attached peptides. They represent an intermediate in one of the intracellular degradation pathways for the cell surface DS-PG-II molecules (11); see below. The generation of this species involves extensive proteolysis of the core protein, which occurs shortly after the proteoglycan is internalized ($t_{1/2}$ 30–60 min) but considerably before ($t_{1/2}$ 3–4 hr) reaching a functional lysosomal compartment, where complete depolymerization of the dermatan sulfate chains occurs.*

The heparan sulfate proteoglycan (HS-PG-I), with a K_d of 0.45 on Sepharose CL-4B, has a structure which is very similar to the cell surface-associated DS-PG-II with similar core protein size and types and amounts of oligosaccharides, but differing in the type of glycosaminoglycan; 5–6 heparan sulfate chains of 30,000 M_r per molecule (8). Its metabolism, described in detail below, is similar in most respects to that for the DS-PG-II molecules.

The smaller heparan sulfate species (HS-II), eluting at K_d 0.7–0.8 on Sepharose CL-4B, are glycosaminoglycan fragments generated as intermediates in one of the degradation pathways for the heparan sulfate proteoglycan (see below). There are two size classes represented in this peak with 10,000 M_r and 5000 M_r (9). Since the parent proteoglycan contains chains of 30,000 M_r, generation of the fragments re-

---

* We define functional lysosome as an intracellular compartment with sufficient exoglycosidases and sulfatases to degrade the glycosaminoglycan chains completely as observed experimentally by the release of free $^{35}SO_4$ from ^{35}S-labeled species.

quires both extensive proteolysis of the core protein and limited, presumably rather specific endoglycosidase cleavage. As for the single-chain dermatan sulfate species (DS-III), the degradation steps leading to these heparan sulfate fragments occur considerably before they reach a functional lysosome compartment where complete depolymerization occurs.

After 20 hr labeling ~60% of the total labeled proteoglycans are in the medium (Fig. 1). Isolation and characterization of these proteoglycans revealed three species: the intact DS-PG-I molecules and the partly processed forms of both the smaller dermatan sulfate proteoglycan (DS-PG-II) and the heparan sulfate proteoglycan (HS-PG-I). Of the intact proteoglycans remaining associated with the cell layer, >90% are present on the cell surface. This was demonstrated by analyzing the trypsin-accessible and trypsin-inaccessible proteoglycans. For the heparan sulfate species (Fig. 3: broken line, left panel), Sepharose CL-4B chromatography showed that ~90% of the intact proteoglycan was removed by trypsin digestion, while the remainder as well as all of the fragments were inside the cell and inaccessible to the trypsin treatment. Similar results were observed for the dermatan sulfate proteoglycans. All of the DS-PG-I and ~90% of the DS-PG-II were trypsin accessible, while all of the DS-III single-chain species and a small amount of DS-PG-II were inaccessible (Fig. 3: right panel). Since the labeling time of 20 hr was sufficient to equilibrate all of the species, the overall distribution indicates that the degradation products are the major intracellular species while most of the intact proteoglycans are on the cell surface.

When 20 hr labeled cultures were chased in medium with heparin, most of the cell-associated DS-PG-I but only 10–15% of the DS-PG-II or HS-PG-I were displaced, indicating that those displaced are probably associated with the cell surface through relatively nonspecific ionic interactions (15). The half-life of the large cell-associated DS-PG-I on the cell surface is 3–4 hr, indicating that the ionic association is relatively long-lived. However, unlike the other two proteoglycans, all of the DS-PG-I eventually is released into the medium.

III.   Metabolism of Granulosa Cell Proteoglycans

A.   Transit Time from Completion of Proteoglycans
in Golgi to Cell Surface

A series of cultures were pulsed for 2 min with $^{35}SO_4$ and chased for different times followed by a 2-min trypsin treatment to remove any labeled proteoglycans on the cell surface at that time. The results (Fig. 4) show a lag of 5–8 min before proteoglycans labeled during the pulse begin to arrive at the cell surface. Over the next 5–10 min there is a rapid rise in the trypsin-accessible molecules to a limit value of ~75% of the total. Thus, 60 min after labeling ~75% of the proteoglycans labeled in the initial 2-min period are on the cell surface. The average transit time from site of sulfate incorporation, the last step in proteoglycan synthesis in the Golgi, to the cell surface was ~13 min. For cultures identically labeled but chased in the continuous presence of low levels of trypsin, ~93% of the labeled molecules were recovered in

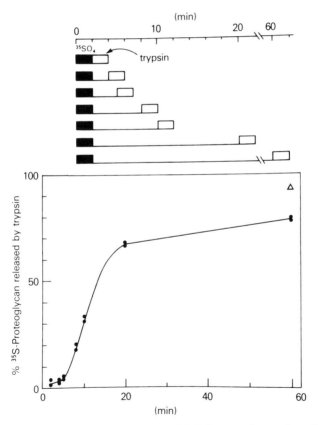

FIG. 4. Measurement of transit time of ^{35}S-labeled proteoglycans from Golgi to cell surface. As summarized in inset, cells were labeled with $^{35}SO_4$ for 2 min and treated with trypsin at different chase times. Amounts of trypsin-released (cell surface) and trypsin-resistant (intracellular) ^{35}S-labeled proteoglycans were quantitated using Sephadex G-50 chromatography. The triangle at 60 min indicates the amount of proteoglycan released by continuous trypsin treatment. Modified from Ref. 11.

the medium (triangle, Fig. 4). This represents all of the molecules which made it to the surface. The remaining 7% were inaccessible to trypsin and directly enter into degradation pathway 2 described below, as shown by their degradation to fragments ~30 min earlier than proteoglycans which travel to the surface before reentering the cells (11). The difference between the total reaching the surface (93%) and the amount on the surface at 60 min (75%) indicates that 18% of the labeled proteoglycans that reached the surface reentered the cell.

B. Intracellular Degradation Pathways
for Heparan Sulfate Proteoglycans

A series of chase experiments was done to define more closely the reentry and subsequent degradation processes for the cell-associated proteoglycans. The experiments focused on the heparan sulfate species because, unlike the dermatan sulfate species, the intact proteoglycan is readily resolved from the fragments and can be quantitated more precisely. One of the protocols used (Fig. 5) was: (a) a 30-min labeling followed by a 30-min chase to clear the secretory pathway, (b) trypsin digestion, and (c) additional chase periods up to 180 min. This focuses initially on those intact proteoglycans which have reached the surface and reentered the cell. The trypsin treatment prevents the reentry of any further labeled proteoglycans from the cell surface and allows the kinetics of the various steps in the degradation pathways to be determined. At each chase time the proteoglycans remaining in the cell layer were extracted, purified by chromatography on Sephadex G-50 and DEAE—Sephacel, and the heparan sulfate peak eluted on Sepharose CL-4B. Selected profiles are shown in Fig. 5A and a summary of the kinetics in Fig. 5B. At the end of the trypsin treatment, 80% of the label inside the cell was present in intact HS-PG-I. With chase time this peak disappeared with an initial slope on a log scale indicative of a half-life of 30 min (Fig. 5B). Some of this proteoglycan being lost from the system appeared in a fragment peak with a K_d equivalent to a glycosaminoglycan chain of ~10,000 M_r (solid vertical line, Fig. 5A). From other similar experiments with shorter labeling pulses, the $t_{1/2}$ for the appearance of the 10,000 M_r fragments from proteoglycans leaving the surface is ~30 min. Approximately 60 min later, the 10,000 M_r fragments are degraded to 5000 M_r fragments (dashed vertical line, Fig. 5A). After the initial increase, the fragments are lost from the cells with log-linear kinetics characteristic of a first-order process having a $t_{1/2}$ of ~3 hr (Fig. 5B). Knowing the initial rate of loss of labeled proteoglycan and the half-life of the fragments,

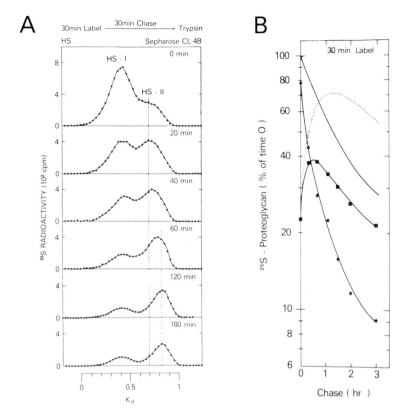

FIG. 5. Measurement of degradation rate of newly internalized proteoglycans. Newly internalized proteoglycans were labeled with a protocol described in the text, and their degradation was analyzed by a chase experiment. (A) Sepharose CL-4B profiles of ^{35}S-labeled HS proteoglycans at different chase times, showing the disappearance of HS-I species and the concomitant increase of HS-II during the earlier chase time and its decrease thereafter. Solid and broken vertical lines indicate the elution positions of HS-II fragments with 10 kDa and 5 kDa, respectively. The areas of peaks are proportional to the total labeled HS proteoglycans in the original extract. (B) The quantitation of HS-I (◆) and HS-II (■) species from (A). Solid line, total. Dotted line indicates the theoretical amounts of HS-II species assuming that all degraded HS-I species give rise to the HS-II. Modified from Ref. 11.

a curve can be calculated for the hypothetical case in which all the internalized HS-PG-I is lost from the system through the fragmentation pathway. This gives the dashed curve in Fig. 5B. The difference between the theoretical and the observed curves indicates that ~60% of the HS proteoglycans are being lost from the cells by some other pathway.

Other experiments have revealed that <5% of the intracellular pro-teoglycans return to the cell surface (11). Further, the net loss of HS-PG-I and HS-II fragments from the system is balanced by the ap-pearance of free $^{35}SO_4$ in the medium, indicating that complete degra-dation in functional lysosomes accounts for the entire loss of the intra-cellular species. This experiment, along with several others (11), indicates that the HS-PG-I molecules reentering the cell are degraded by two separate pathways as outlined in Fig. 6. Approximately 45% of the cell-associated HS-PG-I molecules enter into degradation pathway 1. After internalization, the molecule remains intact with an average half-life of ~25 min before it reaches a functional lysosome compart-

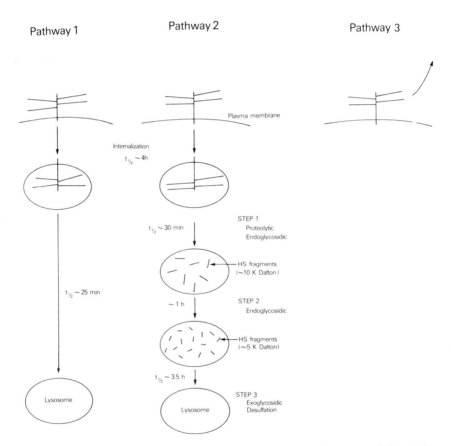

FIG. 6. A schematic model for intracellular degradation (pathways 1 and 2) and the release (pathway 3) of cell surface heparan sulfate (HS) proteoglycans. Modified from Ref. 17.

ment where it is very rapidly degraded with the consequent release of free $^{35}SO_4$. Approximately 25% of the cell-associated HS proteoglycan molecules enter into degradation pathway 2. Approximately 30 min after leaving the surface these molecules are rapidly degraded to the 10,000 M_r fragments (step 1), which involves both proteolysis and an endoglycosidase activity. A further degradation step by a second endo-glycosidase activity, presumably with a different specificity, occurs ~60 min later and produces the 5000 M_r fragments (step 2). These smaller fragments have a long $t_{1/2}$ of ~3 hr before reaching a func-tional lysosomal compartment where they are rapidly depolymerized with the consequent release of $^{35}SO_4$ (step 3). The remaining 30% of the cell-associated HS proteoglycans are released into the medium (pathway 3) with a $t_{1/2}$ of 3 hr, probably by an as yet poorly understood process which partially degrades the molecule via a limited proteolytic cleavage in the ectodomain. Whether this step occurs at the cell sur-face, or inside the cell accompanying a rapid process such as the recy-cling step in receptor-mediated uptake pathways (16), remains to be determined. The latter suggestion, as discussed below, seems likely.

A summary of another label–chase experiment is shown in Fig. 7 to illustrate additional points. In this case a labeling time of 10 min was used and the cultures were not treated with trypsin. As the intact HS-PG-I is lost from the cell layer, one portion appears in HS-II frag-ments, as indicated by the lag in the disappearance of the fragments from the system (Fig. 7A). Another portion is released into the medi-um. There is a greater loss of HS-PG-I from the cells than appears in

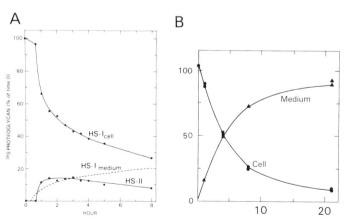

FIG. 7. Turnover of cell surface heparan sulfate (HS) proteoglycan after 10 min $^{35}SO_4$ labeling (A) and dermatan sulfate proteoglycans (DS-PG-I) (B). Modified from Ref. 11.

fragments and is released into medium, reflecting the proportion having been degraded in functional lysosomes. The net loss of the large DS-PG-I from the cell layer and its quantitative appearance as DS-PG-I in the medium is shown in Fig. 7B. Thus, the cells clearly handle the primary secretory proteoglycan (DS-PG-I) differently from the cell-associated species, HS-PG-I (described above) and DS-PG-II (which shows metabolism similar to the HS-PG-I).

As mentioned above, some labeled proteoglycans do not reach the cell surface. These can be studied by using a short labeling pulse and chasing in the continuous presence of trypsin such that any molecules reaching the surface are prevented from reentering the cell. At different chase times these intracellular proteoglycans in the cell layer were isolated and fractionated on DEAE–Sephacel. The heparan sulfate species were then analyzed on Sepharose CL-4B. Representative analyses are shown in Fig. 8, along with analyses from a parallel experiment for cultures not treated with trypsin. It is clear that the HS-PG that do not reach the cell surface enter degradation pathway 2 ~30 min earlier than the HS proteoglycan molecules which go to the surface prior to reentry (Fig. 8B at 1.5-hr chase). This result also validates the use of the trypsin digestion as a means for distinguishing intracellular species, since it rules out significant amounts of cell surface proteoglycans in trypsin-inaccessible sites. Quantitation reveals that little if any of the HS proteoglycan which does not reach the surface is degraded by pathway 1.

C. Metabolism of DS-PG-II

Precise kinetics for analysis of DS-PG-II metabolism are difficult to determine because of the poor resolution of the three DS species. Nevertheless, its metabolism appears to be similar to that for the HS proteoglycans. Approximately 40% are released into the medium with the same $t_{1/2}$, with the remainder being internalized with the same kinetics as for the HS-PG-I. At long labeling time, the DS-III single-chain peptide species is the predominant intracellular dermatan sulfate component, and it is lost from the cells with very similar kinetics as for the HS-II fragments. Finally, a large proportion of DS-PG-II is lost from the cells without appearing in the DS-III chain peptides, suggesting a pathway 1 degradation route (11). Unlike the degradation of HS proteoglycan in pathway 2, there is no endoglycosidase activity directed toward the dermatan sulfate chains, which retain their fully intact size until they reach a functional lysosome.

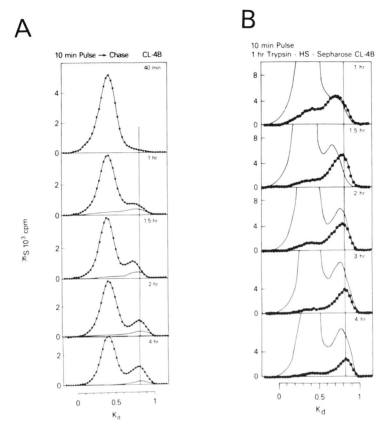

Fig. 8. Sepharose CL-4B profiles for ^{35}S-labeled HS proteoglycans in the cell layer for the 10-min label and chase experiment. This experimental protocol compares the degradation kinetics between the total HS proteoglycan population and those that do not appear on the cell surface. (A) Profiles for the chase experiment of the total heparan sulfate proteoglycan (●) and for that treated with continuous trypsin (——). (B) Same data as (A) but now focusing on the trypsin-treated group (●) drawn at twice the scale. Vertical lines indicate the smallest average size limit for HS-II species at $K_d = 0.82$. The areas of peaks in each panel are proportional to the total labeled HS proteoglycans in the original extract. Modified from Ref. 11.

D. Effects of Chloroquine and Monensin on Intracellular Degradation

A series of experiments were done to test the effects of chloroquine and monensin on the degradation pathways (11, 17). These reagents are known to enter acidic intracellular compartments and to raise their pH. They also affect membrane structure and alter membrane fusion processes. Cultures were labeled with $^{35}SO_4$ for 80 min and chased with or without monensin. At different chase times, the heparan sulfate species in the cell layer and in the intracellular (trypsininaccessible) compartment were isolated and analyzed on Sepharose CL-4B. Some of the results are indicated in Fig. 9. Both chloroquine and monensin showed similar results. Internalization of the membrane-associated proteoglycans was only slowed down slightly. Those HS-PG-I entering into pathway 1, ~60% of those internalized, accumulated as intact molecules inside the cell (Fig. 9). Separate experiments revealed that these molecules still retained the ability to intercalate into micelles (11). However, whether they accumulate in a prelysosomal compartment or the higher pH of the lysosomal compartment effectively blocks any significant degradation remains to be determined. For those HS-PG-I entering pathway 2, ~40% of those internalized, step 1 was only slightly slowed down but proceeded to the 10,000 M_r fragments. This suggests that the protease(s) and endoglycosidase(s) involved in this step do not require an acidic environment. However, step 2 (the generation of 5000 M_r fragments) was totally blocked (compare 8-hr chase results for control and monensin-treated group, Fig. 9, vertical lines). Thus, the 10,000 M_r fragments accumulate in the cell as more cell surface HS-PG-I enter this pathway. This suggests that either step 2 requires an acidic environment or the compartment with the 10,000 M_r fragments is separate and unable to fuse with that responsible for generating the smaller 5000 M_r fragments. Other long-term labeling protocols followed by chase with or without the effector (11) revealed that step 3, the degradation of the 5000 M_r fragments in functional lysosomes, was also totally blocked. This suggests that either the compartment with the 5000 M_r fragments fails to reach a functional lysosome or that step 3 requires an acidic environment. Neither chloroquine nor monensin altered the proportion of HS-PG-I (~30% of the total) released into the medium, although this step was slowed down slightly. The fact that the proportions of internalized HS-PG-I entering the two degradation pathways were apparently not altered by these reagents argues that the pathways are segregated.

E.   Effect of Leupeptin on Degradation Pathways

Leupeptin, an inhibitor of proteases, especially sulfhydryl proteases such as cathepsin B, has been used to perturb steps in the acetylcholine receptor degradation in muscle cells (18). Since proteolytic activity

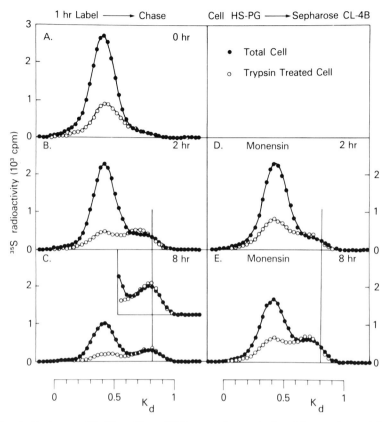

FIG. 9. Sepharose CL-4B profiles of HS proteoglycans isolated from cultures labeled for 60 min and chased with or without monensin. (A–C) Profiles of HS proteoglycans in controls at 0, 2, and 8 hr chase. (D, E) Profiles for monensin-treated cultures at 2 and 8 hr chase. ●, Total cell extract; ○, trypsin-treated cell extract (intracellular) HS proteoglycan. Inset in (C) shows the latter part of the chromatogram in a three times expanded scale. Vertical lines indicate the average size of the smallest HS-II species (K_d = 0.82) in the control cultures. The areas of peaks are proportional to the total labeled HS proteoglycans in the original extract. Modified from Ref. 17.

appears to be involved in the mechanism of release of some of the cell membrane-associated proteoglycans and is a critical part of step 1 in degradation pathway 2, a series of experiments were done with this reagent to test its effect on HS proteoglycan metabolism (19) in the rat granulosa cells. Figure 10A illustrates an experimental protocol to determine the degradation rates of newly internalized proteoglycans (30-min label) and those accumulating in the cell layer after long labeling (20-hr label). Separate experiments showed that the leupeptin

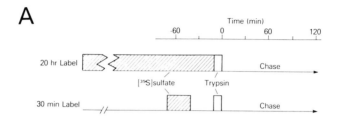

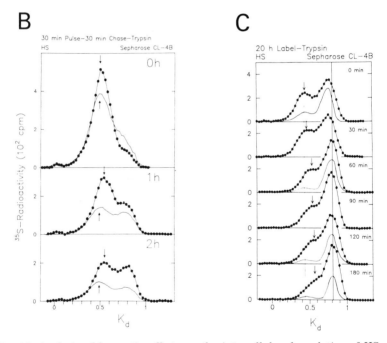

FIG. 10. Analysis of leupeptin effects on the intracellular degradation of HS proteoglycans. (A) An experimental protocol comparing the degradation kinetics of newly
internalized HS proteoglycans (30-min label) and those accumulating after a long label
(20 hr). (B) Sepharose CL-4B profiles of intracellular HS proteoglycans after 30-min
label and their degradation in the presence (●) or absence (——) of leupeptin. (C)
Sepharose CL-4B profiles of intracellular HS proteoglycans after 20-hr label and chase
in the presence (●) or absence (——) of leupeptin. Downward-pointing arrows indicate
peak position for HS proteoglycan at each chase time in leupeptin-treated cultures.
Upward-pointing arrows in (B) and vertical line at $K_d = 0.43$ in (C) indicate the peak
position for HS-PG-I for control cultures. Vertical line at $K_d = 0.82$ indicates the average size of the smallest HS-II species. The areas of peaks in (B) and (C) are proportional
to the total labeled HS proteoglycans in the original extract. Modified from Ref. 19.

did not affect proteoglycan synthesis and transport to the cell surface, nor did it alter the kinetics of internalization of membrane-associated proteoglycans appreciably (19). The release of the medium forms of the HS-PG-I and DS-PG-II into the medium was significantly slowed, but the proportions going this route (pathway 3) remained unchanged. Thus, it is likely that a protease(s) that is partially inhibited by the leupeptin is involved in this release process.

After trypsin digestion to prevent additional internalization, the intracellular HS proteoglycan is lost from the cells with an overall kinetics illustrated in Fig. 11. For example, in the 30-min label protocol, 45% of the HS-PG-I proteoglycan disappears in 2 hr of chase (◆). The $t_{1/2}$ of this loss is somewhat greater than 2 hr, much longer than the control $t_{1/2}$ of 30 min. Only about 15% of these appear in the HS-II fragments (■). Thus, nearly two-thirds are being degraded in functional lysosomes by pathway 1, although the kinetics for the HS proteoglycans

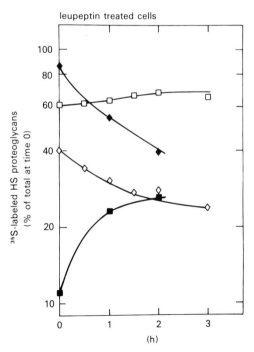

FIG. 11. Degradation of each intracellular HS proteoglycan species labeled for 20 hr or 30 min in the leupeptin-treated cultures. ^{35}S-Labeled HS-I (◆ , ◇) and HS-II species (■ , □) were quantitated from Fig. 10 for leupeptin-treated group. Unfilled and filled symbols show the results for 20-hr labeled culture and 30-min labeled cultures, respectively. Note the logarithmic scale of the ordinate. Modified from Ref. 19.

to reach the functional lysosome compartment are significantly slower than for controls. The effects of leupeptin on pathway 2 were more complex. The generation of the 10,000 M_r fragments in step 1 was significantly slowed, and a probable intermediate in this step was unmasked. This intermediate, indicated by the downward-pointing arrows in Fig. 10, was smaller than the intact HS-PG-I (upward-pointing arrows) but larger than the HS-II fragments. When this intermediate was isolated and degraded with alkaline borohydride, its heparan sulfate chains were intact (i.e., with 30,000 M_r chains identical to those on the intact HS-PG-I). Thus, the intermediate represents HS proteoglycan partially degraded by proteases, which are presumably inhibited sufficiently by the leupeptin to slow this step down significantly. This result also suggests that proteolysis in step 1 must precede the endoglycosidase activity in the generation of the 10,000 M_r fragments, thereby dividing step 1 into two sequential substeps. While difficult to determine precisely, the overall $t_{1/2}$ for generating the 10,000 M_r fragments increases from 30 min in the control to 1–2 hr in the leupeptin-treated cultures.

The subsequent second endoglycosidase cleavage to generate the 5000 M_r fragments proceeds with nearly normal kinetics in the presence of leupeptin (Fig. 10). However, surprisingly, the final degradation of the 5000 M_r fragments in a functional lysosome compartment was totally blocked. This is best observed in the results for the long, 20-hr labeling protocol. At the beginning of the chase ~60% of the label inside the cell was present in fragments, primarily the 5000 M_r form (Figs. 10 and 11). This increased 6% to a constant plateau value during 3 hr of chase (Fig. 11). At the same time the larger HS species (intact HS proteoglycan and the partially degraded intermediate described above) decrease by 15%. As for the short labeling protocol, approximately twice as many intracellular HS-PG-I are lost as appear in the HS-II fragments, reflecting the continued operation of pathway 1.

Why leupeptin blocks step 3 of pathway 2 is not known. However, it is clear that a functional lysosomal compartment still operates, since pathway 1 continues with the complete degradation of the glycosaminoglycan chains. Since there are no apparent chemical differences other than chain length between the 5000 M_r fragments in pathway 2 and the intact chains on HS-PG-I in pathway 1, the blockage of step 3 strongly suggests that the 5000 M_r fragments are in a prelysosomal compartment and that the leupeptin prevents the entry of the 5000 M_r from this compartment into a functional lysosome. In any event, it is most unlikely that the accumulating 5000 M_r fragments in the leupeptin-treated cultures are in the functional lysosomes where the pathway 1 HS proteoglycans are being degraded. As with chloroquine and monensin, the proportions of the molecules entering degradation pathways 1 and 2 were not affected by leupeptin, even though both pathways were perturbed in fundamental ways. This provides further evidence that these two pathways are compartmentalized and reflect different intracellular pathways which involve cell membrane flow.

IV. General Considerations

Proteoglycan metabolism in the granulosa cell system reflects several patterns which will have their counterparts in many other cell systems. The cells synthesize two distinct classes of proteoglycans. One is a secretory product, which is destined to be secreted into the extracellular compartment. In the granulosa cells this is the secretion of DS-PG-I into the follicular fluid, but in other cases it will be secretion to an extracellular matrix, where the molecule will play some specialized role(s) characteristic for that cell or tissue. The other class is proteoglycans (HS-PG-I and DS-PG-II), which are most likely integral membrane components intercalated into the lipid bilayer. These proteoglycans are metabolized quite differently from the secretory ones, participating in several if not all of the complex pathways in the cell involving membrane generation, transport, and degradation. Most, if not all, cells will have representatives of this class of proteoglycan that may be tailored for specialized functions characteristic of the parent cell or tissue. Direct evidence for such proteoglycans dates from the original studies by Kraemer (20) on fibroblasts and has been further refined for hepatocytes (12), capillary endothelial cells (21), and epithelial cells (13), among others.

The current status of research on proteoglycan metabolism in the granulosa cell system has uncovered many new questions which remain to be addressed in future studies. What are the cell morphological and physiological analogs for the two major degradation pathways uncovered by the kinetic experiments discussed above? The kinetics and rather direct route to functional lysosomes for pathway 1 are similar to the kinetics of receptor-mediated uptake pathways worked out elegantly in the hepatocyte (22), while the longer, more complex pathway 2 could reflect membrane flow in a more general, pinocytotic uptake pathway (16). Perhaps a careful kinetic analysis of the uptake and degradation of a specific ligand such as luteinizing hormone, for which these cells have a receptor, could answer this question.

Does the mechanism by which surface-associated proteoglycans are shed involve an intracellular protease cleavage, or does it occur on the

cell surface? In some cases, such as the major chondroitin sulfate proteoglycan synthesized by melanoma cells, this process appears to involve a rather selective cleavage, reducing the core protein size from
M_r 250,000 to M_r 170,000 for the cell-associated and medium forms,
respectively (23). Since very few of the internalized proteoglycans re-
appear on the cell surface or are released into the medium (11), the
involvement of internalization in the processing step would have to be
a rapid one such that the size of the intracellular pool of molecules to
be released remains small relative to other intracellular pools. That
this could be the case is indicated both from the rapid kinetics of
recycling of the asialoglycoprotein receptor to the cell surface in the
hepatocyte, and from the estimate of the proportion of the contents of
the solution which is internalized by this process which reappears to
the surface (24–27). This proportion (30%) is similar to the amount of
HS-PG-I eventually released at the surface. Thus, if receptor-mediated
internalization delivers the contents of a vesicle and its exposed interior surface to an initial sorting location which involves, among other
things, some limited proteolysis, it is possible that partially cleaved
proteoglycans (i.e., the released ectodomain) could recycle rapidly to
the surface where they may remain ionically associated with the surface for eventual release into the medium. The remaining interiorized,
intercalated proteoglycans would be sorted with the membrane com-
partment(s) and enter into the flow culminating in fusion with functional lysosomes.

Are the two degradation pathways segregated entirely, being initiated by interiorization at different locations on the cell surface and
ending in distinctly different lysosomal compartments? Are the HS
proteoglycans which eventually enter pathway 1 different in some key
way from those which enter pathway 2? For example, their core proteins may be closely related but not identical gene products with a
difference in sequence somewhere in the protein targeting the completed proteoglycan for one or another location on the cell surface.
There are many examples of targeting cell surface molecules for different locations, and several of these may involve proteoglycans. For
example, two HS proteoglycans synthesized by capillary endothelial
cells and distinguished by their ability either to bind to antithrombin
or not, may well be segregated in their cell surface localization with
the form that binds to antithrombin located on the luminal surface
(28).

Are the cell surface DS-PG-II molecules interchangeable functionally with their HS-PG-I counterparts? Although more difficult to
determine with the same precision, the metabolism of the DS-PG-II
show similar kinetics and similar proportions entering the different

pathways as for HS-PG-I (11). Furthermore, the overall general structures of these two proteoglycans are very similar (7, 8). How related are the core proteins for these two proteoglycans? It is quite possible that they are derived from the same gene family, but it is also probable that they differ sufficiently such that one is recognized by the Golgi-localized factory for heparan sulfate synthesis and the other by the factory for dermatan sulfate synthesis. Is this distinction built into the sequence of the core proteins? In epithelial cell systems there is increasing evidence for hybrid proteoglycans, that is, single core proteins substituted with both heparan sulfate and dermatan sulfate chains (29, 30). The granulosa cells, however, do not appear to synthesize any appreciable amount of such hybrid species. We can also speculate that the core protein for the secretory DS-PG-I is from a separate gene family than for the cell membrane-associated proteoglycans.

Most cells will quite likely have similar, though modified, degradation pathways for cell membrane-associated proteoglycans. An extreme variation may relate to the metabolism of proteoglycans in neuromuscular synapses. Several laboratories have identified heparan sulfate proteoglycans as major constituents of cholinergic synaptic vesicles (31), and these vesicles are triggered to fuse with the synaptic membrane in response to an action potential. Other investigators have identified a high concentration of HS proteoglycans as a major constituent in the synapse, associated either with the cell membrane or the intervening basement membrane (32). Carlson (31) and associates have also observed reinternalization of some of the proteoglycans as part of the retrograde membrane flow from the synapse eventually to the cell body and have postulated that modifications in the HS proteoglycan structure accompany this process and are possibly involved in the mechanism. The reentry process is quite likely a highly modified degradation pathway involving programmed membrane flow which eventually reaches a functional lysosomal compartment. Perhaps the HS proteoglycan species involved in this critical neurophysiological process are close cousins to the HS proteoglycans in the ovarian granulosa cell. In this case they have been modified and tailored for different functions such as providing a highly anionic matrix for binding the cationic amine neurotransmitters, for organizing the highly specialized extracellular matrix in the synaptic cleft, and for facilitating the reentry of membrane for its eventual degradation in the cell process. How closely analogous the metabolisms of cell membrane-associated proteoglycans are in these two very different biological systems remains to be determined by challenging future experimentation. The likelihood for discovering theme and variation for this class of proteoglycans in different biologically important context seems high.

REFERENCES

1. Richards, J. S., and Midgley, A. R., Jr. (1976). *Biol. Reprod.* **14**, 82–94.
2. Hsueh, A. J. W., Jones, P. B. C., Adashi, E. Y., Wang, C., Zhuang, L.-Z., and Welsh, T. H., Jr. (1983). *J. Reprod. Fertil.* **69**, 325–342.
3. Mueller, P. L., Schreiber, J. R., Lucky, A. W., Schulman, J. D., Rodbard, D., and Ross, G. T. (1978). *Endocrinology* **102**, 824–831.
4. Yanagishita, M., Rodbard, D., and Hascall, V. C. (1979). *J. Biol. Chem.* **254**, 911–920.
5. Yanagishita, M., and Hascall, V. C. (1979). *J. Biol. Chem.* **254**, 12355–12364.
6. Yanagishita, M., Hascall, V. C., and Rodbard, D. (1981). *Endocrinology*, **109**, 1641–1649.
7. Yanagishita, M., and Hascall, V. C. (1983). *J. Biol. Chem.* **258**, 12847–12856.
8. Yanagishita, M., and Hascall, V. C. (1983). *J. Biol. Chem.* **258**, 12857–12864.
9. Yanagishita, M., and Hascall, V. C. (1984). *J. Biol. Chem.* **259**, 10260–10269.
10. Stevens, J. W., Oike, Y., Handley, C. J., Hascall, V. C., Hampton, A., and Caterson, B. (1984). *J. Cell Biochem.* **26**, 247–259.
11. Yanagishita, M., and Hascall, V. C. (1984). *J. Biol. Chem.* **259**, 10270–10283.
12. Kjellén, L., Oldberg, Å., and Höök, M. (1980). *J. Biol. Chem.* **255**, 10407–10413.
13. Rapraeger, A. C. and Bernfield, M. (1983). *J. Biol. Chem.* **258**, 3632–3637.
14. Norling, B., Glimelius, B., and Wasteson, Å. (1981). *Biochem. Biophys. Res. Commun.* **103**, 1265–1272.
15. Höök, M., Kjellén, L., Johansson, S., and Robinson, J. (1984). *Annu. Rev. Biochem.* **53**, 847–869.
16. Steinman, R. M., Mellman, I. S., Muller, W. A., and Cohn, Z. A. (1983). *J. Cell Biol.* **96**, 1–27.
17. Yanagishita, M., and Hascall, V. C. (1985). *J. Biol. Chem.* **260**, 5445–5455.
18. Hyman, C., and Frohner, S. C. (1983). *J. Cell Biol.* **96**, 1316–1324.
19. Yanagishita, M. (1985). *J. Biol. Chem.* **260**, 11075–11082.
20. Kraemer, P. M. (1971). *Biochemistry*, **10**, 1445–1451.
21. Marcum, J. A., and Rosenberg, R. D. (1984). *Biochemistry* **23**, 1730–1737.
22. Wolkoff, A. W., Klausner, R. D., Ashwell, G., and Harford, J. (1984). *J. Cell Biol.* **98**, 375–381.
23. Harper, J. R., Bumol, T. F., and Reisfeld, R. A. (1984). *J. Immunol.* **132**, 2096–2103.
24. Weigel, P. H., and Oka, J. A. (1984). *J. Biol. Chem.* **259**, 1150–1154.
25. Oka, J. A., and Weigel, P. H. (1983). *J. Biol. Chem.* **258**, 10253–10262.
26. Harford, J., Bridges, K., Ashwell, G., and Klausner, R. D. (1983). *J. Biol. Chem.* **258**, 3191–3197.
27. Besterman, J. M., Airhart, J. A., Woodworth, R. C., and Low, R. B. (1981). *J. Cell Biol.* **91**, 716–727.
28. Marcum, J. A., and Rosenberg, R. D. (1985). *Biochem. Biophys. Res. Comm.* **126**, 365–372.
29. Rapraeger, A., Jalkanen, M., Endo, E., Koda, J., and Bernfield, M. (1985). *J. Biol. Chem.* **260**, 11046–11052.
30. David, G., and Van den Berghe, H. (1985). *J. Biol. Chem.* **260**, 11067–11074.
31. Carlson, S. S., and Kelly, R. B. (1983). *J. Biol. Chem.* **258**, 11082–11091.
32. Anderson, M. J., and Fambrough, D. M. (1983). *J. Cell Biol.* **97**, 1396–1411.

Integral Membrane Proteoglycans as Matrix Receptors: Role in Cytoskeleton and Matrix Assembly at the Epithelial Cell Surface

Alan Rapraeger

Department of Pathology, University of Wisconsin School of Medicine, Madison, Wisconsin 53706

and

Markku Jalkanen* and Merton Bernfield

Department of Pediatrics, Stanford University School of Medicine, Stanford, California 94305

I. Introduction

A. Matrix Receptors

The extracellular matrix (ECM) is the substratum for all adherent cells. Stromal or interstitial cells are embedded in an ECM that they produce, and they migrate through it by making and breaking attachments. Epithelial cells, on the other hand, are not surrounded by matrix. Rather, they are bounded at their basal surfaces by a specialized layer of ECM that they produce, the basal lamina. The basal lamina is closely associated with the ECM produced by the stromal cells, form-

* Present address: Department of Medical Biochemistry, Turku University School of Medicine, SF-20520 Turku 52, Finland.

129

ing a basement membrane. The epithelial cells reside on the basement membrane as a stable sheet of cells and do not migrate through it. During development, however, the basement membrane undergoes remodeling directed by these closely apposing tissue types. Indeed, this remodeling has been proposed to direct developmental changes in epithelial morphology (Bernfield *et al.*, 1984). Loss of the basement membrane during development correlates with loss of polarized epithelial morphology and occasionally yields migratory mesenchymal cells. Neoplastic cells also lose their basement membrane, allowing them to invade into the surrounding stroma. Ultimately, in mature and differentiated organs, the specific composition of the basement membrane reflects these prior tissue interactions and provides cues that generate the polarization of epithelial cells and maintains their shape and differentiated behavior. A central question, therefore, is "how do cells detect and respond to matrix components?"

Work from a number of laboratories demonstrates that the interaction of cells with their matrix is mediated by ECM receptors in the plasmalemma. Although only a few have been described, cells may have a battery of such receptors for both epithelial and stromal matrix components. A representative list of such receptors includes a receptor specific for laminin (Rao *et al.*, 1983; Terranova *et al.*, 1983), an adhesive glycoprotein of the basement membrane, collagen receptors on platelets (Chiang and Kang, 1982), chondrocytes (Mollenhauer and Von der Mark, 1983), fibroblasts (Goldberg, 1979), hepatocytes (Rubin *et al.*, 1981) and corneal epithelial cells (Sugrue and Hay, 1981), and a variety of types of fibronectin receptors (see review by Yamada *et al.*, 1985). These receptors may have diverse functions, including roles in matrix secretion and/or assembly and cell/matrix adhesion. Furthermore, different cell types may rely on different molecules, and these may change during embryonic development, wound healing, and neoplastic transformation.

Matrix receptors that are involved in the detection and response to the ECM must differ from hitherto described cell surface receptors such as those involved in chemical signaling of cells or cellular nutrition. In these cases the ligand is soluble or, if particulate as with a virus, is small relative to the size of the cell. The concentration of these ligands can change dramatically and quickly. In contrast, the matrix is composed of molecules that are organized into large insoluble complexes that most often exceed the size of a cell. Also, the matrix is present not transiently, but continuously, at a constant "concentration." Hence, the mechanisms used by matrix receptors to generate a cellular response must accommodate these ECM characteristics. One possible mechanism is that matrix receptors act as matrix anchors, in which binding to the matrix extracellularly causes the receptor to associate with the cytoskeleton intracellularly. This anchoring of the cytoskeleton to the matrix would change or stabilize cell shape, inducing or maintaining a specific differentiative pathway. In this proposed mechanism, cells would sever their association with the matrix either by loss of the matrix or by loss of the receptor itself. For example, when cells change shape, as during morphogenesis, they would shed their matrix receptor.

Many of the cues that influence epithelial cell behavior appear to arise from recognizing matrix products of stromal cells. For example, a substratum of stromal matrix components (interstitial collagens, fibronectin) appears to stabilize epithelial sheets *in vitro* and may accentuate the synthesis of specific differentiated products (Haeuptle *et al.*, 1983; Hay, 1985). Likewise, the acquisition of stable morphology by epithelial sheets *in vivo* correlates with the deposition of stroma-derived interstitial collagens at the interface between these two tissue types (Bernfield *et al.*, 1984). Epithelial cell surface receptors that recognize stromal matrix components may be localized at this interface. Indeed, such receptors may have properties of matrix anchors, as described above. Epithelial sheets display disrupted cytoskeletal architecture when stripped of their ECM (Cohn *et al.*, 1977; Sugrue and Hay, 1981), but rapidly restore a cortical actin network when exposed to even soluble matrix components (Sugrue and Hay, 1981). Such modifications of cell shape have been demonstrated to affect specific cell behavior, including gross control over DNA replication and cell division, and messenger RNA transcription and translation (Folkman and Moscona, 1978; Ben-Ze'ev *et al.*, 1980; Wittelsberger *et al.*, 1981; Haeuptle *et al.*, 1983).

B. Proteoglycans as Membrane-Intercalated Cell Surface Receptors

Cell surface heparan sulfate proteoglycans are excellent candidates to be matrix receptors. They have been implicated in binding several ECM components, including fibronectin (Ruoslahti and Engvall, 1980; Yamada *et al.*, 1980; Stamatoglou and Keller, 1982), laminin (Sakashita *et al.*, 1980; Woodley *et al.*, 1983), interstitial collagens (Stamatoglou and Keller, 1982; Koda *et al.*, 1985), thrombospondin (McKeown-Longo *et al.*, 1984), and other heparan sulfate proteoglycans (Fransson *et al.*, 1983). In addition, cell surface heparan sulfate proteoglycans have been isolated from fibroblast "footpads" (Lark and Culp, 1984) and "adherons" of neural retinal cells (Cole *et al.*, 1985), further implicating them in cell–substratum adhesion.

Cell surface proteoglycans are found on most, if not all, cell types. These include chondroitin sulfate (Oldberg *et al.*, 1981; Bumol *et al.*, 1983; Hedman *et al.*, 1983), dermatan sulfate (Yanagishita and Hascall, 1984), and heparan sulfate forms (see review by Höök *et al.*, 1984). Although some cell surface heparan sulfate proteoglycans are proposed to be peripheral cell surface components, bound to the cell surface via receptors specific for their glycosaminoglycan (GAG) chains, most have lipophilic properties and are believed to be anchored directly into the plasma membrane.

The cell surface heparan sulfate proteoglycans exist as diverse families. Their core proteins range from 35 kDa in hepatocytes (Höök *et al.*, 1984) to 240 kDa in colon carcinoma (Iozzo, 1984). Similar diversity characterizes the number and length of the GAG chains and suggests that the characteristics and/or functions of these molecules or their core proteins may differ, perhaps based on the type of cell from which they are derived.

C. Cell Surface Heparan Sulfate Proteoglycan of Mouse Mammary Epithelial Cells

This chapter will focus on a proteoglycan isolated from the surfaces of cultured NMuMG mouse mammary epithelial cells (Rapraeger and Bernfield, 1985). Immunostaining with a monoclonal antibody directed against its core protein indicates that the proteoglycan is limited to only certain epithelia, suggesting that it belongs to a distinct family of epithelial proteoglycans. The proteoglycan bears chondroitin sulfate and heparan sulfate on the same core protein, and binds via the latter to interstitial collagens and fibronectin. On confluent NMuMG cells *in vitro* and on various simple epithelia *in vivo*, the proteoglycan is sequestered to the basolateral plasmalemma, apparently in close association with the actin-containing cytoskeleton.

The orientation of this receptor on the epithelial cell suggests a role for the proteoglycan as a matrix anchor. The proteoglycan at the basal cell surface may serve to bind and organize interstitial collagens or fibronectin deposited by the adjacent stromal tissue. In a similar manner, the proteoglycan may serve as a focal site at the cytoplasmic face of the plasmalemma for the binding and assembly of the cytoskeleton, similar to mechanisms leading to biogenesis of the erythroid membrane-bound cytoskeleton (Moon and Lazarides, 1984). Binding of the proteoglycan to the matrix may stabilize, if not enhance, cytoskeleton binding to the receptor, thus providing a mechanism for anchorage in which the matrix influences cell behavior.

II. CHARACTERISTICS OF MAMMARY EPITHELIAL
CELL SURFACE PROTEOGLYCAN

A. *Integral Membrane Proteoglycan Is the Predominant Form in NMuMG Cell Cultures*

Mild treatment with trypsin releases proteoglycan from suspended mouse mammary epithelial cells. In contrast, treatment with GAG-degrading enzymes or heparin, known to displace peripheral proteoglycan from Chinese hamster ovary (CHO) cells (Kraemer, 1977) and hepatocytes (Kjellén *et al.*, 1980), fails to remove it and suggests that the molecule is an integral membrane protein. This suggestion is supported by estimates, based on liposome intercalation (Fig. 1), that at least two-thirds of the cellular proteoglycan of these cells is lipophilic and the majority of this lipophilic fraction is cleaved by the cell surface protease treatment (Rapraeger and Bernfield, 1985). The extracellular portion (ectodomain) released by the protease is only slightly smaller than the intact proteoglycan but fails to intercalate into liposomes, suggesting that it has been cleaved from the lipophilic domain (Rapraeger and Bernfield, 1983). The nature of the latter domain, presumably a membrane-anchored portion of the core protein, has not been elucidated.

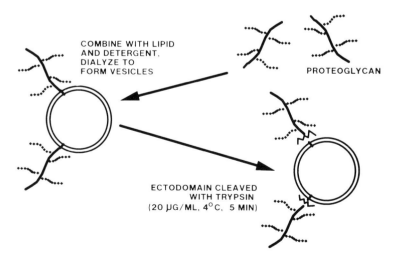

FIG. 1. Cell surface proteoglycan intercalates into liposomes. Isolated cell surface proteoglycan intercalates quantitatively into the lipid bilayer of liposomes (Rapraeger and Bernfield, 1983, 1985). Treatment with trypsin releases the extracellular domain (ectodomain) from a putative membrane-anchored domain. [Reprinted from Rapraeger *et al.* (1987). © 1987 Marcel Dekker, Inc.]

Quantification of the $^{35}SO_4$ in cells labeled to the steady state demonstrates that ~80 to 90% of the proteoglycan in newly confluent NMuMG monolayers is associated with the cells (Rapraeger and Bernfield, 1985). In addition to cell-associated proteoglycan, the NMuMG monolayers have at least two classes of extracellular proteoglycans at their basal surfaces (see Table I). These classes are distinguished from one another by their size and/or buoyant density and are not lipophilic. One of these classes, a low-density (~1.25 g/ml) fraction, is recognized by antibodies specific for the basement membrane heparan sulfate proteoglycan of the EHS tumor (BM-1 antigen, Hassell *et al.*, 1980), antibodies that recognize basement membrane proteoglycan in a wide variety of tissues. However, the BM-1 antibody

TABLE I

MAJOR HEPARAN SULFATE-RICH PROTEOGLYCANS PRODUCED BY NMuMG CELLS

Location	Size (K_{av} on Sepharose 4B)	Buoyant density (g/ml)	Other features	Probable function
Cellular				
1. Cell surface	0.30	1.55	MnAb 281 positive; lipophilic	Matrix receptor
2. Intracellular	0.30	1.55	Lipophilic	Precursor for 1 (?)
3. Intracellular	0.33	1.60	Not lipophilic	Derived from 1 (?) within endosomes
Extracellular				
1. Basal	0.10–0.15	1.25	BM-1 positive; MnAb 281 negative	Basement membrane proteoglycan
2. Basal	0.20–0.25	1.38	BM-1 negative; MnAb 281 negative; collagenase-susceptible	(?)
3. Culture medium	0.32	1.55	MnAb 281 positive; not lipophilic	Derived from 1
4. Culture medium	0.40	1.60	MnAb 281 negative; all heparan sulfate	Degradation product

fails to recognize the cell surface proteoglycan (Jalkanen *et al.*, 1987a). Thus, the cell surface and extracellular heparan sulfate proteoglycans made by these cells may represent distinct gene products and even discrete functions.

B. *Metabolism of Cell Surface Proteoglycan*

Radiolabeling studies indicate that the cell surface proteoglycan of newly confluent cells has a limited lifetime at the cell surface. The proteoglycan disappears from the plasma membrane with a half-life of ~6 hr (Rapraeger and Bernfield, unpublished) concomitant with the appearance of intracellular heparan sulfate GAG fragments, apparently generated by an endoglycosidase either in endosomes or lysosomes. Thus, much of the intracellular proteoglycan, which constitutes a third of the cell-associated proteoglycan, may be the cell surface form en route to or from the cell surface (Table I). In addition, proteoglycan is apparently shed from the cell surface into the culture medium (Jalkanen *et al.*, 1987b), though as a nonlipophilic molecule. A similar molecule, namely a nonlipophilic form, is released from the surface of cells rounded with EDTA, which apparently promotes shedding of the cell surface proteoglycan. These findings suggest that shedding may be a mechanism of rapid removal when the cells change shape, possibly to move or divide, similar to the loss reported by Kraemer and Tobey (1972) during cytokinesis of CHO cells.

C. *Physical Characteristics of the NMuMG Cell Surface Proteoglycan*

The structural features of the cell surface proteoglycan from NMuMG cells have been explored by biochemical and immuno-chemical approaches. Analysis of $^{35}SO_4$-labeled GAG released from the isolated ectodomain indicates that it is a hybrid molecule bearing predominantly heparan sulfate but also some chondroitin sulfate (Rapraeger *et al.*, 1985). Estimation of the size of the molecule, detected with a monoclonal antibody (MnAb 281, Jalkanen *et al.*, 1985) on Western transfers from polyacrylamide gel electrophoresis, demonstrates that (1) the proteoglycan exhibits a wide size distribution, with a relative modal mass of 200–260 kDa, (2) it is reduced to a more uniform size by removal of heparan sulfate, suggesting that a variable number of heparan sulfate chains are attached to the core protein, and (3) the moiety remaining after removal of heparan sulfate is reduced in size again by removal of chondroitin sulfate, yielding a denuded

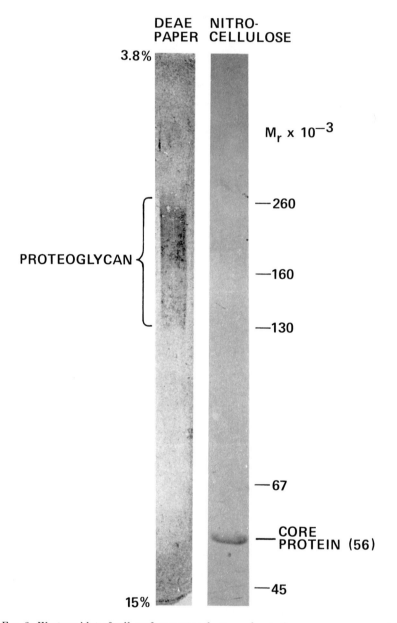

FIG. 2. Western blot of cell surface proteoglycan and putative precursor core protein. Immunoassay of MnAb 281 antigen in NMuMG cell extracts transferred to DEAE–cellulose (A) or nitrocellulose (B) following polyacrylamide gel electrophoresis detects proteoglycan with a modal mass of 200–260 kDa and a cytoplasmic protein of 56 kDa, respectively.

core protein, presumably lacking the lipophilic domain, of ~53 kDa (Rapraeger *et al.*, 1985). In addition, a putative precursor protein of ~56 kDa is detected in cytoplasmic extracts of the NMuMG cells (Fig. 2).

Despite the variability in GAG chain length and number, an estimate of the general proportions of the proteoglycan can be made (see Fig. 3). Approximately two chondroitin sulfate chains (~17 kDa) and four heparan sulfate chains (~36 kDa) are bound to a core protein estimated to be 56 kDa. Similar molecular sizes are estimated from preliminary rotary shadowing data of the core protein (~130 nm long, assuming that it is in an extended configuration) and of the GAG chains (~65 nm).

The presence of chondroitin sulfate on the heparan sulfate proteoglycan may imply a unique type of regulatory control. For example, the proportion of each GAG type may vary, even to the exclusion of one or the other, depending on substratum type, cell cycle, developmental stage, and other factors. Insight into the significance of these possibilities will come with a further understanding of the role of each GAG type on the proteoglycan. Reports of heparan sulfate–chondroitin sulfate proteoglycan hybrids include the cell surface form of NMuMG cells (David and Van den Berghe, 1985; Rapraeger *et al.*, 1985), a form present in intracellular granules of rat basophilic leuke-

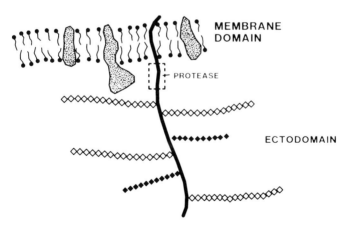

FIG. 3. Model of NMuMG cell surface proteoglycan. The core protein consists of membrane and extracellular (ectodomain) domains that can be separated by cleavage with trypsin or chymotrypsin at a protease-susceptible site. The intact protein is ~56 kDa, while the protease-released ectodomain is ~53 kDa. A typical ectodomain bears both heparan sulfate (four chains of ~36 kDa) and chondroitin sulfate (two chains of ~17 kDa), although the number of the GAG chains is variable.

mia (RBL-1) cells, which contain predominantly chondroitin sulfate (Seldin *et al.*, 1985), and a form secreted by rat Sertoli cells (Skinner and Fritz, 1985). Because heparan and chondroitin sulfate are thought to be linked to core proteins by an identical linkage region, the occurrence of these hybrids suggests that the polypeptide chain contains information for designating which GAG chain is present.

III. DISTRIBUTION OF EPITHELIAL
CELL SURFACE PROTEOGLYCAN

*A. Immunolocalization of Cell Surface Proteoglycan
on NMuMG Cells*

The cell surface proteoglycan is detected on the surface of the NMuMG epithelial cells by a monoclonal antibody (MnAb 281; Jalkanen *et al.*, 1985). Indirect immunofluorescent staining of islands of subconfluent cells demonstrates proteoglycan at the apical cell surface and cell periphery, especially where the cells are contiguous. If the cells are rounded by a brief treatment with EDTA prior to staining, the antibody stains the entire cell surface. Although staining is unaffected by GAG removal, enzymatic removal of the proteoglycan abolishes the staining, as would be expected.

Although proteoglycan can be detected on the entire surface of cells grown at subconfluence, it vanishes from the apical surface of confluent cells (Rapraeger *et al.*, 1986). Treatment of the apical surface of such monolayers with trypsin indicates that only 4% of the cell surface proteoglycan is accessible to the protease. However, when these monolayers are detached with EDTA (Fig. 4) or extracted with detergent, stain is readily demonstrated on the basolateral cell surface. Thus, polarization of the proteoglycan's distribution appears to accompany the formation of a stable epithelial monolayer.

Localization of proteoglycan to the basolateral cell surface is consistent with a functional role at this site, the site of anchorage of the epithelium to the basement membrane *in vivo*. Two mechanisms are likely to be responsible for this restricted localization: (1) a cellular process in which the proteoglycan is preferentially displayed at basolateral surfaces and (2) an extracellular process in which the proteoglycan is immobilized at basolateral surfaces by anchorage to the ECM or adjacent cell surfaces. Some insight into the first mechanism is derived from culture of the NMuMG cells *in vitro,* a circumstance where a complicating ECM is minimized. This is demonstrated by

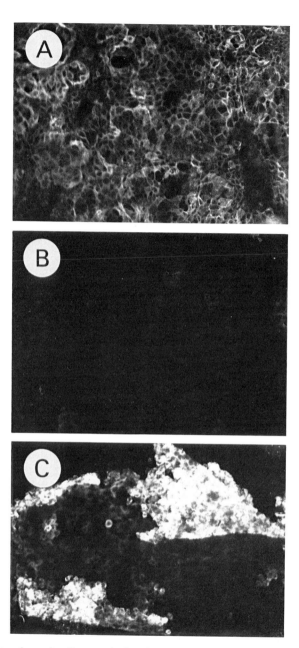

FIG. 4. Proteoglycan localizes at the basolateral surface of confluent NMuMG mono-layers. Indirect immunofluorescence with MnAb 281 detects proteoglycan on the apical surface of newly confluent NMuMG monolayers (A), but not on monolayers confluent for a week (B). However, if such confluent monolayers are briefly treated with EDTA, some regions detach and fold back on themselves, exposing the basolateral cell surface. Staining (C) reveals the proteoglycan in these monolayers exclusively at the basolateral cell surfaces (After Rapraeger *et al.*, 1986).

staining with the BM-1 antibody (Jalkanen *et al.*, 1987a), for example, which detects basement membrane proteoglycan only as sparse aggregates beneath the NMuMG cells rather than as an organized matrix (Fig. 5). At the same time, the cell surface proteoglycan, detected by MnAb 281, is present at the basal cell surface in an intricate pattern that colocalizes with components of the cytoskeleton (see Section V,B). This suggests that, at least *in vitro,* the localization of the cell surface proteoglycan can be attributed to its cellular, possibly cytoskeletal, interactions.

Cell polarization, not matrix accumulation, appears to constitute a sufficient signal for the restricted distribution of the cell surface proteoglycan at the basolateral plasma membrane. The presence of proteoglycan on the apical surface of subconfluent cells may be explained by its diffusion in the plane of the membrane from the sites of insertion on the basolateral surface. When the cells establish intercellular junctions in confluent cultures, this diffusion is prevented and, with time, the preexisting proteoglycan is shed from the apical surface. Alternatively, proteoglycan may be inserted randomly into the plasma membrane of subconfluent cells, resulting in an apical as well as basolateral distribution, but an intracellular pathway leading only to the basolateral membrane may be established in confluent cells together with the acquisition of more pronounced cell polarity.

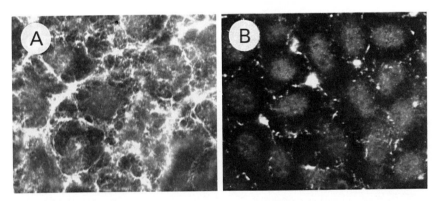

Fig. 5. Cell surface proteoglycan and basement membrane proteoglycan have distinct distributions at the basal surface of NMuMG monolayers. Triton-extracted residues of NMuMG monolayers are stained with MnAb 281 to localize cell surface proteoglycan (A) and with antibody directed against the basement membrane (BM-1) proteoglycan (B), then visualized by indirect immunofluorescence.

B. The Cell Surface Proteoglycan Is Restricted
to Epithelial Cells

Immunolocalization of the proteoglycan with MnAb 281 shows that the antigen is restricted to epithelial tissues; no stain is noted in various mesenchymal tissues. There appear to be four distinct patterns of staining of epithelia. Stratified epithelia such as the epidermis, esophagus, cornea, vagina, and various transitional epithelia of the urogenital tract show intense staining. The entire surface of the cells within the stratified layer stain heavily. Interestingly, the apical cells of the bladder as well as the mucified cells of the vagina show no stain, although these cells are thought to be derived from the same precursors as the stratified cells. Various cuboidal and columnar simple epithelia stain less intensely, and these cells stain only at their basolateral surfaces. This pattern is seen in the epithelia of the trachea, bronchiole, stomach, duodenum, jejunum, rectum, uterus, and the ducts but not the acini of the pancreas, epididymus, prostate, and mammary glands. The hepatic cords of the liver show modest staining only at the sinusoidal surfaces. No staining is seen in various endocrine epithelia (thyroid, adrenal, pancreatic islets), renal tubules, endothelia, or mesothelia (Table II). The significance of these staining patterns is not clear. The antigen is present solely at the cell surface in every tissue. On the other hand, its proposed function as a matrix receptor is consistent only with its localization on simple epithelia and hepatocytes. The possibility exists,

TABLE II

IMMUNOLOCALIZATION OF MnAb 281 ANTIGEN AT EPITHELIAL CELL SURFACES[a]

Type of epithelium	Cell surface stain
I. Stratified squamous (epidermis, vagina, esophagus, cornea); transitional (bladder, ureter)	Intense; entire surface of stratified cells; absent from differentiated apical cells
II. Cuboidal (mammary, prostatic, pancreatic ducts); columnar (trachea, bronchiole, stomach, intestine, uterus, prostate)	Moderate; basolateral surfaces only
III. Hepatocytes	Modest; sinusoidal surfaces only
IV. Endothelia, mesothelia; mammary, prostatic, pancreatic acini; thyroid, adrenal, pancreatic islets; renal tubules	None

[a]After Hayashi et al. (1987).

however, that the antigen differs between the stratified and simple epithelia. Although the Mnab 281 recognizes the core protein, the antibody detects only a single determinant and this may be shared between otherwise dissimilar core proteins of the stratified and simple epithelia. On the other hand, the core protein of the two epithelial types may be identical, but the proteoglycan molecules may differ in the number or type of GAG chains. Indeed, the core protein may be at the cell surface without GAG chains.

C. Developmental Regulation of the Cell Surface Proteoglycan

Staining of epithelia with MnAb 281 shows similar changes in distribution in cultured and developing cells. As noted above, stain surrounds mammary epithelial cells in early cultures, but with time after confluence, stain is lost from the apical surfaces and remains solely basolaterally. Reactivity surrounds epithelial cells of early embryonic organs, such as the epithelia of the lung and uterus, but becomes localized solely at the basolateral surfaces of these epithelia in mature tissues. These distributions of MnAb 281 reactivity differ from those of the BM-1 antigen, which recognizes an extracellular heparan sulfate proteoglycan in basement membranes (Hassell *et al.*, 1980). A significant difference in the distributions of these antigens, for example, is seen in the embryonic submandibular gland. Reactivity with MnAb 281 surrounds each epithelial cell, whereas that for the BM-1 antibody is localized to the epithelial–mesenchymal junction along the stalk and deep within the interlobular clefts. These results indicate that the cell surface and basement membrane heparan sulfate proteoglycans are antigenically distinct and that these proteoglycans, though made by the same cells, are localized differently.

IV. Heparan Sulfate Proteoglycan Binds Stromal Matrix Components

A. Ectodomain of the Cell Surface Proteoglycan Binds Interstitial Collagens

The isolated ectodomain of the cell surface proteoglycan, isolated from suspended cells by mild treatment with trypsin, binds to native collagen fibrils with high affinity and specificity, assessed in cosedimentation assays, binding to matrix-coated wells or collagen affinity columns (Koda *et al.*, 1985). At physiological ionic strength, the proteoglycan will bind to type I collagen fibrils but will not recognize either collagen monomers or heat-denatured collagen. Scatchard analysis indicates a high-affinity interaction with a K_d of ~10^{-9} M. In addition, the proteoglycan binds interstitial collagen types III and V. However, it fails to bind types II or IV. These findings demonstrate that the proteoglycan requires a native collagen fibril for binding. Although type II collagen is fibrillar, the failure of proteoglycan binding to this collagen may be explained by the difficulty in forming stable fibrils under physiological conditions *in vitro*.

Binding of the proteoglycan to collagen is via the heparan sulfate chains. The isolated chains bind to collagen, although binding is less stable than proteoglycan binding and occurs only at subphysiological salt concentrations. In addition, proteoglycan is displaced from type I collagen affinity columns by a select group of polyanions with characteristics similar to heparan sulfate. For example, although high concentrations of chondroitin sulfate and dermatan sulfate fail to displace the proteoglycan, a 100-fold lower concentration of heparin or dextran sulfate is highly effective (Koda *et al.*, 1985). An explanation for this effectiveness may lie in the sulfation pattern of these polyanions. Whereas the ineffective chondroitin sulfates are sulfated uniformly and rarely have more than one sulfate per disaccharide, heparin, dextran sulfate, and heparan sulfate can bear up to three sulfates per disaccharide, often as highly sulfated blocks interrupted by regions of low sulfation. These block regions may recognize specific sites on the collagen fibril.

These data suggest that heparan sulfate GAG chains on the cell surface proteoglycan mediate binding to fibrillar collagen. The increased affinity of proteoglycan binding compared to that of isolated GAG chains indicates that the GAG chains may be acting cooperatively to bind to the collagen fibril. Removal of the chondroitin sulfate chains does not abolish the binding. These findings suggest that collagen types I, III, and V, in their native configurations, contain a binding region selective for heparan sulfate-like GAG.

B. *Proteoglycan Binds NMuMG Cells to Collagen*

The specificity of ectodomain binding to collagens is duplicated by EDTA-suspended NMuMG cells bearing the cell surface proteoglycan (Fig. 6; Koda *et al.*, 1985). Binding under physiological ionic strength conditions to polyvinylchloride (PVC) wells coated with collagens demonstrates that (1) the cells bind collagens I, III, and V, but not II or IV, (2) a native collagen fibril is required, and (3) competition with hep-

Collagen Type I

II

III

IV

V

Fibronectin

Laminin

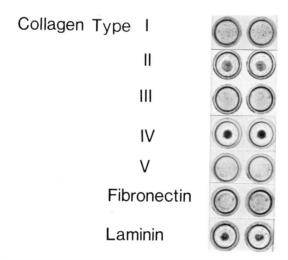

FIG. 6. Binding of NMuMG cells to matrix-coated wells. PVC wells are incubated with 1 μg/ml protein, washed, then incubated in the presence or absence of heparin with cells suspended by scraping into phosphate-buffered saline containing 0.02% EDTA at 0°C. After 30 min incubation, the wells are centrifuged and the cells are fixed and stained. Failure to bind the matrix results in the cells sedimenting into a button at the bottom of the well (cf. denatured collagen) as described by Koda *et al.* (1985).

arin or digestion of the heparan sulfate chains with heparitinase reduces the binding. Therefore, proteoglycan in its native state (i.e., embedded in the plasma membrane) acts as a receptor for interstitial collagens.

C. *Proteoglycan Binds NMuMG Cells to Fibronectin*

The NMuMG cells also bind to fibronectin, a major component of the interstitial matrix (Fig. 6). However, unlike several fibroblastic cell types which bind solely to the amino acid sequence Arg-Gly-Asp (RGD) within fibronectin, the mammary epithelial cells also bind via their cell surface proteoglycan to the C-terminal heparin-binding domain of fibronectin. These two types of binding are distinguished by treating the NMuMG cells with trypsin; trypsin treatment removes the cell surface proteoglycan and cell binding is solely to the RGD sequence, while EDTA-suspended cells retain the cell surface proteoglycan and bind to both the RGD sequence and the heparin-binding domain. Cell binding to fibronectin via the cell surface proteoglycan can be duplicated by liposomes containing the intact purified proteoglycan (Saunders *et al.*, 1986).

D. Significance of Proteoglycan-Mediated Cell
Attachment to Interstitial Matrix Components

The binding of the cell surface proteoglycan and of cells via this molecule is highly specific for interstitial matrix components. For example, at physiological ionic strength, there is no detectable binding to laminin (Fig. 6) or type IV collagen, authentic components of the basal lamina, while there is substantial binding to interstitial collagens and fibronectin. It is surprising because laminin and type IV collagen are deposited by the epithelium at sites where they would be encountered by the cell surface proteoglycan. The interstitial matrix molecules, on the other hand, are deposited predominantly as components of the stromal matrix distal to the epithelial basal lamina. However, assessment of the physical dimensions of the proteoglycan (see Section II,C) indicates that it should be capable of spanning the basal lamina and interacting with the stromal components. Interpretation of these findings alone suggest that the proteoglycan interacts preferentially with components deposited by the stroma rather than the epithelium itself.

V. Cell Surface Proteoglycan as a Matrix
Anchor to the Cytoskeleton

A. Cell Surface Proteoglycan Resists
Extraction with Detergents

Cell surface proteoglycan remains bound via an ionic linkage to the cytoskeleton-rich residue following extraction of NMuMG monolayers with nonionic detergents (Rapraeger et al., 1986). The resistant fraction comprises a third of the cell surface fraction when the extraction is at pH 7.5 and accounts for all of the cell surface proteoglycan when at pH 5.0. In contrast, all of the nonlipophilic cellular proteoglycans and the bulk of the extracellular proteoglycans in these monolayers are released. Therefore, a property peculiar to the cell surface form anchors it to the residue.

The resistance of the cell surface proteoglycan to extraction is apparently due to binding via its membrane-intercalated domain to components in the residue, possibly to another integral membrane protein or components of the filamentous cytoskeleton. The proteoglycan can be displaced from its binding site by 0.5 M KCl, but only after extraction of the cell with Triton X-100. Even a treatment with KCl combined with saponin, which disrupts cholesterol-rich regions of the membrane and presumably allows KCl entry to the cell interior, fails to release the proteoglycan without prior detergent extraction. This suggests that the ionic site is protected by membrane lipid and may even reside within the membrane itself.

A similar pH-dependent extraction has been described for other membrane proteins, attributed to pH-dependent binding or aggregation with cytoskeletal components. Band 3, the cytoskeleton-bound anion channel of erythrocytes, has an increased resistance to detergent extraction at pH 5.0 (Yu *et al.*, 1973), apparently due to isoelectric aggregation of spectrin (pK_a of 4.8) at the cytoplasmic face of the membrane (Nicolson, 1973; Golan and Veatch, 1980). Binding of actin to this complex is also stabilized in this pH range.

B. Detergent-Insoluble Proteoglycan Colocalizes with F-Actin

The ionic linkage that enables the cell surface proteoglycan to resist extraction by nonionic detergents may be part of an assembly linking the proteoglycan to the intracellular filamentous cytoskeleton. Immunolocalization of proteoglycan at the basal cell surface following detergent extraction reveals a filamentous distribution that mimics the pattern of actin in the residue (Rapraeger *et al.*, 1986). Actin, detected by a fluorescent phalloidin probe, is present as fibers in the spreading margins of single NMuMG epithelial cells, as irregularly arranged filament bundles in newly confluent cells and as highly organized parallel fibers at the basal surface of cells in established monolayers. Double staining with MnAb 281 demonstrates that the cell surface proteoglycan mirrors this actin distribution (Fig. 7). In addition, F-actin and proteoglycan are found at the lateral borders of the cell and, if examined prior to polarization of the proteoglycan, on the apical surface.

Several lines of evidence are consistent with the notion that the cell surface proteoglycan is bound via its membrane-intercalated domain to components of the cytoskeleton. In fibroblastic cells, cell surface heparan sulfate proteoglycan localizes with actin filament bundles to cell attachment sites (Lark and Culp, 1987) and with concentrations of actin in stress fibers (Woods *et al.*, 1984). Similar colocalizations are seen with NMuMG cells. In addition, treating these cells with cytochalasin D, which disrupts actin filament organization and collapses the actin network into numerous aggregates, also disrupts the arrangement of the cell surface proteoglycan and causes it to localize to the actin aggregates (Rapraeger *et al.*, 1986). Indeed, a fraction of the intact cell surface proteoglycan, but not its isolated ectodomain, cosediments with F-actin in a centrifugation assay (Rapraeger and Bernfield, 1982).

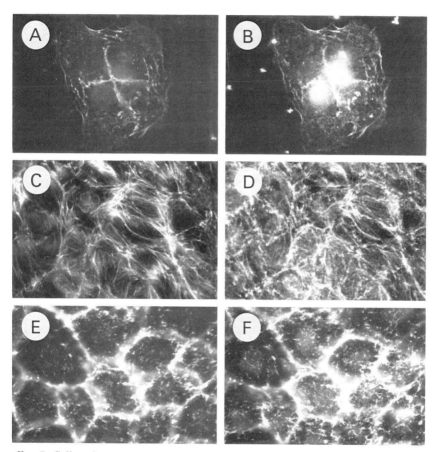

Fig. 7. Cell surface proteoglycan colocalizes with F-actin. F-Actin, detected by FITC-conjugated phalloidin (A, C, E), and cell surface proteoglycan, detected by MnAb 281 (B, D, F), are localized in Triton (pH 7.5)-extracted residues of subconfluent (A, B) or confluent (1 week) monolayers (C–F) of NMuMG cells. Photographs are taken in a focal plane at the basal cell surface. In (E, F), the cells have been treated for 30 min with cytochalasin D (2 μM) prior to extraction and staining (After Rapraeger *et al.*, 1986).

C. Cytoskeletal Anchorage Is Promoted
by Ligand Binding

Interaction of cell surface proteoglycan with interstitial matrix ligands may have a profound influence on the epithelial cell. One activity of matrix-bound proteoglycan may be to generate or stabilize cytoskeleton binding sites at the cytoplasmic face of the plasmalemma potentially as a direct response to the proteoglycan becoming cross-linked by its matrix ligands. Work with integral membrane receptors, for example, concanavalin A receptors of *Dictyostelium* (Condeelis, 1979; Goodloe-Holland and Luna, 1984), surface IgG on lymphocytes (Flanagan and Koch, 1978), and *N*-formyl chemotactic peptide receptors on granulocytes (Jesaitis *et al.*, 1984), has led to the suggestion that multivalent ligands (i.e., those capable of cross-linking receptors on the cell surface) promote an association between the receptor and the cytoskeleton.

Cross-linking of receptors at the cell surface, leading to cluster or aggregate formation, is often achieved artificially by the use of antibodies or lectins. For example, with hormone receptors, such as receptors for epidermal growth factor or insulin, biological activity similar to native hormone binding is elicited by those antibodies capable of cross-linking the receptor (Schlessinger *et al.*, 1978; Schreiber *et al.*, 1983).

When proteoglycan on the apical surface of the NMuMG cells is treated with cross-linking antibodies, it is assimilated into clusters (Fig. 8; Rapraeger *et al.*, 1986). This assimilation occurs independently of energy. Cross-linking can be achieved using the monoclonal antibody (MnAb 281) together with a second anti-MnAb 281 IgG. The monoclonal alone, presumably due to its recognition of only a single antigenic site on the proteoglycan, is an ineffective cross-linking ligand. Once assimilated into clusters, the clusters are aggregated and lost by shedding or endocytosis, similar to the capping behavior of integral membrane receptors reported for suspended cells. However, this behavior is dependent on energy and is prevented by cytochalasin D. Thus, although the proteoglycan on the apical cell surface appears freely mobile, allowing rapid assimilation into clusters even at 4°C, the clusters appear immobilized if cytoskeletal function is impaired, suggesting that the initial assimilation of proteoglycan into clusters promotes binding to the cytoskeleton.

Although similar data for other antigens have been subject to alternate interpretations, such as antigen displacement by membrane flow (Oliver and Berlin, 1982), another line of evidence supports the idea of

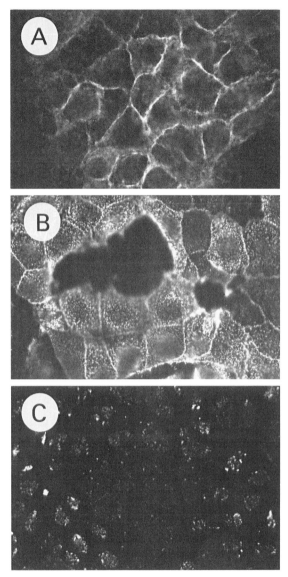

FIG. 8. Cross-linked cell surface proteoglycan is assimilated into clusters. Living NMuMG cells are treated with MnAb 281, then (A) fixed, and MnAb 281 is detected by a second FITC-conjugated IgG, (B) stained directly with the second IgG at 4°C prior to fixation, and (C) stained with the second IgG, then incubated for 30 min at 37°C prior to fixation (After Rapraeger et al., 1986).

cytoskeletal anchorage. Extraction with nonionic detergents at neutral pH, a treatment that normally removes most of the proteoglycan detected on the apical cell surface, fails to remove proteoglycan assimilated into clusters (Rapraeger *et al.*, 1986). This suggests that crosslinked proteoglycans are anchored, possibly by cooperative binding, to the cytoskeleton at the cytoplasmic face of the plasmalemma.

VI. CELL SURFACE PROTEOGLYCAN IS A MATRIX RECEPTOR

Interaction of the proteoglycan with a matrix ligand can lead to a sequence of events that affect cell behavior. Importantly, matrix ligands exist as highly insoluble multicomponent complexes. Thus, the cell surface proteoglycan, mobile in the plane of the membrane, may (1) bind and coalesce around its appropriate ligand in the matrix, (2) become anchored as a cluster at this site, and (3) evoke a cytoplasmic response, part of which may be localized assembly of the cytoskeleton. Moreover, because the matrix is present continuously over a large number of cells, this mechanism would account for coordinate regulation of a sheet of cells in response to matrix deposition. Clearly, mechanisms such as those active during epithelial–mesenchymal interactions that lead to localized matrix degradation and remodeling would free some receptor clusters, leading to their shedding or endocytosis, and anchor others, leading to changes in epithelial cell behavior.

Initial attempts to study this model indicate that proteoglycan in clusters at the epithelial cell surface are immobilized by a closely apposing collagen matrix (Fig. 9). Adding a layer of type I collagen fibrils to the apical surface of NMuMG monolayers after proteoglycan clusters have been generated causes the clusters to become immobilized. In contrast, the clusters aggregate and are lost in the absence of the collagen layer or when the cell surface GAG chains are removed. Thus, immobilization of the clusters is due to proteoglycan binding to the collagen fibrils. This binding is mechanically stable because the collagen layer cannot be physically separated from the cells. However, if the GAG chains are removed, the cells appear to lack other means of binding the collagen and remain bound to the glass coverslip when the collagen layer is removed. Thus, in the presence of the collagen fibrils, the proteoglycan clusters represent direct, stable anchorage foci between the matrix, the cell surface, and, presumably, the cytoskeleton.

VII. SUMMARY

During morphogenesis of embryonic organs, changes in tissue form are paralleled by changes in the extracellular matrix. Indeed, remodel-

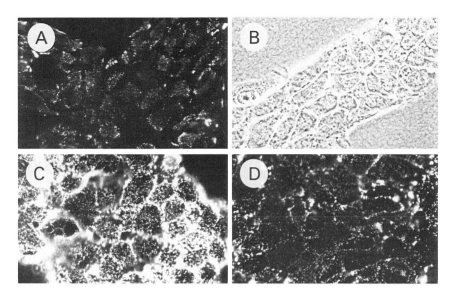

FIG. 9. Cell surface proteoglycan clusters are immobilized by a collagen matrix. NMuMG monolayers were treated with MnAb 281 followed by a second, FITC-conjugated IgG to generate proteoglycan clusters. The monolayers were then overlaid at 4°C with type I collagen (1 mg/ml) in phosphate-buffered saline. When these monolayers are incubated at 37°C for 60 min, the collagen forms a fibrillar matrix while the proteoglycan clusters can aggregate in the plane of the membrane. (A) Indirect immunofluorescence of a monolayer incubated in the absence of collagen; (B) phase micrograph of a monolayer overlaid with collagen; (C) indirect immunofluorescence of a monolayer overlaid with collagen; (D) monolayer treated with heparitinase to remove the heparan sulfate chains from the proteoglycan and then overlaid with collagen.

ing of the matrix is thought to be a mechanism by which one tissue can influence the morphogenesis of other tissues. The cellular response to changes in the matrix must involve cell surface molecules that recognize matrix components extracellularly and initiate a transmembrane signal intracellularly. The cell surface proteoglycan of NMuMG mouse mammary epithelial cells is such a molecule, a matrix receptor.

The cell surface proteoglycan behaves as a matrix receptor involved in epithelial–stromal interactions. The proteoglycan consists of at least two functional domains: (1) a membrane-intercalated domain which anchors the proteoglycan to the cell and (2) an ectodomain which bears both heparan sulfate and chondroitin sulfate chains. The proteoglycan is present solely on epithelial cell surfaces: on simple epithelia, both *in vitro* and *in situ*, the proteoglycan localizes to the basolateral cell surface, the interface between epithelium and stroma. The ectodomain binds with specificity and high affinity to components of the interstitial matrix, providing a means by which the stromal cells that produce this matrix can influence epithelial behavior. The cell surface proteoglycan appears to function as a matrix receptor that is involved in translating the composition, organization, and stability of the interstitial matrix into the behavior of epithelial sheets. Binding of the proteoglycan to matrix ligands at the cell surface promotes the association of the proteoglycan to the cytoskeleton, potentially stabilizing the epithelium. The expression of the proteoglycan during the development of epithelial organs also suggests a role in generating stable morphology. The proteoglycan is shed when the cells round up, suggesting a potential mechanism by which the cells may alter the association of the proteoglycan with the matrix and promote changes in epithelial morphology.

The function of the epithelial cell surface proteoglycan, as well as other matrix receptors, may change with alterations in the stability of the matrix. They may modify cellular behavior during development or with neoplastic invasion, when the matrix is labile, but play a different role in mature tissues, when the matrix is metabolically and physically more stable. An understanding of these roles for the cell surface proteoglycan awaits further definition of its structure and its molecular interactions.

Acknowledgments

The authors thank Shib Banerjee, William Benitz, Joy Koda, and Scott Saunders for helpful discussions, and are grateful to Margareta Svensson-Rosenberg and Hung Nguyen for their expert technical assistance.

This work has been supported by a NIH New Investigator Research Award (HD17146 and HD21881) to A.R., Cystic Fibrosis Foundation fellowship (F030-5-01) to M.J., and grants from the NIH (HD06763 and CA28735) and support from the Weingart Foundation to M.B.

References

Ben-Ze'ev, A., Farmer, S. R., and Penman, S. (1980). *Cell* **21,** 365–372.
Bernfield, M. (1970). *Dev. Biol.* **22,** 213–231.
Bernfield, M., Banerjee, S. D., Koda, J. E., and Rapraeger, A. (1984). *In* "The Role of Extracellular Matrix in Development" (R. L. Trelstad, ed.), pp. 545–572. Liss, New York.
Bumol, T. F., Wang, Q. C., Reisfeld, R. A., and Kaplan, N. O. (1983). *Proc. Natl. Acad. Sci. U.S.A.* **80,** 529–533.
Chen, W.-T., Hasegawa, E., Hasegawa, T., Weinstock, C., and Yamada, K. M. (1985). *J. Cell Biol.* **100,** 1103–1114.
Chiang, R. M., and Kang, A. H. (1982). *J. Biol. Chem.* **257,** 7581–7586.
Cohn, R. H., Banerjee, S. D., and Bernfield, M. (1977). *J. Cell Biol.* **73,** 464–478.
Cole, G. J., Schubert, D., and Glaser, L. (1985). *J. Cell Biol.* **100,** 1192–1199.
Condeelis, J. (1979). *J. Cell Biol.* **80,** 751–758.

David, G., and Van den Berghe, H. (1985). *J. Biol. Chem.* **260**, 11067–11074.

Flanagan, J., and Koch, G. (1978). *Nature (London)* **273**, 278–280.

Folkman, J., and Moscona, A. (1978). *Nature (London)* **273**, 345–349.

Fransson, L.-A., Carlstedt, I., Coster, L., and Malmstrom, A. (1983). *J. Biol. Chem.* **258**, 14342–14345.

Golan, D. E., and Veatch, W. (1980). *Proc. Natl. Acad. Sci. USA* **77**, 2537–2541.

Goldberg, B. (1979). *Cell* **16**, 265–275.

Goodloe-Holland, C. M., and Luna, E. (1984). *J. Cell Biol.* **99**, 71–78.

Haeuptle, M. T., Suard, Y. L. M., Bogenmann, E., Reggio, H., Racine, L., and Kraehenbuhl, J. P. (1983). *J. Cell Biol.* **96**, 1405–1414.

Hassell, J., Robey, P., Barrach, H., Wilczek, J., Rennard, S., and Martin, G. (1980). *Proc. Natl. Acad. Sci. U.S.A.* **77**, 4494–4498.

Hay, E. D. (1985). *J. Cell. Biochem.* **27**, 143–156.

Hayashi, K., Hayashi, M., Jalkanen, M., Firestone, J., Trelstad, R. L., and Bernfield, M. (1987). Submitted.

Hedman, K., Christner, J., Julkunen, I., and Vaheri, A. (1983). *J. Cell Biol.* **97**, 1288–1293.

Höök, M., Kjellén, L., Johansson, S., and Robinson, J. (1984). *Annu. Rev. Biochem.* **53**, 847–869.

Iozzo, R. (1984). *J. Cell Biol.* **99**, 403–417.

Jalkanen, M., Nguyen, H., Rapraeger, A., Kurn, N., and Bernfield, M. (1985). *J. Cell Biol.* **101**, 976–984.

Jalkanen, M., Rapraeger, A., Hassell, J., and Bernfield, M. (1987a). *J. Cell Biol.*, submitted.

Jalkanen, M., Rapraeger, A., Saunders, S., and Bernfield, M. (1987b). *J. Cell Biol.*, submitted.

Jesaitis, A. J., Naemura, J. R., Sklar, L. A., Cochrane, C. G., and Painter, R. C. (1984). *J. Cell Biol.* **98**, 1378–1387.

Kjellén, L., Oldberg, A., and Höök, M. (1980). *J. Biol. Chem.* **255**, 407–413.

Koda, J., Rapraeger, A., and Bernfield, M. (1985). *J. Biol. Chem.* **260**, 8147–8162.

Kraemer, P. (1977). *Biochem. Biophys. Res. Comm.* **78**, 1334–1340.

Kraemer, P., and Tobey, R. (1972). *J. Cell Biol.* **55**, 713–717.

Lark, M., and Culp, L. (1984). *J. Biol. Chem.* **259**, 6773–6783.

Lark, M., and Culp, L. (1987). *In* "Diseases of Connective Tissue: Pathology of the Extracellular Matrix" (J. Uitto and A. Perejda, eds.), pp. 185–211. Dekker, New York.

Laurie, G. W., Leblond, C. P., and Martin, G. R. (1982). *J. Cell Biol.* **95**, 340–344.

Lindahl, U., and Höök, M. (1978). *Annu. Rev. Biochem.* **47**, 385–417.

McKeown-Longo, P. J., Hanning, R., and Mosher, D. F. (1984). *J. Cell Biol.* **98**, 22–28.

Mollenhauer, J., and Von der Mark, K. (1983). *EMBO J.* **2**, 45–50.

Moon, R. T., and Lazarides, E. (1984). *J. Cell Biol.* **98**, 1899–1904.

Nicolson, G. L. (1973). *J. Cell Biol.* **57**, 373–387.

Oldberg, A., Hayman, E. G., and Ruoslahti, E. (1981). *J. Biol. Chem.* **256**, 10847–10852.

Oliver, J., and Berlin, R. (1982). *Int. Rev. Cytol.* **74**, 55–94.

Rao, N. C., Barsky, S. H., Terranova, V. P., and Liotta, L. (1983). *Biochem. Biophys. Res. Comm.* **111**, 804–808.

Rapraeger, A., and Bernfield, M. (1982). *In* "The Extracellular Matrix" (S. Hawkes and J. Wang, eds.), pp. 265–269. Academic Press, New York.

Rapraeger, A., and Bernfield, M. (1983). *J. Biol. Chem.* **258**, 3632–3636.

Rapraeger, A., and Bernfield, M. (1985). *J. Biol. Chem.* **260**, 4103–4109.

Rapraeger, A., Jalkanen, M., Endo, E., Koda, J., and Bernfield, M. (1985). *J. Biol. Chem.* **260**, 11046–11052.

Rapraeger, A., Jalkanen, M., and Bernfield, M. (1986). *J. Cell Biol.*, **103**, 2683–2696.

Rapraeger, A., Koda, J. E., and Bernfield, M. (1987). In "Diseases of Connective Tissue: Pathology of the Extracellular Matrix" (J. Vitto and A. Perejda, eds.), pp. 213–231. Dekker, New York.

Ruoslahti, E., and Engvall, E. (1980). *Biochim. Biophys. Acta* **631**, 350–358.

Rubin, K., Höök, M., Obrink, B., and Timple, R. (1981). *Cell* **24**, 463–470.

Sakashita, S., Engvall, E., and Ruoslahti, E. (1980). *FEBS Lett.* **116**, 243–246.

Saunders, S., Jalkanen, M., and Bernfield, M. (1986). *J. Cell Biol.* **103**, 533a.

Schlessinger, J., Shechter, Y., Willingham, M., and Pastan, I. (1978). *Proc. Natl. Acad. Sci. U.S.A.* **75**, 2659–2663.

Schreiber, A., Libermann, T., Lax, I., Yarden, Y., and Schlessinger, J. (1983). *J. Biol. Chem.* **258**, 846–853.

Seldin, D. C., Austen, K. F., and Stevens, R. L. (1985). *J. Biol. Chem.* **260**, 11131–11139.

Skinner, M. K., and Fritz, I. B. (1985). *J. Biol. Chem.* **260**, 11874–11883.

Stamatoglou, S., and Keller, J. (1982). *Biochim. Biophys. Acta* **719**, 90–97.

Sugrue, S., and Hay, E. D. (1981). *J. Cell Biol.* **91**, 45–54.

Terranova, V. P., Rao, C. N., Kalebic, T., Margulies, I. M., and Liotta, L. (1983). *Proc. Natl. Acad. Sci. U.S.A.* **80**, 444–448.

Wittelsberger, S. C., Kleene, K., and Penman, S. (1981). *Cell* **24**, 859–866.

Woodley, D. T., Rao, C. N., Hassell, J. R., Liotta, L. A., Martin, G. R., and Kleinman, H. K. (1983). *Biochim. Biophys. Acta* **761**, 278–283.

Woods, A., Höök, M., Kjellan, L., Smith, C. G., and Rees, D. A. (1984). *J. Cell Biol.* **99**, 1743–1753.

Yamada, K. M., Kennedy, D. W. Kimata, K., and Pratt, R. M. (1980). *J. Biol. Chem.* **255**, 6055–6063.

Yamada, K., Akiyama, S. K., Hasegawa, T., Hasegawa, E., Humphries, M. J., Kennedy, D. W., Nagata, K., Urushihara, H., Olden, K., and Chen, W.-T. (1985). *J. Cell. Biochem.* **28**, 79–97.

Yanagishita, M., and Hascall, V. C. (1984). *J. Biol. Chem.* **259**, 10270–10283.

Yu, J., Fischman, D. A., and Steck, T. L. (1973). *J. Supramol. Struct.* **3**, 233–248.

Structural Organization of Proteoglycans in Cartilage

Ernst B. Hunziker and Robert K. Schenk

Department of Anatomy, University of Bern, 3000 Bern 9, Switzerland

I. Introduction

The intercellular substance of connective tissue, called "matrix" in cartilage tissue, is composed of two phases. One of these, consisting of a variety of insoluble fibers, can be discerned upon routine light- or electron-microscopic examination, while the other, often referred to as the amorphous ground substance, is optically homogeneous and transparent, special preparative techniques being required to reveal its components. Proteoglycans have been found to constitute a major proportion of this phase in cartilage matrix, where they represent up to 95% of the nonfibrous dry-weight matrix material (1). Tremendous progress has been made in the last 10–20 years in the study of isolated PG.* Detailed analyses of their structural characteristics and elucidation of functional implications have considerably improved our understanding of these molecules. Unfortunately, the progress made does not fully correlate to our understanding of the *in vivo* situation, since the morphological methods for PG preservation *in situ* have not developed to a similar degree. Morphological information regarding tissue PG is, in most instances, still obtained from tissue fixed in conventional media, possibly supplemented with cationic dyes for the precipitation of PG *in situ*. Under these conditions the destruction of PG integrity is unavoidable, and hence no reliable information regarding their physiological structure *in vivo* nor of their interaction with other matrix components may be obtained.

Is it reasonable to assume that contemporary morphological methods are powerful enough to resolve such molecules in their native state at all? The biochemical data obtained on isolated molecules describe them as polymers of dimensions far exceeding the resolution limits of

---

* Abbreviations used in this chapter are as follows: PG, proteoglycans; PG-S, proteoglycan subunit or proteoglycan monomer; PG-A, proteoglycan aggregate.

conventional transmission electron microscopy (Fig. 3). Indeed, these dimensions even lie in the range of the light microscope.

It is the aim of this chapter to present recent progress in morphological methods for improved cartilage PG preservation and to summarize the currently used conventional procedures. An understanding of PG structure *in vivo* is pertinent to an understanding of cartilage function, since the properties of this tissue reflect the characteristics of matrix molecules and their interactions with one another. Recent findings relating to the role played by morphology in increasing our understanding of cartilage function are therefore discussed in the light of recent findings relating to PG structure *in vivo*.

II. Macromolecular Structure of Isolated Proteoglycans Based on Electron-Microscopic Analysis

Early investigations of PG structure were mainly biochemical in nature, rapid progress in this field being achieved after the development of efficient PG extraction procedures by Sajdera and Hascall (2). More detailed information about the structural characteristics of PG molecules was gained step by step within a relatively short period of time (1965–1985) by the contributions of many investigators (see Refs. 3–18 for original articles and reviews). During this time, the first electron-microscopic studies of isolated PG were undertaken by Rosenberg *et al.* (19, 20), who were able to visualize the structure of PG-A; subsequent electron-microscopic investigations (21–26) confirmed these initial findings (Fig. 1). More recently, Wiedemann *et al.* (27) have been able to visualize for the first time details of the PG-S (Fig. 2), using rotatory shadowing. It is important to realize, however, that the results obtained by the preparation techniques used for electron-microscopic investigation of isolated PG, and the electron micrograph per se, yield a two-dimensional spreading and projection of these molecules and provide no direct information relating to three-dimensional structure.

PG are polymeric molecules, the subunits of which consist basically of a core protein to which various kinds of (polyanionic) carbohydrate chains are linked. In most instances, these are long glycosaminoglycans (e.g., chondroitin sulfate and keratan sulfate) or various kinds of oligosaccharides (N-linked or O-linked). Various numbers of PG-S can be linked to hyaluronic acid residues under the stabilizing effect of link proteins to form PG-A (see Refs. 3–18). The dimensions of the various structural elements (core protein, chondroitin sulfate chains,

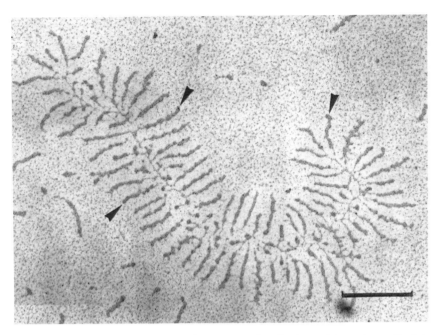

Fɪɢ. 1. Electron micrograph of a PG aggregate from bovine epiphyseal growth plate cartilage, prepared as a PG–cytochrome c film which was subsequently shadowed with platinum–palladium (for details see Ref. 21). Arrow, PG-S. ×38,000. Bar = 500 nm. By courtesy of Dr. J. A. Buckwalter.

etc.) as well as the dimensional relationships between such elements for cartilage PG are summarized in Fig. 3. Determinations of the dimensions given have been based on electron-microscopic (21–27) and biochemical (3–18) studies of isolated PG-S and PG-A.

III. Preservation Methods for Proteoglycans in Tissue

A. Peculiarities of Proteoglycan Structure Requiring Special Consideration

The major portion of a PG molecule is represented by the glycosaminoglycans. These carbohydrate polymers are characterized by the presence of a considerable number of negative charges ($-SO_3^-$, COO^-, etc.) along their chains, this being the reason why PG behave as polyanionic compounds under physiological conditions. These chemical

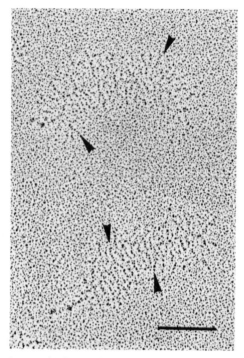

Fig. 2. Electron micrograph of two PG subunits from bovine nasal cartilage, revealed by rotatory shadowing with platinum–carbon. Arrows, Glycosaminoglycan chains. ×160,000; bar = 100 nm. Reprinted with permission from the *Biochemical Journal* [1985 (Vol.) **224**, (p) 332 (facing page)], copyright © 1985, the Biochemical Society, London, and the author.

characteristics have important consequences for PG, namely, that they are extremely hydrophilic and tend to expand their structure completely in space (aqueous phase) due to the electrical repulsive forces acting among the innumerable fixed negative charges present on the polymeric carbohydrate chains. PG in aqueous solution thus tend to occupy a huge space (domain) in relation to their molecular mass (5, 7). Indeed, the very high concentration of PG within cartilage matrix (~20 to 40 mg/ml) is achieved by virtue of their existence in an underhydrated state. Even so, the organic molecular material present per unit of matrix volume is still very small (water content, 70–80%; for review, see Ref. 28). Removal of PG from cartilage matrix leads to an immediate increase in the water domains of these molecules by a factor of 5 to 7 (4). The high internal pressure (29, 46) existing within cartilage, and the inherent tendency of tissue slices to swell in any

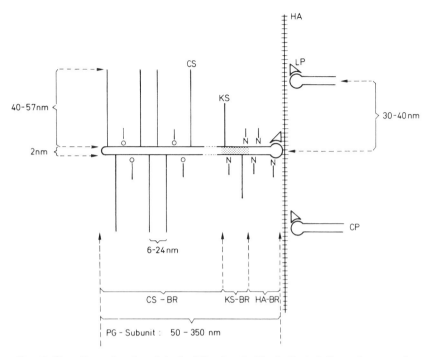

F ɪ ɢ. 3. Two-dimensional model of a PG subunit. The indicated dimensions are those
determined by Buckwalter (21, 22), Hascall (4), Heinegard (23), Kimura (24), Rosenberg
(19, 20), and Thyberg (25, 26). BR, Binding region of core protein (CP); CS, chondroitin
sulfate chains; HA, hyaluronic acid residue; KS, keratan sulfate chains; LP, link protein; —N, N-linked oligosaccharides; ——O, O-linked oligosaccharides. Reproduced
with permission from the author and the *Journal of Cell Biology* 1984, **98**, 277–282, by
copyright permission from the Rockefeller University Press.

aqueous phase, may be attributed mainly to the presence of PG in an
underhydrated state. When this swelling phenomenon occurs, PG,
upon taking up water, being to diffuse out of tissue blocks with a
rapidity suggestive of aqueous extraction. All these processes pose
major problems for aqueous chemical fixation of cartilage tissue.
Apart from these specific peculiarities of structure which cause difficulties in preservation, there are others that render morphological
visualization a problem. Their very low organic mass relative to the
large domain that they occupy results in such a low mass density that
they become transparent. This particular problem may, however, be
overcome by adopting special procedures of structure enhancement,
and these will be described later in this chapter.

B. Preparation Procedures for Light-Microscopic Analysis

Thin slices of fresh and native cartilage tissue may be examined directly in the light microscope. The matrix then appears transparent and completely unstructured. This phenomenon relates to the old term homogeneous "matrix" for cartilage or to the "amorphous ground substance" for connective tissue (for review, see Ref. 30). In growth plate cartilage the matrix is composed mainly of PG, and under such conditions collagen fibrils cannot be visualized. This is a problem relating to the fact that the "index of refraction" for both of these components is identical rather than to a "masking" phenomenon by PG.

A very reliable procedure to preserve cartilage PG *in situ* is cryofixation (31). Simple immersion fixation in a coolant (e.g., liquid nitrogen, propane, Freon) may be adopted prior to cutting cryosections and their subsequent drying. Slow cryofixation in a freezer or even storage of tissue blocks between $-30°$ and $-70°C$ prior to cryosectioning is not recommended because of the possible occurrence of recrystallization, ice sublimation, and PG concentration effects in the matrix. After drying of cryosections, the polyanionic PG may be stained using a variety of cationic dyes for their visualization in the light microscope (for review, see Ref. 32).

In many laboratories, processing of cartilage tissue on a routine basis is performed by conventional methods of fixation (i.e., immersion in aldehyde solutions). After dehydration, embedding, and sectioning, the section may be stained with suitable dyes for gross analysis of PG distribution. Since aldehyde fixation does not chemically preserve carbohydrate compounds, it is not surprising that during this procedure PG are lost to an uncontrollable degree (32–35). An alternative method for aqueous chemical fixation would be a medium containing 70% (v/v) ethanol, in which PG are insoluble. This is the only alternative for aqueous chemical cartilage fixation capable of preventing significant PG loss (32).

C. Preparation Procedures for Electron-Microscopic Analysis

When cartilage tissue is processed using conventional chemical fixation techniques (aqueous aldehydic media), a number of problems arise with respect to PG preservation within the matrix (as explained in Sections III,A and III,B). For electron-microscopic examination, tissue is not only prefixed (usually in a buffered glutaraldehyde solution), but also postfixed in a buffered osmium tetroxide medium. Such media contribute significantly to PG loss, the result being that cartilage matrix is depleted of ~65% (±10%) of its physiological PG content (33). The few PG remnants usually collapse completely during the subsequent step of dehydration. Smaller loss of PG is incurred when postfixation is omitted (33), as is the case when using alcohol fixation solutions. Such media are, however, not recommended for electron-microscopic studies, since alcohols effect complete cell destruction caused by extraction of all lipids (membranes, lipid droplets, etc.). The possibility of fixation using aqueous aldehyde and OsO_4 solutions without concomitant loss of PG was suggested by Luft in 1965 (36). He included the cationic dye ruthenium red in both primary- and secondary-fixation media. In this way, precipitation of PG in $situ$ reduced their loss to a considerable degree. Although a variety of cationic dyes is available, the selection of an appropriate one for cartilage fixation deserves careful evaluation (see Section IV,C).

A newly developed cryotechnical tissue processing method is now available as an alternative to aqueous aldehydic fixation (37). Cartilage tissue blocks are cryofixed immediately after dissection by rapid high-pressure freezing (alternative freezing techniques for native tissue may be used, such as propane jet freezing for cartilage cell cultures; cf. Ref. 31). By this first step of tissue freezing, all matrix components are immobilized in $situ$ under conditions avoiding the formation of ice crystals within the resolution limits of transmission electron microscopy.

In a second step, the solidified tissue water is re-solved, and replaced by an organic solvent at $-90°C$ (e.g., methanol). This freeze substitution process represents at the same time a tissue dehydration step. Its performance at a very low temperature ($-90°C$) is important for preventing molecular collapse (of proteins, carbohydrates, etc.), which begins to occur at $-30°C$ (±10°C) and above in pure organic solvents (38). Execution of freeze substitution at temperatures lower than $-80°C$ is also necessary to avoid recrystallization processes within the solidified interstitial water. Uranyl acetate is added to the substitution medium for block contrast, and glutaraldehyde to assure immobilization of matrix components when warming up for embedding and polymerization. These steps are performed at $-35°C$ with Lowicryl K4M (or HM 20, at $-50°C$). The preservation of PG without loss, and in the native (extended) state can thus be achieved by the principles of (1) rapid cryofixation, (2) strict avoidance of any aqueous immersion steps by substituting tissue water for organic solvents within which

PG are insoluble, and (3) processing at very low temperatures for avoidance for PG collapse and dislocation. Details of this procedure have been given elsewhere (37).

IV. MORPHOLOGY OF PROTEOGLYCANS IN CARTILAGE

A. Appearance of Proteoglycans in the Light Microscope

Staining of cartilage for light-microscopic examination is generally performed on tissue sections (see Section III,B). All dyes used are characterized by the presence of cationic moieties, these sites reacting with the anionic groups along the carbohydrate chains of PG. They will, however, also react with any other anionic groups (e.g., on proteins) present in the cartilage matrix (1). Despite this, the staining profiles obtained are generally considered to be representative for PG, since ~95% of the noncollagenous dry-weight material of cartilage consists of these components.

In the history of cartilage histology, it was the staining patterns observed on tissue sections that led to the definitions of the various matrix compartments (30). Considerable variation in such staining profiles can, however, be obtained depending on the chemical preparation and staining conditions (e.g., pH, temperature of staining solution, composition of the chemical fixation medium; for review, see Refs. 30, 32). It is thus not surprising that the definition and description of cartilage matrix compartments led to a confusing terminology. A categorization of cartilage matrix compartments independent of any staining procedure, and based exclusively on structural parameters, has been made recently for the growth plate (39).

The intensity of staining in various cartilage zones or matrix compartments has been used for a semiquantitative assessment of PG content. The prerequisites for such quantitative estimations are, however, the use of appropriate dyes [i.e., having a single net positive charge per dye molecule and a stoichiometric interaction with glycosaminoglycans (40)] and the use of specific fixation methods (as explained in Section III,B). Semiquantitative assessment of PG content in cartilage is a powerful tool in the study of, for example, diseased states of cartilage (41).

B. Proteoglycan Structure in the Electron Microscope after Cryotechnical Tissue Processing

After thorough cryotechnical tissue processing (as described in Section III,C), cartilage PG are preserved in their native (extended) state

(42). Upon electron-microscopic examination at low power, the PG appear as a dense reticular network filling the interfibrillar collagen spaces of the cartilage matrix (Fig. 4). This filamentous network is present at an even density throughout the various matrix compartments. At higher magnifications (Fig. 5), the network is found to be heterogeneous in composition. It consists of heavily contrasted thick filaments (T) and less intensely stained, finer branches (F). On the basis of their reaction with uranyl acetate, the T filaments (intensely stained) are thought to represent the core proteins of the PG subunits, since proteins are known to have a considerably higher affinity for this stain than do carbohydrates (43–45). The faintly stained fine filaments may thus be identified with the glycosaminoglycan chains of the PG.

Studies on isolated cartilage PG have previously shown that a single PG subunit consists of one core protein (Figs. 2 and 3), to which are attached ~250 carbohydrate chains. For every PG molecule within cartilage matrix, ~100 chondroitin sulfate and 50 keratan sulfate chains would be expected to be visible (about 40 to 60 nm in length, Fig. 3). The oligosaccharides (~100 per core protein) most likely are too short to be identifiable. The much higher numerical density of

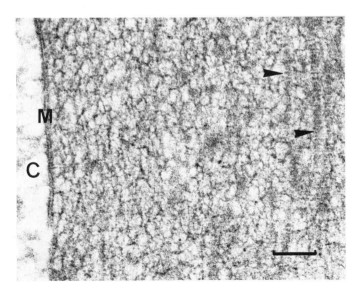

FIG. 4. Electron micrograph of rat growth plate cartilage processed by high-pressure freezing, freeze substitution, and low-temperature embedding. The PG are discernible in the form of a fine reticular network, filling evenly the interstices between collagen fibrils (arrows) and cells (C). M, Plasmalemma. ×118,000; bar = 0.1 μm.

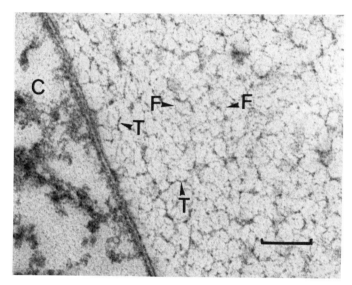

FIG. 5. High-power electron micrograph of rat growth plate cartilage processed by high-pressure freezing, freeze substitution, and low-temperature embedding. Within the cartilage matrix, fine (F), faintly stained and thick (T), darkly stained threads of the PG network are apparent. C, Cells. The F threads have been identified with the glycosaminoglycan components and the T threads with the core proteins of PG. ×140,000; bar = 0.1 μm.

carbohydrate chains relative to the number of core proteins is reflected in the greater number of F filaments relative to T strands, supporting this interpretation. Further support for this description is obtained by analysis of the dimensional relationships encountered. The lengths of the glycosaminoglycan chains determined on isolated molecules lie within the range 40–57 nm, this being consistent with that found for the lengths of F filaments measured on electron micrographs (40–60 nm, Fig. 6). Core proteins have been previously found to be around 250 nm (±50 nm) in length (22–25). Hence, within the volume occupied by the thin sections (35 ± 5 nm thick) analyzed in the electron microscope, it is possible to detect only portions (Fig. 7) of the core protein threads. The lengths of these observable portions will be a function of the angle between the direction of the core protein axis (traversing the section thickness) and the section surface. Since any angle may be possible, a tremendous variability in length of the visible portion of a core protein axis (i.e., T strands) would be expected (Fig. 7). The lengths of T strands were found to cover a considerable range, thus providing further evidence for the suggested identifications.

In preparations of isolated PG, glycosaminoglycan chains have pre-

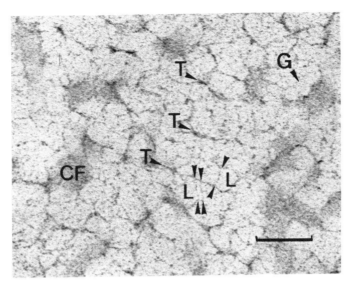

FIG. 6. High-power electron micrograph of rat growth plate cartilage processed by high-pressure freezing, freeze substitution, and low-temperature embedding. F-strand lengths (L) lie within the range of 40–60 nm, and the distance separating these structures varies between 5 and 15 nm. T strands (T) are separated by a distance of between 60 and 80 nm. CF, Cross-sectioned collagen fibril; G, background granularity. ×149,000; bar = 0.1 μm. Reproduced with permission from the *Journal of Cell Biology* 1984, **98**, 277–282, by copyright permission from the Rockefeller University Press.

viously been found to be separated laterally by distances of between 6 and 24 nm (21, 23–25). The theoretical lateral distances between two core proteins lying parallel in the same plane lies between 80 and 114 nm (i.e., twice the length of a glycosaminoglycan chain—assuming that interdigitation of these chains does not occur). In native cartilage tissue, however, the PG molecules exist in an underhydrated state and their spatial domains are reduced in volume by a factor of approximately 4 to 6 (4). As a consequence of this, the above-mentioned PG components would be packed more closely together. If the volume of any three-dimensional structure is reduced by a factor x, then its linear parameters (diameters, lengths, etc.) are shortened by a factor of $x^{-1/3}$. The lateral distances determinable between F filaments (5–15 nm) and T strands (50–80 nm) would thus correlate with PG molecules being underhydrated by a factor of between 3 and 6. A schematic comparison of lateral distances between PG core proteins in tissue, and those of isolated molecules (being just in contact), as well as the dimensions of their domains, is made in Fig. 8.

An estimation of the underhydration factor may be obtained for the

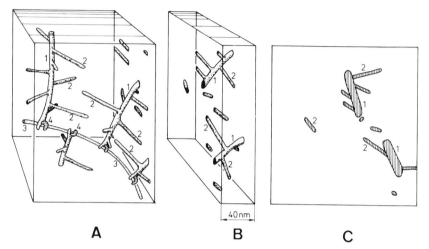

A   B   C

FIG. 7. (A) Schematic representation of a cartilage tissue block containing parts of PG subunits and aggregates. All components are drawn to scale (according to the dimensions given in Fig. 3). 1, PG core protein; 2, glycosaminoglycan chain; 3, hyaluronic acid residue; 4, link protein. (B) A thin (40 nm) section cut from the tissue block contains fragments of PG subunits and aggregates within the section volume. (C) An electron micrograph of the thin section (B) does not yield a two-dimensional profile of the PG subunit within the section, but rather a two-dimensional projection of the structures contained in the section volume. The part of the core protein length (1) actually visible is a function of the angle between the section surface and the core protein axis, and hence this parameter varies considerably. As the angle approaches 90°, the core protein projections become more dotlike in appearance. As the angle approaches 0°, the filament will become longer.

rat growth plate cartilage tissue presented in Figs. 4–6. Assuming that the lateral resolution of the transmission electron micrographs is at best 5 nm, the maximal lateral distances between two T strands would be 75 ± 5 nm (Fig. 6). Given a section thickness of 35 ± 5 nm (Fig. 9), this would yield minimal lateral distances of 66 ± 5 nm (α ~28°). Thus, the theoretical range of 61–80 nm lies within the numerical range measured (60–80 nm). These linear parameters correspond to a spatial underhydration factor for PG domains of 3.5–4 (in rat growth cartilage).

C. Proteoglycan Ultrastructure after Fixation of Cartilage Tissue in the Presence of Cationic Dyes

The high susceptibility of cartilage matrix PG to solubilization in aqueous media reflects a property of these molecules which has been largely responsible for the slow progress made in elucidating their

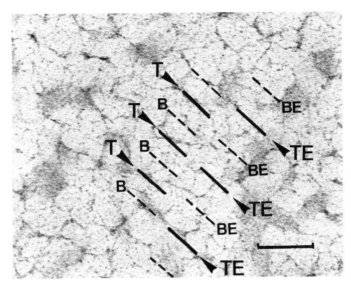

FIG. 8. High-power electron micrograph of rat growth plate cartilage processed by high-pressure freezing, freeze substitution, and low-temperature embedding. The lateral distances between three T threads (T) may be easily measured; they lie within a range of 60–80 nm. On the basis of these data, an underhydration factor of 4 can be calculated. If PG subunits were present in a completely expanded state, their core protein (TE) would be separated by a distance of approximately 80–120 nm. Borders between underhydrated (B) and completely expanded (BE) PG subunits are indicated. ×149,000; bar = 0.1 μm.

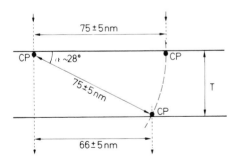

FIG. 9. Schematic view of a thin section (thickness, $T = 35 \pm 5$ nm) containing two core proteins (●, CP) separated by a minimal distance of 75 ± 5 nm (⋯→ direction of observation—projection). When both core proteins lie within the same plane near the section surface, then the lateral separation distance in the projected image is again 75 ± 5 nm. If the two core proteins were lying in different planes—for example, on opposite surfaces of the section (producing a maximum angle of ∼28° relative to the surface)— then they would appear to be separated by a distance of only 66 ± 5 nm in the projected plane.

ultrastructure in native tissue. In 1965, Luft (36) introduced the idea of including a cationic dye into cartilage tissue fixation media in an attempt to overcome this problem. These agents interact with PG causing their precipitation and effective insolubility in aqueous solution. Although no cationic dye has been found to prevent PG loss completely, there are those which may reduce the losses considerably (see below).

A disadvantage associated with the use of these dyes is that the precipitation of PG leads inevitably to molecular collapse, the extent of which depends on the nature of the dye used. This is a problem which, incidentally, cannot be entirely avoided even in the absence of a precipitating agent when conventional chemical processing protocols are adopted; PG undergo collapse during tissue dehydration in organic solvents within which they are insoluble.

The principal criteria on which the choice of a particular cationic dye is based are that it should penetrate cartilage tissue rapidly and homogeneously, and react stoichiometrically with PG in the matrix (48). Under physiological conditions, the diffusion of a substance through cartilage matrix obeys specific diffusion laws [mainly of the Gibbs–Donnan Type (47)], and it is important to remember that during the initial stages of fixation these laws are also applicable. In accordance with these, a cationic dye should have a low molecular weight and specific volume, and yet carry a high positive charge density (48).

The structural form of chondrocytes may be used as an indicator of the penetration qualities of a particular dye. Under physiological conditions the plasma membrane of these cells is in intimate contact with the pericellular matrix PG (37, 49). Under conditions where a dye fails to penetrate thoroughly, PG will be lost preferentially from this compartment (49) where a supportive collagen network is absent (39). As a consequence, an empty lacunar space will develop around the cells, which will themselves be collapsed or shrunken in appearance (48, 49). A total tissue PG loss of between 3 and 5% is sufficient to deplete the pericellular matrix and produce this effect. Hence, the occurrence of collapsed and/or shrunken cells surrounded by "lacunar spaces" in normal cartilage may be used as a simple but very sensitive indicator of PG extraction resulting from poor dye penetration (for comparisons see Figs. 10–13).

Ruthenium hexamine trichloride (RHT) appears to fulfill the requirements of a cationic dye most adequately, and its inclusion in fixation media results in homogeneous preservation of cartilage matrix and chondrocytes throughout tissue blocks (Fig. 10). Ruthenium

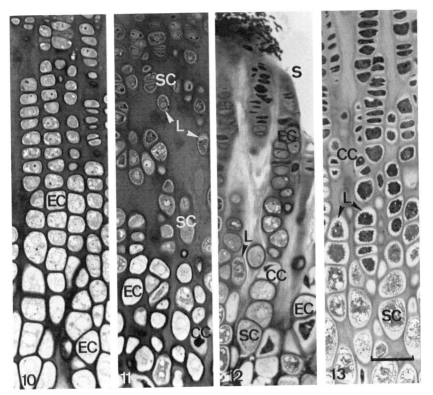

Figs. 10–13. Light micrographs of rat growth plate cartilage. Thick (1 μm) sections, stained with toluidine blue O. All photographs are taken at the same magnification: ×238; bar = 50 μm. Abbreviations used: EC, expanded chondrocyte, filling its lacunar space; CC, collapsed chondrocyte; SC, partially shrunken chondrocyte; L, lacuna.

Fig. 10. Fixation in the presence of RHT. Rapid and homogeneous penetration of fixative and dye assures preservation of all chrondrocytes in an expanded state. Since pericellular matrix PG are not lost, no artificial lacunar spaces develop.

Fig. 11. Fixation in the presence of ruthenium red. Dye penetration during fixation is unhomogeneous, the result being that chondrocytes are shrunken (SC) and collapsed (CC), and surrounded by empty lacunae (L) (due to PG loss into the fixation medium). A few chondrocytes (EC), mainly in the hypertrophic zone, are preserved in an expanded state.

Fig. 12. Fixation in the presence of safranin O. S, Tissue block surface. A reasonable proportion of optimally preserved cells (EC) is observed only along the tissue block surface, up to a depth of ~100 μm. All other cells, especially those deep within the tissue block, appear shrunken or collapsed (SC/CC). A high loss of pericellular matrix PG has led to the formation of optically empty artifact spaces around the cells. L, Lacunae.

Fig. 13. Fixation in the presence of alcian blue. The quality of cell preservation is not different from that obtained after fixation in the absence of a cationic dye, demonstrating that this precipitating agent does not penetrate into cartilage tissue blocks more than a few microns. All cells appear either shrunken (SC) or collapsed (CC), and are surrounded by artifact lacunar spaces (L), as a result of extensive PG loss during fixation.

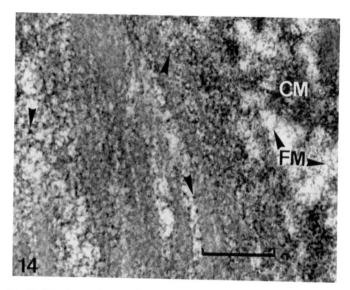

FIGS. 14–20. Electron micrographs of growth plate cartilage matrix. All pictures are taken at the same magnification (×40,000; bar = 0.5 μm) and originate from corresponding matrix compartments. Abbreviations used: arrow, matrix granule; CM, fused matrix granules; FM, fine filamentous material.

FIG. 14. Chemical fixation in the presence of RHT. Having a high positive charge density, RHT is capable of effective PG precipitation. These molecules (or their fragments) appear as small, dense matrix granules. Fusions (FM) between matrix granules may also occur. Fine filamentous material (FM) is interspersed irregularly between the granules.

FIG. 15. Tissue processed by high-pressure freezing, freeze substitution, and low-temperature embedding. The PG are preserved in an extended state and appear as a dense network between the collagen fibrils.

FIG. 16. Tissue fixed in the presence of toluidine blue O. The "matrix granules" (arrows) appear as bulky or coarse, interconnecting (fused?) tubes, forming a very coarse network.

FIG. 17. Fixation in the presence of alcian blue. The PG appear as fine, slender tubes (arrows), frequently possessing pointed ends; they are rarely interconnected.

FIG. 18. Fixation in the presence of safranin O. The PG appear as bulky, short rods, often curved, frequently having pointed or irregular ends.

FIG. 19. Fixation in the presence of ruthenium red. The PG appear as clumps (arrows) with variable shapes and sizes. The irregularity of the "empty" spaces between the matrix granules illustrates well the directional irregularity of the precipitating process per se.

FIG. 20. Fixation in the presence of cuprolinic blue. The PG are precipitated in an irregular pattern; they may appear as clumps of various shapes and sizes or as short tubes, these sometimes having quite pointed ends.

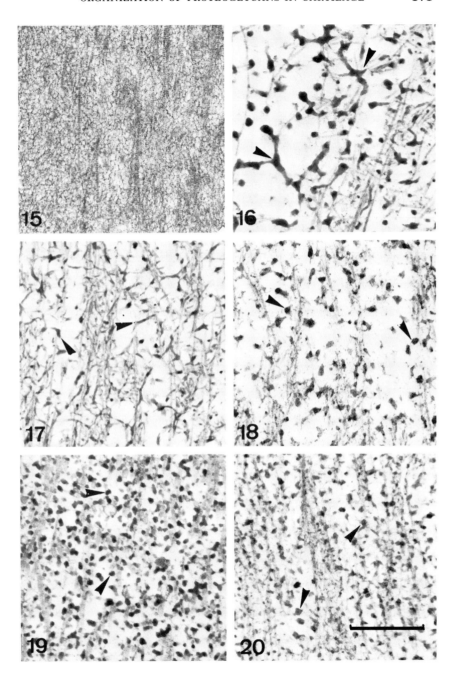

red, which penetrates in an uneven manner, yields only scattered optimal PG precipitation and chondrocyte preservation (Fig. 11). Safranin O possesses poor penetration qualities, and is capable of preserving cells optimally only along the tissue block surfaces (Fig. 12). Alcian blue penetrates only a few microns into cartilage tissue blocks, preserving only a very superficial layer sufficiently well for morphological matrix analysis.

While RHT preserves cartilage matrix most effectively in content, its high precipitative power effects very severe PG collapse, these molecules appearing as condensed "matrix granules." The extent of PG collapse produced by cationic dyes can be easily assessed by comparing their appearance after such treatment with isolated molecules (50; Fig. 24) or with cryotechnically processed material (Figs. 4–6, 15, 25).

The term "matrix granule" is a fairly unspecific term used to describe a variety of structures, the nature of which depends on the cationic dye used. Ruthenium red, for example, yields clumps of variable shapes (Figs. 19, 24) interconnected with filamentous material. Toluidine blue O and, to a lesser extent, alcian blue (Figs. 16, 17) produce quite different effects, PG appearing as bulky "tubes," rather than granules, with irregular shapes. After fixation in the presence of safranin O, PG are precipitated as curved, bulky "rods" with pointed or irregular ends (Fig. 18); cuprolinic blue produces a similar, although less pronounced, pattern of precipitation (Fig. 20). The manner in which PG are precipitated depends also on the coordinates of these molecules with respect to the penetrating dye front. Hence irregularities in the size and distribution of PG precipitates will result irrespective of the penetration qualities of the dye (Figs. 10–13).

The highly variable appearance of a unique structure obtained using cationic dyes makes interpretation of structure difficult. Furthermore, it is not possible to ascertain with certainty precisely what individual components are represented in PG precipitates (see Figs. 21, 22).

The resolution obtainable from these precipitates is, at maximum, twice their lateral diameter (for explanation see legend to Fig. 23). Assuming that the PG within cartilage matrix are present in a highly condensed state (i.e., underhydrated by a factor of $\sim$4), the double lateral dimension of a PG precipitate would lie between 120 and 160 nm (Fig. 23). Considering that with the conventional transmission electron microscope, lateral dimensions of 3–5 nm may be obtained under optimal conditions, the lateral resolution range for PG precipitated in the presence of cationic dyes is quite low (Fig. 24). Discrete relationships existing between PG and collagen fibrils, for example, can thus not be analyzed with any degree of accuracy (Fig. 24).

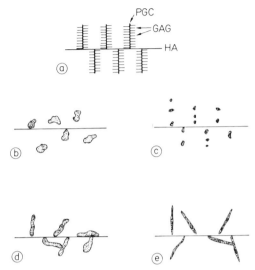

Fig. 21. (a) Schematic two-dimensional representation of a PG aggregate. HA, Hyaluronic acid residue; PGC, proteoglycan core protein; GAG, glycosaminoglycan chain. All subsequently drawn models (b–e) relate to this figure. (b–e) Schematic representation of PG-S precipitation patterns obtained using different cationic dyes. (b) Retraction of whole PG-S to granulelike structures (that may be induced by a cationic dye such as ruthenium red). (c) Extreme retraction of either whole or fragmented PG-S to "tight" granules with a high electron absorbency. Cationic dyes with a high positive charge, such as RHT, produce this pattern of precipitation. (d) Retraction of glycosaminoglycan chains toward the PG core protein produces "tubelike" structures which may fuse with one another. Both toluidine blue and safranin O precipitate PG in this manner. (e) Extreme retraction of glycosaminoglycan chains toward the PG core protein yielding "spindlelike" structures which may fuse together. This pattern of precipitation is produced by alcian blue.

D. Identification of Different Proteoglycan Species

By application of the critical electrolyte concentration principle (51, 52), it has been possible to discriminate between various types of PG (e.g., keratan sulfate and chondroitin sulfate) in cartilage tissue. Although this technique has permitted identification of individual PG species, elucidation of fine-structural relationships such as those existing between these molecules and collagen fibrils has not been possible (53). The main reason for this is that the success of the technique relies on the use of very high osmolarities, to which cartilage matrix structure is particularly susceptible (48). Under such conditions delicate interactions existing between molecules are destroyed (54, 55).

The use of highly purified antibodies raised against different types of PG (or their components) in immunohistochemical investigations

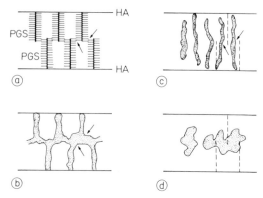

Fig. 22. (a) Two neighboring PG aggregates are represented (in two dimensions). The PG subunits (PGS) of each are drawn on one side only. Arrows indicate locations of PGS overlap (interdigitation). Subsequently drawn models (b–d) relate to this figure. (b–d) Schematic representation of PG aggregate precipitation patterns obtained using different cationic dyes. (b) Precipitation of glycosaminoglycan chains by a cationic dye of low strength; individual units may fuse together at points of interdigitation (arrows) to form a continuous network. (c) Precipitated PG subunits, dislocated from their original positions, may fuse together at points of interdigitation (arrows). In such an instance, a single "tubelike" "matrix granule" would represent two PG subunits from neighboring aggregates (- - - - original positions of PG core proteins). Unlike the precipitated units in (b), these "tubelike" elements will not form a true three-dimensional network. (d) Two or more neighboring PG subunits may coalesce under precipitation to form "matrix granules" clumps of variable sizes and shapes.

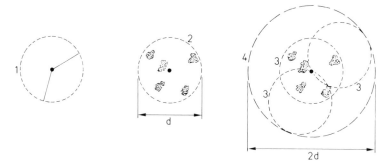

Fig. 23. (1) Schematic drawing of a cross-sectioned proteoglycan subunit, the domainal border (- - - -) of which is represented as a circle (60–80 nm in diameter) in the (underhydrated) native state. Solid lines represent glycosaminoglycan chain. (2) Upon precipitation, the forming matrix granule (represented by irregularly shaped, shaded areas) may be located anywhere within the domain of the native PG subunit. (3, 4) Each of the depicted matrix granules may represent the position either of the original central core protein axis (●) or of the one that has been maximally (within the diameter, d, of its own domain) dislocated (3). In a transmission electron micrograph of such a precipitated PG-S it is not possible to distinguish between these possibilities. Hence, the lateral resolution for locating the original position of the core protein axis is twice the diameter of its native domain (i.e., approximately 120–160 nm) (4).

will probably prove to be a more promising approach for the future (56, 57). At the level of the light microscope, tissue PG have already been localized by immunofluorescence (58, 59). Improvements in methods of tissue preservation will, however, be required before the information gleaned from immunoelectron-microscopic studies can be accepted with assurance.

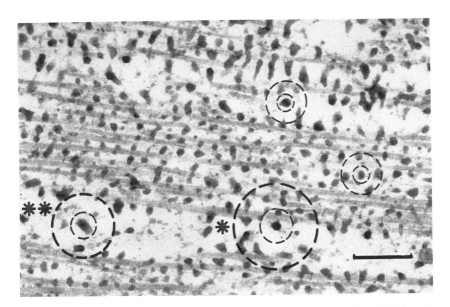

FIG. 24. Electron micrograph of cartilage matrix after precipitation of PG with ruthenium red. The two concentric circles marked by ** around a matrix granule represent the domainal borders of a PG-S when present in a completely hydrated state. The inner circle describes the cross-sectional range (80–120 nm in diameter), and the outer circle the longitudinal domain (250–350 nm). The two pairs of unmarked circles around matrix granules delineate the cross-sectional (60–80 nm in diameter) and longitudinal (160–220 nm) domains of a PG-S in an underhydrated state of approximately four. The two concentric circles marked by * around a matrix granule represent the best lateral resolution possible for localizing the original position of PG-S. If the matrix granule represents a PG-S in cross section, then the core protein axis may have been originally located at any point within the inner circle (120–160 nm in diameter). If the matrix granule represents a longitudinal section through a PG-S, then the extremities of the core protein axis could originally have been located at any position within the outer circle (220–440 nm) (for explanation see Fig. 23). The range of optimal lateral resolution possible for localizing the original position of the PG-S from a single matrix granule cannot be determined exactly, since the orientation of the core protein axis within the section plane, and the specific precipitation pattern for PG-S by the cationic dye, are unknown. ×62,000; bar = 0.25 μm.

V. SPATIAL ARRANGEMENT OF PROTEOGLYCANS
IN NATIVE TISSUE

A. The Problems of Image Analysis

Structural details of PG precipitated in the form of "matrix granules" (cf. Figs. 14–20) are lost during the preparative procedure and can be neither detected nor identified. The precipitation process per se leads to a limitation in the resolution limits, these being at best in the order of 120–160 nm (for explanation see Section IV,C). Using such material it is thus not possible to ascertain with any degree of certainty the nature of fine (within the range of a few nanometers) relationships existing between structural molecules. Cryotechnically processed material, however, allows identification of PG components in the dimensional order of 5–15 nm (for explanation see Section III,B). Since these and other structures (e.g., collagen fibrils) have dimensions within the range of the section thickness (~40 nm), electron micrographs can no longer be considered as being representative two-dimensional profiles of these objects (as, for example, is acceptable for thin sections through nuclei or mitochondria). The images need to be analyzed as projections of structures present in a three-dimensional section volume (Fig. 7). Three-dimensional relationships between structures cannot be judged unambiguously by examining electron micrographs taken at a single projection angle. A series of pictures taken at various angles is required for this and indeed then is possible only if the structures of interest are of simple geometric configurations (60).

For a study of the topological relationships existing between structures, it is important that like structures are contrasted with equal intensity throughout the section volume. This may be achieved most satisfactorily by staining tissue *en bloc* prior to embedding. Contrasting thin sections is more likely to yield inhomogeneities due to problems of stain penetration through plastics (61). If contrasting is performed on ultrathin sections, then care should be taken that the stains and plastics are compatible with respect to hydrophilicity. [Lowicryl K4M is an example of a hydrophilic resin which permits good penetration of uranyl acetate (61, 62).]

The contrast intensity of an object or part of one as projected on the screen of an electron microscope is related not only to the intensity of staining of the tissue per se but also to its orientation within the three-dimensional tissue volume.

A filament, for example, with low contrast intensity may be only just apparent or even invisible (truncation phenomenon) when lying parallel to the section surface (Fig. 26). The visibility of this structure

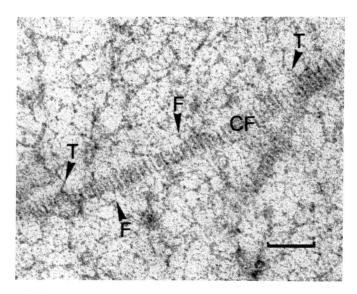

Fig. 25. High-power electron micrograph of rat growth plate cartilage matrix processed by high-pressure freezing, freeze substitution, and low-temperature embedding. A collagen fibril (CF) running parallel to the section surface and somewhere within the section volume can be identified. Due to the possibility of superimposition of PG components (e.g., fine, F, and thick, T, threads) upon the collagen fibril, it is not possible to determine whether contact sites really exist or not. (For detailed explanation see Figs. 26 and 27.) ×130,000; bar = 0.1 μm.

will, however, increase as the angle of orientation with respect to the surface increases reaching a maximum when perpendicular (see Fig. 26). The reason for the change in apparent intensity is a function of the thickness of electron-absorbent material presented to the electron beam when traversing the structure. This will increase to a maximum as the angle of orientation is increased from 0 to 90° with respect to the surface.

The contours of a projected object will also change with respect to its orientation in the tissue. A filament, for example, when aligned at right angles to the section surface will appear as a mere point on an electron micrograph. If the diameter of this point lies below the resolution limits of the transmission electron microscope (~5 nm), then it will not be detectable at all (Fig. 26). The carbohydrate chains of PG, for example, cannot be discriminated from the background granularity when orientated at an angle of 90° to the section surface (see Fig. 6).

A further difficulty encountered when attempting to interpret images of small objects whose dimensions lie within the section volume is

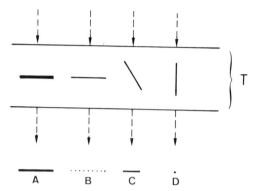

FIG. 26. Truncation phenomenon. (A) A thick filament lying parallel to the section surface, within a section volume (T), is clearly identifiable upon projection by the electron beam. (B) The same filament as in (A), having an identical position becomes invisible (truncation), when less intensely stained. (C) If this filament could be rotated through the section volume, there would be a critical point at which it becomes visible, this position being a function of the minimal mass of electron-absorbing material required for projection in the electron beam. (D) In the extreme case, with the filament rotated through 90° (from its position parallel to the section surface), it will be projected as a point, and as such possibly no longer distinguishable from the background granularity (see Fig. 6).

that of object superimposition. The dimensions of PG components and the thickness of collagen fibrils (for example), both lie within the depth of the section. When these components lie at different levels within this depth, they may appear to be in contact with one another if their positions are overlapping in the plane of the section (Figs. 25, 27). An electron micrograph would in this case suggest the existence of PG–collagen contact sites where none exist.

To overcome these problems of image analysis, a set of pictures taken from various angles need to be combined mathematically (Fourier synthesis) to build up a three-dimensional map of the structures (60). Stereo-electron microscopy may also be of value in this respect; the main drawbacks of this technique are, however, that data from pairs of corresponding points are combined visually, and thus on a subjective basis. For simple three-dimensional objects such as tubes or fibers it remains nonetheless a powerful tool.

B. Structural Relationships between Proteoglycans and Collagen Fibrils

When certain spatial geometric arrangements are satisfied, it is possible to analyze structural relationships within a single electron

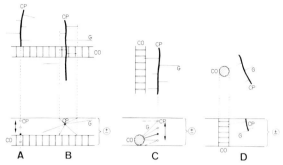

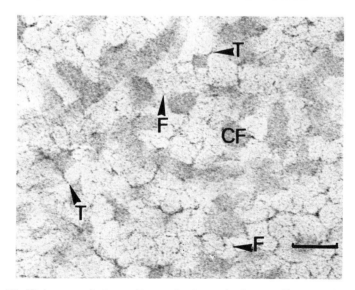

Fig. 27. Each figure (A–D) consists of a corresponding pair (upper and lower). The upper drawings represent projections of PG and collagen fibrils, as may be seen in electron micrographs. The lower drawings reveal the spatial organization of the same structures within the section volume. Abbreviations: CP, core protein of a PG; G, glycosaminoglycan chain of a PG; CO, collagen fibril; ± indicates that the spatial organization may also occur in reverse without affecting the projected image. (A, upper) The core protein apparently contacts a collagen fibril in the projected image. (A, lower) Within the section volume, the core protein may actually lie above (or below) the collagen fibril. (B, upper) Core protein or glycosaminoglycan chains appear to contact a collagen fibril. (B, lower) The core protein may lie above (or below) this fibril and the glycosaminoglycan chain may end freely within the section volume. (C, upper) A glycosaminoglycan chain appears to contact a collagen fibril from a lateral position. (C, lower) This glycosaminoglycan chain may contact the collagen fibril from various planes within the section volume; it may also end freely without establishing a contact (- - - -). (D, upper) A collagen fibril, running perpendicular to the section surface, is contacted by a glycosaminoglycan chain. (D, lower) Providing that the collagen fibril runs through the whole depth of the section volume, this contact really does exist.

Fig. 28. High-power electron micrograph of growth plate cartilage matrix processed by high-pressure freezing, freeze substitution, and low-temperature embedding. Most of the collagen fibrils (CF) run perpendicularly through the section plane and are in contact with fine (F) and thick (T) components of PG. ×118,000; bar = 0.1 μm.

micrograph (42). This case is illustrated with specific reference to PG and collagen fibrils. A collagen fibril lying perpendicularly to the section surface (i.e., cross-sectioned in a two-dimensional projection) runs parallel with the electron beam; providing that the length of this fibril exceeds the thickness of the section, any PG contact apparent may be interpreted as a genuine one, since in this case superimposition of one structure upon another is not possible (Figs. 27, 28). Under these conditions, it is not possible to assess whether these contact sites exhibit a periodicity along the length of the collagen fibril. The necessarily subjective analysis of stereo-pair electron micrographs does, however, indicate that no periodicity exists, contact sites between PG components and the collagen fibril surface being randomly distributed (Fig. 29).

C. Structural Relationships between Proteoglycans and Cell Membranes

It has been previously suggested that chondroitin sulfate-rich PG may be intercalated in the chondrocyte plasma membrane, and that these may interact with similar molecules in the pericellular matrix

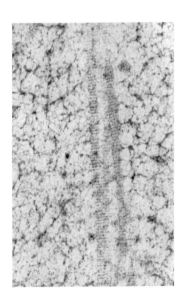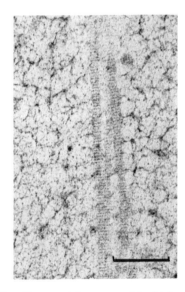

FIG. 29. Stereo-pair electron micrograph of cartilage matrix processed by high-pressure freezing, freeze substitution, and low-temperature embedding (tilt angle 7°). Fine and thick threads of PG contact type II collagen in a random manner, without any relationship to the cross-banded structure of these fibrils. ×77,000; bar = 0.25 μm.

(49). These structural relationships have not, however, been directly demonstrated, and their existence was proposed as being the most plausible explanation to account for the morphological appearance of chondrocytes after chemical fixation in the presence of RHT (48). Such material cannot be used to investigate the existence of such discrete relationships because of the poor lateral resolution obtained on electron micrographs after treatment of tissue with precipitating agents (see Figs. 19, 24).

In cryotechnically processed material, the lateral resolution is improved to such an extent that a morphological analysis becomes realistic (37). However, to avoid misinterpretations arising from superimposition of structures, only those plasmalemmata that are cut perpendicularly should be analyzed (42).

Examination of tissue processed under these conditions reveals the existence of numerous points of contact between chondrocyte plasmalemmata and PG components within the pericellular matrix (Fig. 30).

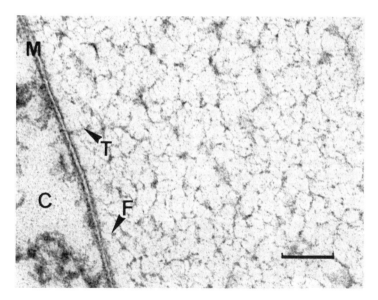

FIG. 30. High-power electron micrograph of growth plate cartilage processed by high-pressure freezing, freeze substitution, and low-temperature embedding. The plasmalemma (M) of a chondrocyte is cut perpendicularly. Intimate contact between the pericellular matrix and the cell is established, this being a manifestation of structural relationships between fine (F) and thick (T) filaments of PG and the outer (extraplasmatic) cell membrane surface. C, Cell cytoplasm. ×138,000; bar = 0.1 μm.

D. Spatial Organization of Proteoglycans in vivo

Using chemical procedures it is not possible to preserve PG both *in situ* and in their native state, due to the numerous technical difficulties intrinsic to this method of processing (e.g., PG precipitation, dislocation, and loss). The spatial organization of PG within the cartilage matrix of such tissue can thus not be analyzed with sufficient accuracy (see Section V,A). When cartilage is processed using cryotechniques these difficulties may be avoided or overcome; such tissue would thus serve as a reliable basis on which to undertake a structural analysis. To date, no work in this direction has been undertaken, and the ideas relating to PG organization presented in this section are purely speculative. It is hoped that their inclusion will stimulate interest in this area of research.

PG occupy the interstices within a structural framework of collagen fibrils and chondrocytes, and their existence in a considerably underhydrated state suggests that they are densely packed in a restricted space. Optimal spatial packing of PG may be achieved by their organization into a "liquid" crystalline form whose units are capable of sliding with respect to one another, thus ensuring flexibility and reversible deformity of the structure as a whole. In such a system, a crystal unit could be visualized as a polytope consisting of several and possibly different kinds of PG. The number of geometric arrangements possible for a crystal unit in such a maximal space-filling system is by no means unlimited but is indeed rather small (63, 64). On the other hand, PG could be distributed completely at random; in such a situation, a dense spatial packing would still be conceivable providing that the structural units were highly flexible.

Although these ideas have been conceived intuitively on the basis that a PG molecule is a geometric body with a clearly delineated surface, this, in fact, may not be the case. The domainal border of a PG molecule may end as a zone or coat region characterized by a gradient of decreasing repulsive force (negative charges), and units of this type could be arranged in an interdigitating fashion (interdigitating "surface zones") without loss of "liquid" phase characteristics (i.e., the ability of PG to move with respect to one another).

The limited freedom of movement of PG within cartilage matrix may be an important factor contributing to the elasticity and compressibility of the tissue as a whole. Hence any model of PG organization should be able to account for this phenomenon. Each of the three "models" proposed above can do so, not least the last mentioned which, to the authors, seems the most attractive.

E. The Role of Proteoglycans in Cartilage

Within cartilage matrix, PG are present at a very high concentration (i.e., they exist in an underhydrated state), and, as such, they exert a considerable pressure, assuring resiliency of the tissue to external compressive forces. Their main function is thus a biomechanical one (65). At the same time, however, they also control the diffusion and transport of ions through cartilage matrix (46), and may also influence the mineralization process (66, 67). Some classes of PG have been shown to be involved in cell attachment (68, 69), and if this holds true in cartilage, then these molecules may be involved in integrating chondrocytes into the matrix. At cell surfaces, PG have been demonstrated as having receptor functions. Developmental processes and fibrillogenesis (of collagen) may also be partially controlled by PG (70, 71). It is now being realized that these molecules play a key role in many biological processes of a general character—as, for example, in the pathogenesis of degenerative and inflammatory osteoarthritis (41, 72).

ACKNOWLEDGMENTS

The authors are indebted to Ceri England for the English correction of the manuscript. They are very grateful to Wolfgang Herrman, Karl Babl, and Peter Eggli for their help with the artwork, and to Regina Channi for typing the manuscript.

This work was supported by the Swiss National Science Foundation grant number 3.058.-0.84.

REFERENCES

1. Paulsson, M., and Heinegard, D. (1981). *Biochem. J.* **197**, 367.
2. Sajdera, S. W., and Hascall, V. C. (1969). *J. Biol. Chem.* **244**, 77.
3. Hardingham, T. (1981). *Biochem. Soc. Trans.* **9**(6), 489.
4. Hascall, V. C., and Lowther, D. A. (1982). *In* "Biological Mineralization and Demineralization" (G. H. Nancollas, ed.), pp. 179. Springer Verlag, Berlin and New York.
5. Hascall, V. C. (1977). *J. Supramol. Struct.* **7**, 101.
6. Hascall, V. C. (1982). *In* "Limb Development and Regeneration" (E. O. Kelley, P. F. Goetinck, and J. A. Mac Cabe, eds.), Part B, p. 3. Liss, New York.
7. Hascall, V. C. (1981). *In* "Biology of Carbohydrates" (V. Ginsburg and P. Robbins, eds.), Vol. 1, p. 1.
8. Heinegard, D., and Hascall, V. (1970). *J. Biol. Chem.* **245**, 4920.
9. Heinegard, D. (1977). *J. Biol. Chem.* **252**, 1980.
10. Kimura, J. H. *et al.* (1981). *J. Biol. Chem.* **256**, 7890.
11. Lohmander, L. S., *et al.* (1980). *J. Biol. Chem.* **255**(13), 6084.
12. Lohmander, L. S., *et al.* (1983). *J. Biol. Chem.* **258**(20), 12280.
13. Mason, R. M. (1981). *Progr. Clin. Biol. Res.* **54**, 87.
14. Muir, H. (1977). *Soc. Gen. Physiol. Ser.* **32**, 87.

15. Oegema, T. R., et al. (1975). J. Biol. Chem. 250, 6151.
16. Oegema, T. R., et al. (1984). J. Biol. Chem. 259(63), 1720.
17. Rosenberg, L., et al. (1983). In "Limb Development and Regeneration" (E. O. Kelley, P. F. Goetinck, J. A. Mac Cabe, eds.), Part B, p. 67. Liss, New York.
18. Thonar, E. J-M. A., et al. (1979). Arch Biochem. Biophys. 194(1), 179.
19. Rosenberg, L., et al. (1970). J. Biol. Chem. 245(16), 4123.
20. Rosenberg, L., et al. (1975). J. Biol. Chem. 250(5), 1877.
21. Buckwalter, J. A., and Rosenberg, L. C. (1982). J. Biol. Chem. 257(16), 9830.
22. Buckwalter, J. A. (1983). Clin. Orthop. 172, 207.
23. Heinegard, D., et al. (1978). Biochem. J. 175, 913.
24. Kimura, J. H., et al. (1978). J. Biol. Chem. 253(13), 4721.
25. Thyberg, J., et al. (1975). Biochem. J. 151, 157.
26. Thyberg, J. (1977). Histochem. J. 9, 259.
27. Wiedemann, H., et al. (1984). Biochem. J. 224, 331.
28. Muir, H. M. (1980). In "The Joints and Synovial Fluids II" (L. Sokoloff, ed.), p. 28. Academic Press, New York.
29. Urban, J. P., et al. (1979). Biorheology 16, 447.
30. Szirmai, J. A. (1969). In "Thule International Symposium on Aging of Connective and Skeletal Tissue" (A. Engel and T. Larsson, eds.), p. 163. Nordiska Bokhandelns, Stockholm.
31. Hunziker, E. B., and Schenk, R. K. (1984). In "Methods of Calcified Tissue Preparation" (G. R. Dickson, ed.), p. 199. Elsevier, Amsterdam.
32. Szirmai, J. A. (1963). J. Histochem. Cytochem. 11, 24.
33. Engfeldt, B., and Hjertquist, S. O. (1968). Virchows Arch. B Zellpathol. 1, 222.
34. Jozsa, L., and Szederkényi, G. (1967). Acta Histochem. 26, 255.
35. Jubb, R. W., and Eggert, F. M. (1981). Histochemistry 73, 391.
36. Luft, J. H. (1965). J. Cell Biol. 27a, 61.
37. Hunziker, E. B., et al. (1984). J. Cell Biol. 98, 267.
38. Mac Kenzie, A. P. (1972). Scanning Electron Microsc. Part II, 273.
39. Eggli, P., et al. (1985). Anat. Rec. 211, 246.
40. Rosenberg, L. (1971). J. Bone Jt. Surg. 53A, 69.
41. Mankin, H. J., et al. (1971). J. Bone Jt. Surg. 53A, 523.
42. Hunziker, E. R., and Schenk, R. K. (1984). J. Cell Biol. 98, 277.
43. Garavito, R. M., et al., (1982). J. Ultrastruct. Res. 80, 344.
44. Hascall, G. K. (1980). J. Ultrastruct. Res. 70, 369.
45. Smith, J. W. (1970). J. Cell Sci. 6, 843.
46. Maroudas, A., and Urban, J. P. G. (1980). In "Studies in Joint Disease" (A. Maroudas and A. J. Holborow, eds.), Vol. 1, p. 87. Pitman, London.
47. Maroudas, A. (1980). In "The Joints and Synovial Fluid II" (L. Sokoloff, ed.), Vol. 2, p. 239. Academic Press, New York.
48. Hunziker, E. B., et al. (1982). J. Ultrastruct. Res. 81, 1.
49. Hunziker, E. B., et al. (1983). J. Histochem. Cytochem. 31, 717.
50. Hascall, G. K. (1980). J. Ultrastruct. Res. 70, 369.
51. Scott, J. E. (1974). In "Normal and Osteoarthrotic Articular Cartilage" (S. Y. Ali, M. W. Elves, and D. H. Leaback, eds.), p. 19. Institute of Orthopaedics, University of London.
52. Scott, J. E., and Hughes, E. W. (1982). J. Microsc. (Oxford) 129(2), 209.
53. Goldberg, M., and Escaig, F. (1984). J. Microsc. (Oxford) 134(2), 161.
54. Koda, J. E., and Bernfield, M. (1984). J. Biol. Chem. 259(19), 11763.
55. Shimabayashi, S., et al. (1984). Chem. Pharm. Bull. 32(9), 3337.

56. Christner, J. E., *et al.* (1980). *J. Biol. Chem.* **255**(15), 7102.
57. Poole, A. R., *et al.* (1982). *J. Cell Biol.* **93**, 921.
58. Caterson, B., *et al.* (1982). *J. Invest. Dermatol.* **79**, 455.
59. Couchman, J. R., *et al.* (1984). *Nature (London)* **307**, 650.
60. Rosier de, D. J., and Klug, A. (1968). *Nature (London)* **217**, 130.
61. Horobin, R. W. (1983). *J. Microsc. (Oxford)* **131**(2), 173.
62. Carlemalm, E. R. (1982). *J. Microsc. (Oxford)* **126**, 123.
63. Coxeter, H. S. M. (1963). "Regular Polytopes." Macmillan, New York.
64. Mc Mullen, P. (1981). *In* "The Geometric Vein (The Coxeter Festschrift)" (C. Davis, B. Grünbaum, and F. A. Sherk, eds.), p. 123. Springer-Verlag, Berlin and New York.
65. Maroudas, A., *et al.* (1980). *Ann. Rheum. Dis.* **39**, 514.
66. Pal, S., *et al.* (1981). *Coll. Rel. Res.* **1**, 151.
67. Vittur, L., *et al.* (1977). *Bull. Mol. Biol. Med.* **2**, 189.
68. Woods, A., *et al.* (1984). *J. Cell Biol.* **99**, 1743.
69. Turley, E. A. (1984). *Cancer Metastasis Rev.* **3**, 325.
70. Chandrasekhar, S., *et al.* (1984). *Coll. Rel. Res.* **4**, 323.
71. Solursh, M., *et al.* (1984). *Dev. Biol.* **105**, 451.
72. Mc Devitt, C. (1981). *In* "Tissue Repair and Regeneration" (L. E. Glynn, ed.), Vol. 3, p. 111. Elsevier/North Holland, Oxford.

Proteoglycans, Chondrocalcin, and the Calcification of Cartilage Matrix in Endochondrial Ossification

A. Robin Poole

Joint Diseases Laboratory, Shriners Hospital for Crippled Children, Department of Experimental Surgery, McGill University, Montreal H3G 1A6, Quebec, Canada

and

Lawrence C. Rosenberg

Orthopaedic Research Laboratories, Montefiore Hospital and Medical Center, Bronx, New York 10467

I. Anatomical Organization of the Growth Plate

In order to understand clearly the events that occur within the growth plate and, in particular, the calcification of cartilage matrix, it is essential that the reader has a clear perspective of how the growth plate develops. Figure 1 provides a diagrammatic representation of the various zones of the primary growth plate. In the proliferative zone, chondrocytes divide at a rapid rate and there is synthesis of new matrix including proteoglycan (Greulich and Leblond, 1953; Bélanger, 1954; Hjertquist, 1961), collagen, and noncollagenous proteins. In the primary growth plate in particular, where the growth in length of long bones occurs, these cells are organized into columns of cells along the longitudinal axis of the developing bone.

Early in the development of the longitudinal septa, in the upper proliferative zone, matrix vesicles are first seen, being budded off from the plasma membranes of cells into the longitudinal septa (Cecil and Anderson, 1978; Hunziker *et al.*, 1981; Ali, 1983; Morris *et al.*, 1983; Anderson, 1985). It is in these matrix vesicles, which were first

187

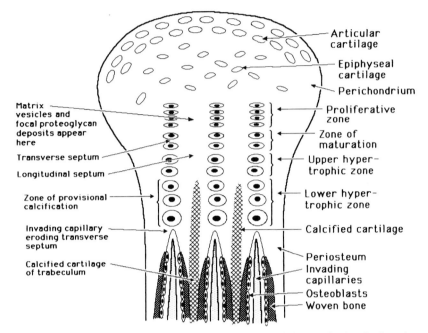

FIG. 1. A diagrammatic representation of the zones of the epiphysis of a long bone showing the primary growth plate. The large focal deposits of proteoglycan and link protein detected by antibody in cartilage treated with EGTA prior to fixation appear in the upper proliferative zone, as described in the text. They are very difficult to detect with antibody in the lower hypertrophic zone when calcification is initiated but can then be detected by dye binding or by microprobe analysis.

described by Bonucci (1967) and by Anderson (1967), that the earliest sign of mineral formation, in the form of apatite crystals, has been detected by electron microscopy (Anderson, 1969). It is not until the lower hypertrophic zone (also known as the zone of provisional calcification) that significant and plentiful calcification occurs: this is mainly restricted to the longitudinal septa. Only the earliest signs of mineral formation, detectable by careful ultrastructural analysis, are seen prior to the lower hypertrophic region. Some workers have claimed that the early presence of mineral in matrix vesicles is an artifact of tissue preparation (Landis et al., 1977), although others disagree (Schraer and Gay, 1977; Ali et al., 1977; Cecil and Anderson; 1978; Hunziker et al., 1981; Morris et al., 1983).

In the zone of maturation, the more flattened cells of the proliferative zone start to enlarge. In the lower hypertrophic zone this

enlargement is essentially complete. The volume of matrix decreases
as the cells enlarge (Buckwalter et al., 1986). Autoradiographic studies
of matrix synthesis in rats using $H[^{14}C]O_3$ (Greulich and Leblond,
1953) and $^{35}SO_4$ (Bélanger, 1954) have revealed that this is most pronounced in the cells of the maturing and upper hypertrophic zones.
Cells in the lower hypertrophic zone where calcification is occurring
showed no signs of synthesis. More recently, however, Hjertquist
(1961) reported that $^{35}SO_4$ labeling in rats was present throughout the
growth plate and most concentrated in cells of the proliferative zone.
The matrix vesicles persist in the hypertrophic zone but become progressively more fragmented, particularly at the time when calcification occurs in the zone of provisional calcification. Calcification is
usually centered in the longitudinal septa in the territorial and interterritorial matrix. Staining of bovine growth plate with antibodies to
link protein clearly reveals the presence of distinct pericellular, territorial, and interterritorial matrix in the longitudinal septa (Poole et
al., 1982a). The transverse septa usually remain uncalcified and these,
as well as the longitudinal septa, stain intensely for both proteoglycan
and link protein. Later the transverse septa are eroded by invading
capillaries, and associated chondroclastic cells, coming from the
metaphysis.

Calcification in the lower hypertrophic zone is centered in multiple
discrete foci consisting of large (>500 nm diameter) spherical mineral
clusters in the longitudinal septa, which contain a calcium-binding
molecule called chondrocalcin (Poole et al., 1984; Fig. 6) which has a
high affinity for hydroxyapatite (Choi et al., 1983). Mineral grows out
from these calcific nuclei into the surrounding matrix (Poole et al.,
1984). Finally, at the junction with the metaphysis, the longitudinal
septa become almost totally calcified. With the selective erosion of the
transverse septa and the survival of about half of the calcified longitudinal septa (Schenk et al., 1967), osteoblasts settle upon these calcified
trabecular extensions of the growth plate and deposit osteoid which
calcifies to form woven bone (Fig. 1). Thus a bony shell is formed on a
calcified cartilaginous core which acts as a framework directing the
assembly of the forming bone.

The discrete focal calcification that is centered in the lower hypertrophic zone occurs in the immediate vicinity of both intact and fragmented matrix vesicles (Thyberg, 1974; Poole et al., 1984). Some have
argued that this calcification is always centered within and originates
from the matrix vesicles (Anderson, 1969, 1985; Bonucci, 1970), but
evidence for this is inconclusive. The large, spherical mineral clusters
are clearly observed in close proximity to matrix vesicles. Thyberg

(1974) reported that mineral is initially deposited in the matrix and secondarily in the vesicles. The highest density of matrix vesicles has been found in the proliferative zone and the lowest in the zone of provisional calcification (Reinholt et al., 1982; Shepard and Mitchell, 1985). This may result from disruption of matrix vesicles as they calcify.

Thus within the growth plate we have the development of cells which synthesize an extracellular matrix which will later calcify. This is a very special series of events which include extracellular changes involving proteoglycans and chondrocalcin which lead to the calcification of the cartilage matrix.

II. COMPOSITION AND ORGANIZATION OF THE CARTILAGE MATRIX OF GROWTH PLATE

The growth plate represents a highly specialized hyaline cartilage committed to the calcification, in a highly ordered manner, of part of its extracellular matrix prior to the invasion of capillaries from the metaphysis. The chemistry of this matrix and the biology of its cells inevitably reflect the fact that this is a tissue much of which is destined to calcify.

A. Collagen

Like all hyaline cartilages, the matrix of growth plate cartilage contains a backbone of type II collagen fibrils which constitute at least half of the dry weight of this matrix. In addition it contains other collagens. The 1α, 2α, 3α collagen chains (Burgeson and Hollister, 1979) are present in pericellular sites throughout the growth plate (Ricard-Blum et al., 1982). Type IX collagen (van der Rest et al., 1985), previously referred to as HMW and LMW collagen or type M, is probably also present in growth plate (Reese and Mayne, 1981; von der Mark et al., 1982). In other cartilages it has also been found in pericellular sites (Duance et al., 1982). Type IX is an intriguing molecule in that, besides being a collagen, it is also by definition a proteoglycan. It is composed of three different α-chains (van der Rest et al., 1985), to one of which dermatan sulfate is bound (Noro et al., 1983; Vaughan et al. 1985).

Type X collagen is also present within the growth plate (Gibson et al., 1981; Gibson et al., 1982; Schmid and Conrad, 1982; Grant et al., 1985). It appears around hypertrophic chondrocytes (Schmid and Linsenmayer, 1985) just before calcification is detectable. Other than type

II collagen, which provides tensile strength (Kempson, 1980), we have no clear idea of the functions of the other collagens. Their presence in cartilage matrix suggests that they play a structural role and, in the case of type X collagen, a role in its calcification. Biosynthetic studies also indicate that type X collage is a major synthetic product of hypertrophic cells (Schmid and Conrad, 1982).

A collagenase capable of cleaving triple-helical collagen has been detected in the hypertrophic zone of the growth plate cartilage (Dean *et al.*, 1985). This may be involved in the controlled remodeling of the growth plate matrix as it decreases in volume during maturation from the proliferating zone to the hypertrophic zone; it may also be involved in degradative changes in the organic matrix associated with calcification, since collagen degradation has been detected in calcifying cartilage (Vittur *et al.*, 1971; Wuthier, 1969).

B. Proteoglycan

The proteoglycans of growth plate cartilage are, as far as we are aware, structurally closely related to those found in other hyaline cartilages such as nasal and articular. Our knowledge of these molecules is limited by the fact that proteoglycans of the growth plate have been very little studied. Proteoglycans have been isolated from the developing epiphyseal cartilage and from the metaphysis, which includes the calcified trabeculae derived from the longitudinal septa of the growth plate. But proteoglycans have not yet been isolated and biochemically characterized specifically from the individual proliferative, maturing, and hypertrophic zones of the growth plate itself. This has been mainly due to the technical difficulties encountered in working with such small pieces of tissue and the great need for a rigorous histological characterization of what is being studied. Such work has recently been initiated in our laboratories.

If we base our judgments of the proteoglycans of growth plate cartilage on those of other hyaline cartilages, we can arrive at the structure shown in Fig. 2. This depicts a large proteoglycan aggregate isolated from bovine fetal epiphyseal cartilage. These proteoglycan monomers of fetal or immature cartilage are larger than those found in adult cartilages (Roughley and White, 1980; Pal *et al.*, 1981; Buckwalter and Rosenberg, 1982). The majority aggregate with hyaluronic acid. Hyaluronic acid-based proteoglycan aggregates have been detected *in situ* in articular cartilage (Poole *et al.*, 1982c) but not in growth plate cartilage. Only circumstantial data for their presence *in situ* has been obtained (see below). The aggregation with hyaluronate

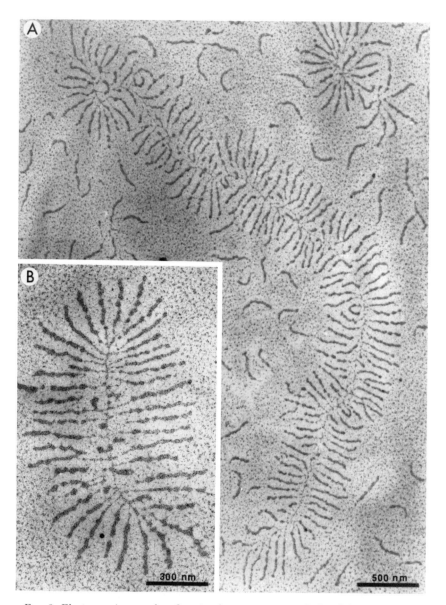

Fig. 2. Electron micrographs of proteoglycan aggregates isolated from bovine fetal epiphyseal cartilage. (A) Large aggregate: the aggregate consists of multiple closely spaced monomers bound to a central hyaluronic acid filament. Free monomers and small clusters of monomers surround the aggregate. (B) Small aggregate: monomers consist of a thin segment and a thick segment. The thin segment attaches directly to the central hyaluronic acid filament, while the thick segment extends peripherally. Reproduced from Buckwalter and Rosenberg (1982), with permission.

is stabilized by link protein. Link protein has been detected immu-
nohistochemically in growth plate cartilage (Poole *et al.*, 1982a) and in
hypertrophic cartilage formed by injection of decalcified bone matrix
(Poole *et al.*, 1982b). It has also been detected in fluid containing proteoglycan aggregates which has been aspirated from rat growth plate
(Pita *et al.*, 1979). The work of Axelsson *et al.* (1983) has demonstrated
the presence of proteoglycan aggregates in normal rabbit growth plate
cartilage. Using gel chromatography and electron microscopy, they
noticed that aggregates in the uncalcified cartilage were smaller than
those seen in articular cartilage and contained larger proteoglycan
monomers. There were at least two recognizable populations of aggregated proteoglycan in uncalcified cartilage. Earlier analyses of extracellular fluid aspirated from rachitic rat growth plate also indicated
the presence of two aggregate populations (Pita *et al.*, 1979). As the
authors stated, this may be due to the presence of two distinct populations of hyaluronic acid.

It has been reported that rat growth plate proteoglycans contain
keratan sulfate in view of the high content of glucosamine (Reinholt *et
al.*, 1985). The presence of a significant amount of keratan sulfate in a
fetal cartilage is, however, uncharacteristic, since this molecule is normally absent or almost absent from fetal cartilages (Roughley and
White, 1980; Pal *et al.*, 1981; Webber *et al.*, 1985). Immunohistochemical studies with a monoclonal antibody to this molecule have also
revealed that it cannot be detected within the fetal growth plate (Poole
et al., 1986). Thus the glucosamine may be present in oligosaccharides
rather than in keratan sulfate.

A small dermatan sulfate-rich proteoglycan has been isolated from
adult bovine articular cartilage (Rosenberg *et al.*, 1985). This molecule
has also been shown to be present in fetal cartilage, but there appears
to be relatively little of this molecule present in growth plate cartilage
(Poole *et al.*, 1985).

The structural organization of proteoglycans within growth plate
cartilage has been studied at the ultrastructural level using both dye-
binding techniques (Thyberg *et al.*, 1973; Thyberg, 1974; Shepard and
Mitchell, 1976, 1977) and immunohistochemical techniques (Takagi *et
al.*, 1983; Poole *et al.*, 1982c). Proteoglycan monomers, identified as
discrete granular staining (Hascall, 1980), have been observed in association with collagen fibrils throughout the cartilage matrix using dye
binding (Thyberg *et al.*, 1973; Shepard and Mitchell, 1976, 1977). Similar structures have been seen using antibodies (Poole *et al.*, 1982c;
Takagi *et al.*, 1983) and are also considered to represent collapsed
proteoglycan monomers (Poole *et al.*, 1982c). Similar staining for link

protein has also been seen (Poole *et al.*, 1982c). More recently, using special preparative techniques, much larger focal proteoglycan deposits have been seen in the calcifying longitudinal septa of growth plate cartilages using both dye binding to detect proteoglycan (Shepard and Mitchell, 1985) (Fig. 3) and immunolocalization for proteoglycan and link protein (A. R. Poole, I. Pidoux, N. Mitchell, and L. C. Rosenberg, manuscript in preparation) (Fig. 4). These structures are only visualized by dye binding in calcified cartilage when proteoglycans are precipitated with acridine orange and then demineralized. Using antibodies, focal proteoglycan concentrations were observed in growth plate cartilage prior to calcification but only when the tissue was treated with ethylenediamine tetraacetic acid (EDTA) or 1,2-di(2-aminoethoxy)ethane-*NNN′N′*-tetra-acetic acid (EGTA), prior to fixation in formaldehyde. In these cases, we are producing some kind of artifact. These focal concentrations of proteoglycan, detected by antibodies directed against both proteoglycan monomer and link protein, are spread throughout the uncalcified longitudinal septa of the pro-

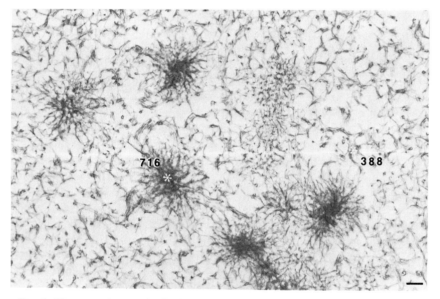

FIG. 3. Electron micrograph of a demineralized section of the lower hypertrophic zone of rat growth plate to show the rosettelike radial formation of proteoglycan fixed in acridine orange and glutaraldehyde. The sulfur peaks for proteoglycan determined by electron microprobe analysis revealed that the sulfur concentrations over these clusters were almost twice those of the remaining matrix. Bar = 200 nm. From Shepard and Mitchell (1985), with permission.

liferative, maturing, and upper hypertrophic zones. When calcification occurs they are hard to detect with antibody, yet can then be clearly seen with dye-binding methods. They clearly represent the focal sites where the major calcification of the cartilage matrix is initiated and indicate that the distribution and organization of proteoglycans in extracellular matrix seems to change and they become focally concentrated as calcification occurs: this change can be prematurely induced by decalcifying the tissue. Focal concentrations of calcium and sulfur in calcifying sites have also been demonstrated using elemental analysis in studies of rat (Shepard and Mitchell, 1985; Fig. 3) and bovine growth plates (A. R. Poole, I. Pidoux, N. Mitchell, and L. C. Rosenberg, manuscript in preparation). Independent studies employing electron-spectroscopic imaging have also revealed the focal concentration of sulfur coincident with the initial sites of calcium accumulation in the hypertrophic matrix (Arsenault and Ottensmeyer, 1983).

For calcification to occur where there are high focal concentrations of proteoglycans (which are highly sulfated), these studies indicate that there must be a redistribution of a preexisting organization of proteoglycan. The inability to detect with antibodies these structures in calcified cartilage as easily as in uncalcified growth plate cartilage is thought to be related to a masking of proteoglycan and link protein epitopes from reaction with their respective antibodies, possibly by chondrocalcin. The apparent presence of link protein within these structures in view of the near-identical staining observed for proteoglycan and link protein in terms of distribution (size and shape) suggests that they represent proteoglycan monomers aggregated with hyaluronic acid, with the interaction stabilized by link protein. Since these large structures which stain for proteoglycan monomer and link protein have a mean diameter of up to 500 nm in bovine growth plate cartilage compared with diameters from 12 to 30 nm for the individual collapsed proteoglycan monomers, observed in articular cartilage (Poole *et al.*, 1982c), we believe that they represent a "conglomeration" of aggregated proteoglycan monomers which have collapsed from a spread aggregate organization to a condensed structure. For this to occur it is considered that the hyaluronic acid must be mobilized. Normally it appears bound to collagen fibrils in articular cartilage (Poole *et al.*, 1982c; and see Fig. 5A). Since these focal deposits of proteoglycan can be detected only by pretreatment with EDTA or EGTA (A. R. Poole, I. Pidoux, N. Mitchell, and L. C. Rosenberg, manuscript in preparation), the hyaluronic acid may be bound to collagen by calcium and be artificially mobilized by EDTA or EGTA to produce a collapse of spread aggregates. The concept of this collapse is shown diagram-

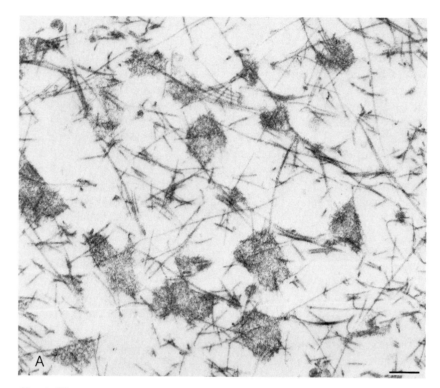

Fɪɢ. 4. Electron micrographs of the proliferative zone to show antibody staining (with the peroxidase–antiperoxidase method) of (A) proteoglycan and (B) link protein in longitudinal septa of the bovine growth plate. These focal deposits are only detectable when

matically in Fig. 5B and C. This may naturally also result from the activity of a hyaluronidase, since a shortening of hyaluronic acid in calcifying cartilage (suggestive of cleavage) has been observed (Buckwalter and Rosenberg, 1986; Table I).

C. Chondrocalcin

Calcification sites rich in proteoglycan occur exactly when and where the molecule we have called chondrocalcin (Choi *et al.*, 1983; Poole *et al.*, 1984) appears in the matrix in relatively high concentrations (Fig. 6). We now know that this molecule is very similar or identical to the C-propeptide of type II collagen (van der Rest *et al.*, 1986), which is cleaved from the procollagen by a specific metallo C-propeptidase (Leung *et al.*, 1979). Sequence analyses have revealed the

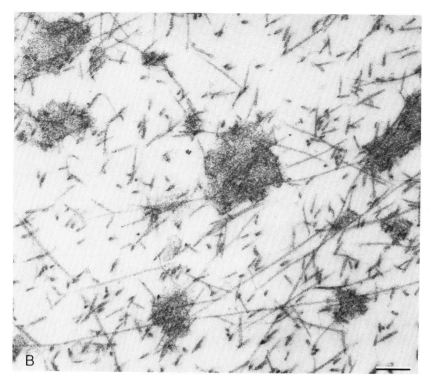

tissue is treated with EDTA prior to fixation. They are not clearly seen when calcification occurs, even after demineralization. Bars = 200 nm.

identity of the N terminus of chondrocalcin with the N-terminal sequences of the purified chicken and human C-propeptides of type II collagen and the sequence deduced from the cDNA of the bovine C-propeptide (van der Rest *et al.*, 1986); there is also identity between tryptic peptides isolated from chondrocalcin and the human and chick molecules. The only difference would appear to be that whereas we only observe dimers of a MW 35,000 subunit* (Choi *et al.*, 1983), a trimer would be expected for the C-propeptide of type II collagen. The concept of this molecule being involved in the calcification of cartilage matrix in addition to its established role in the formation of collagen fibrils (Schofield *et al.*, 1974; Uitto and Prockop, 1974; Rosenbloom *et*

* More recent unpublished studies by L. C. Rosenberg and co-workers of the molecular size of chondrocalcin using sedimentation equilibrium and HPLC have produced molecular weights of 39,000 and 78,000 for the subunit and dimer, respectively.

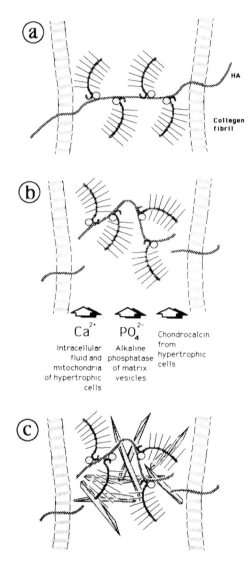

Fig. 5. Diagrammatic representation of the changes in proteoglycan organization during the calcification of cartilage matrix. (a) Natural organization of proteoglycan aggregate. (b) There are focal collapses of the aggregated proteoglycan. These can be artificially produced prior to calcification with EDTA extraction prior to fixation. They may occur naturally as a result of hyaluronidase cutting the hyaluronic acid (HA) backbone. The artificially collapsed proteoglycan aggregate can be detected with antibodies to proteoglycan and link protein as shown in Fig. 4. When collapse occurs commensurate with calcification, the collapsed aggregate may be detected in acridine orange–glutaraldehyde prepared tissue as shown in Fig. 3. (c) Mineral deposition on aggregate.

TABLE I

CHANGES IN THE DIMENSIONS OF PROTEOGLYCANS AND HYALURONIC ACID
FROM DIFFERENT ZONES OF BOVINE RIB GROWTH PLATE[a]

Slice number[b]	Hyaluronate chain length (nm ± SD)	Numbers of monomers per aggregate	Length of aggregated monomers (nm ± SD)	Length of nonaggregated monomers (nm ± SD)	Aggregated monomers (%)
1	772 ± 560	23 ± 12	309 ± 71	217 ± 118	31
2	676 ± 568	20 ± 9	302 ± 65	211 ± 82	28
3	641 ± 509	21 ± 12	304 ± 74	236 ± 84	28
4	484 ± 533	14 ± 8	308 ± 74	243 ± 88	14
5	300 ± 205	12 ± 5	336 ± 74	229 ± 92	8

[a] From J. A. Buckwalter and L. C. Rosenberg, unpublished results.

[b] The following 1-mm-thick zones were identified: 1 and 2, epiphyseal cartilage; 3, upper proliferative zone; 4, upper one-third of hypertrophic zone; 5, lower two-thirds of hypertrophic zone.

al., 1976) raises a fascinating and intriguing bifunctional role for the C-propeptide in hypertrophic cartilage. Clearly low levels of the molecule are detectable as weak to moderate staining in forming epiphyseal, nasal, and growth plate matrix in the proliferating and upper hypertrophic zones (Poole *et al.*, 1984) where the molecule is no doubt involved in the fibrillogenesis of collagen. But in the lower hypertrophic zone this staining is reduced or absent before intense staining appears in focal calcifying tissue, suggesting that here it is selectively involved in calcification.

Chondrocalcin appears to be intimately associated with the mineral phase (Poole *et al.*, 1984) as well as with the proteoglycan. Chondrocalcin is a calcium-binding protein (P. Hauschka and L. C. Rosenberg, unpublished observations) which is codistributed with the focally concentrated proteoglycan exactly where and when cartilage calcifies. It may thus play an important role in the calcification of cartilage matrix. In its native state it is also a very basic molecule with several isoforms exhibiting a range of pI from 8.2 to 8.8 (J. S. Mort, L. C. Rosenberg, and A. R. Poole, unpublished results), although in the presence of urea it behaves as a more acidic molecule, as observed previously (Choi *et al.*, 1983). In its native state it may therefore bind ionically to proteoglycan, neutralizing (in part at least) its highly anionic character. This may lead to a removal of the capacity of proteoglycan to inhibit mineral formation which has been observed *in vitro* (see below). Alternatively or in addition, it may play an important role in the accretion of mineral once mineralization has been

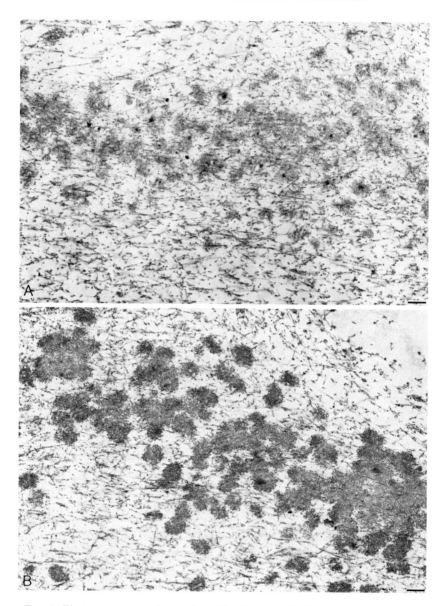

Fig. 6. Electron micrographs to show the close association between mineral and chondrocalcin (B) in the lower hypertrophic zone in a bovine fetus. Fixed, undecalcified tissue was stained in (B) with an antibody to chondrocalcin using the peroxidase–antiperoxidase technique. (A) Tissue treated with nonimmune immunoglobulin, showing only the mineral deposits. Bars = 500 nm. (From Poole *et al.*, 1984, with the permission of the Editors.)

initiated within the matrix, in view of its ability to bind to hydroxy-apatite.

Chondrocalcin is synthesized by chondrocytes from noncalcifying and calcifying cartilages (Hinek *et al.*, 1985a), but compared with cultures of nasal and epiphyseal chondrocytes it accumulates in increased amount (per DNA and per hydroxylysine) in cultures of growth plate chondrocytes. When plated at high density, the latter cells elaborate an extracellular matrix which calcifies, as shown by incorporation of ^{45}Ca, von Kossa staining, and electron microscopy. The secretion of the protein is regulated by vitamin D: in vitamin D deficiency it is absent from the uncalcified lower hypertrophic matrix of the growth plate. 24,25-dihydroxycholecalciferol stimulates synthesis in vitamin D deficient growth plate cartilage (Hinek and Poole, 1985b). Chondrocalcin can be detected within the cisternae of the endoplasmic reticulum, in the Golgi apparatus, and in condensing vacuoles of hypertrophic cells. In the fully enlarged cell in the lower hypertrophic zone it is seen mainly in very enlarged cisternae of the endoplasmic reticulum (E. Lee and A. R. Poole, unpublished results). Chondrocalcin has not been detected in the extracellular matrix of the lower hypertrophic zone in hypophosphatasia in humans (A. R. Poole, F. Glorieux, L. C. Rosenberg, and D. Rimoin, unpublished results). This condition is characterized by a deficiency of alkaline phosphatase and a gross lack of mineralization of the hypertrophic matrix (Jaffe, 1975). The selective presence of this molecule in epiphyseal cartilage in hypophosphatasia and its continuing presence in epiphyseal cartilage in vitamin D deficiency, together with its selective absence from the epiphyseal cartilage matrix in Kniest dysplasia but its presence in the calcifying lower hypertrophic zone (A. R. Poole, L. Murray, L. C. Rosenberg, and D. Rimoin, manuscript in preparation), suggests that the synthesis and/or secretion of this molecule in developing epiphyseal and growth plate cartilages is differently regulated.

III. CHANGES IN THE PROTEOGLYCANS OF THE GROWTH
PLATE ASSOCIATED WITH CALCIFICATION

Let us now examine in more detail other evidence for changes in the organization and the removal of proteoglycans in calcifying cartilage matrix. This has been a very controversial area of research where many workers using biochemical, histochemical, immunohistochemical, and electron-microscopic and electron microprobe analyses have produced widely divergent conclusions regarding changes in the amount and organization of proteoglycans in the calcifying growth plate.

Biochemical studies have concentrated upon analyses of the proteoglycan content in extracts of different parts of the growth plate cartilage. The greatest variation in results may be due to sampling problems which involve the extraction of quite different tissues such as the more cellular regions of the hypertrophic zone and metaphysis and the less cellular proliferative zone (Buckwalter et al., 1986). Some have reported that there is no evidence for a loss of chondroitin sulfate down to and including the hypertrophic zone where calcification occurs, although in the metaphysis, the amount of chondroitin sulfate is abruptly decreased (Weatherell and Wiedman, 1963). Here the transverse septa have been removed together with about one-half of the calcified longitudinal septa (Schenk et al., 1967). Thus, in the metaphysis, the chondroitin sulfate content does fall, but this occurs after mineralization. Others have also failed to detect a loss and, in fact, observed an increase in glycosaminoglycan in the calcifying hypertrophic zone (Lindenbaum and Kuettner, 1967; Vittur et al., 1971). One biochemical study, however, has recorded a net loss of proteoglycan in calcifying cartilage (Wuthier, 1969). Collagen loss has also been recorded (Wuthier, 1969; Vittur et al., 1971). All these observations were made with normal growth plates.

More detailed studies have examined the monomeric size and aggregatability of proteoglycans in the growth plates of vitamin D-deficient rats. Vitamin D deficiency induces an arrest in calcification in the hypertrophic zone, which elongates considerably, permitting an easier analysis of the upper and lower growth plate. Addition of vitamin D leads to a rapid "healing" of the rickets which involves a restoration of calcification in the elongated hypertrophic cartilage and the rapid shortening of this zone to normal with capillary ingrowth from the metaphysis. Examinations of these vitamin D-deficient rat growth plates have revealed that proteoglycans exhibit a decreased aggregatability in healing rachitic cartilage (Lohmander and Hjerpe, 1975; Howell and Pita, 1977; Pita et al., 1979; Reinholt et al., 1985). This is accompanied by a reduction in the content of galactosamine (as a measure of chondroitin sulfate), glucosamine (as a measure of keratan sulfate and/or oligosaccharide and hyaluronic acid) (Lohmander and Hjerpe, 1975; Reinholt et al., 1985) and of uronic acid (as a measure of chondroitin sulfate) (Reinholt et al., 1985). Proteoglycan degradation was not observed and only one main polydisperse population of proteoglycan monomer was identified (Reinholt et al., 1985). If cleavage occurred in the vicinity of the hyaluronic acid-binding region (leading to the observed loss of aggregation), degradation may not be detected by a reduction of size. A loss of aggregation with little sign of a change in size is commonly observed when proteoglycans are degraded in articular cartilage (see Tyler, 1985, and her discussion of related data). A loss of proteoglycan aggregation in normal growth plate has in fact been found to be associated with calcification (Franzén et al., 1982).

Healing rickets may be said to provide a very sensitive indicator of matrix changes in mineralization, but it must be remembered that the rapid extensive calcification of this hypertrophic zone coupled with its rapid destruction in healing rickets is in itself an abnormal process compared with what happens naturally. In vitamin D deficiency, calcification has been arrested as well as capillary invasion. Suddenly it is all restored. Thus we must accept such data with caution. It may reveal naturally occurring mechanisms that in a healing rickets are grossly magnified. But whether such degradation and change occur naturally remains to be determined.

Buckwalter and Rosenberg (1986) compared the dimensions of proteoglycan aggregates from epiphyseal cartilage, growth plate cartilage, and metaphyseal calcified cartilage from the distal femurs and proximal tibiae. In the growth plate cartilage and metaphysis there was no decrease in the length of aggregated monomers, but the sizes of aggregates were smaller, as indicated by decreased hyaluronate chain length and numbers of monomers per aggregate, the differences being greatest in the metaphysis. Nonaggregating monomers were not examined. However, the data indicate that there may be a cleavage of hyaluronate central filaments (or a change in chain length synthesis) during mineralization. In a later, as yet unpublished study, of transverse slices through the different zones of bovine fetal rib growth plate, Buckwalter and Rosenberg have observed a shortening of the length of proteoglycan aggregates in the hypertrophic zone, caused by a decrease in the hyaluronate central filament length (Table I). This agrees with the decrease in aggregate size in calcifying cartilage reported elsewhere (Pita et al., 1979). A sharp increase in the percentage of nonaggregating monomers was also detected in the hypertrophic zone (Table I). Moreover, they noticed that the nonaggregated monomers were shorter. Together these observations are in accord with the concept of a cleavage of proteoglycan monomers with the removal of the hyaluronic acid-binding region suggested earlier (Lohmander and Hjerpe, 1975; Howell and Pita, 1977; Pita et al., 1979; Reinholt et al., 1985) and the cleavage of hyaluronic acid in association with the mineralization of cartilage.

Earlier immunohistochemical studies of proteoglycans in the growth plate suggested that there was a gross removal of these molecules during calcification (Hirschman and Dziewiatkowski, 1966). More recent studies of growth plate (Poole et al., 1982a) and of other examples of endochondral bone formation (Poole et al., 1982b) have failed to confirm these findings: in fact, the latter authors could find no evidence for a net loss of proteoglycan and link protein during the calcification of cartilage: there may have been some loss of proteoglycan but this was not detectable. The differences in these two sets of observations may be due to how the earlier workers prepared their tissue for immunohistochemistry using methods which may have led to the loss of proteoglycan and/or damage to antigenic sites (epitopes) during tissue preparation.

Histochemical studies of proteoglycans at the ultrastructural level have employed metachromatic dyes to localize these molecules. Individual staining particles (usually called "matrix granules") are considered to represent proteoglycan monomers which have been collapsed by dye binding to their glycosaminoglycan chains (Hascall, 1980). Subsequent dehydration for electron microscopy probably leads to further collapse. Using this approach, several groups have come to the conclusion that proteoglycan monomers decrease in concentration when, or just before, calcification occurs (Lohmander and Hjerpe, 1975; Matukas and Krikos, 1968; Thyberg, 1974; Thyberg et al., 1973). Others have detected proteoglycan monomers (as matrix granules) within dense globular bodies in calcifying cartilage (Takagi et al., 1983). These "bodies" were detected using tannic acid–ferric chloride staining and may well correspond to the sites of calcification where chondrocalcin is deposited (Poole et al., 1984). During calcification these matrix granules were found to decrease in size as observed by others (Thyberg et al., 1973). This may indicate limited degradation of proteoglycan during calcification or a reduction in dye binding to these structures caused by the interaction of other molecules such as chondrocalcin. Ultrastructural studies employing positively charged colloidal thorium dioxide to detect proteoglycans have concluded that there is no evidence for proteoglycan removal during or after calcification except in noncalcified areas outside the lamina limitans (Scherft and Moskalewski, 1984).

Employing electron microprobe analyses, some workers have measured sulfur as an indicator of the presence of the highly sulfated proteoglycans, and have found no evidence for a loss of sulfur when cartilage calcifies (Howell and Carlson, 1968), although others disagree (Boyde and Shapiro, 1980). As we discussed in the preceding section, there is clear evidence to indicate that proteoglycans are condensed and concentrated in spherical loci at the primary sites of calcification (Barckhaus *et al.*, 1981; Arsenault and Ottensmeyer, 1983; Shepard and Mitchell, 1985; A. R. Poole, I. Pidoux, N. Mitchell and L. C. Rosenberg, manuscript in preparation) or are at least retained (Davis *et al.*, 1982).

Differences in the above findings could be reconciled by the fact that calcification may occur in focal loci where proteoglycans are more concentrated, possibly by cleavage of hyaluronic acid and of proteoglycans close to the hyaluronic acid-binding region. Then, as calcification proceeds outward from the calcific nucleus, proteoglycans may be further degraded after calcification has been initiated; this degradation may also extend to the surrounding tissue to permit calcification to spread throughout the matrix. This would reconcile data that argues that proteoglycans can inhibit calcification (see next section). Evidence in support of this has come from the work of Mitchell *et al.* (1982), who showed that once mineralization is fully established, no microbe signal for sulfate or brominated toluidine blue was detectable. Similar results also using microprobe analyses were reported by Barckhaus *et al.* (1981), who observed a reduction in sulfur content after calcification was initiated. The observed progressive reduction in the proportion by volume of extracellular matrix from the lower proliferative zone to the lower hypertrophic zone (Buckwalter *et al.*, 1986) may result from a net loss of proteoglycan during matrix remodeling and calcification.

IV. PURIFIED PROTEOGLYCANS AND GLYCOSAMINOGLYCANS CAN BIND CALCIUM AND INHIBIT CALCIFICATION *in Vitro*

There have been a number of studies which suggest that proteoglycans can bind calcium (Bowness, 1968; Woodward and Davidson, 1968; Németh-Csóka and Sarközi, 1982). Woodward and Davidson showed that chondroitin 4-sulfate was much less efficient at binding calcium than the intact proteoglycan: highly concentrated proteoglycans may bind calcium even better. This would fit in well with the observations described above, which have identified high concentrations of proteoglycans in calcifying centers in cartilage matrix.

In vitro studies of purified chondroitin sulfate (Bowness and Lee, 1967) and proteoglycans have revealed that these molecules alone can arrest the precipitation of calcium phosphate and can inhibit mineral growth *in vitro* (Weinstein *et al.*, 1963; Di Salvo and Schubert, 1967). Di Salvo and Schubert, however, also showed that *in vitro* a unique complex of proteoglycan and $Ca_3(PO_4)_2$ is formed. Digestion of the proteoglycan complex with either trypsin or hyaluronidase removed the so-called inhibitory effect of the proteoglycan, permitting precipitation of calcium phosphate to proceed. Others have shown that proteoglycan aggregates are more inhibitory than proteoglycan monomers (Blumenthal *et al.*, 1979), further indicating that a high concentration of negative charge, as is found in aggregate, is more effective. Even high molecular weight dextran can prevent hydroxyapatite formation (Blumenthal *et al.*, 1979; Chen *et al.*, 1984), indicating that this effect cannot be simply caused by high negative charge density but retardation of ion movement in solution may also be sufficient to delay mineral formation. Aspirates removed from the growth plate also inhibit the formation of amorphous calcium phosphate, and removal of proteoglycan can remove this inhibitory effect (Howell *et al.*, 1969; Cuervo *et al.*, 1973; Howell and Pita, 1976).

Although these *in vitro* observations demonstrate an inhibition of mineral formation, they also indicate that calcium and phosphate are bound to proteoglycan and thus not available for mineral growth. Thus these observations should be reevaluated. In a biological situation where different molecules and organizations are involved, matters cannot be so simple. In addition to proteoglycan being in calcifying sites, we have noted that chondrocalcin is also present (Poole *et al.*, 1984). This calcium-binding protein may act as an important mediator in the calcification process, binding to proteoglycan, neutralizing its highly anionic character, and yet retaining calcium-binding capacity within the complex: it may also or alternatively be an essential adjunct for mineral growth to occur once calcification has been initiated in the matrix. Degradation of surrounding proteoglycan may be a requirement for growth of the mineral clusters, if they truly have an inhibitory effect *in vivo*.

V. A Unifying Hypothesis to Explain the Proteoglycan Changes That May Occur in Calcifying Growth Plate Cartilage

We have examined the literature describing the many changes that take place in growth plate cartilage which are directly or indirectly associated with its calcification. Since there are widely differing conclusions, it is necessary to arrive at a working hypothesis that can ideally unify the many observations we have discussed. We would like to propose the following based on a careful assessment of these differing observations.

First, although proteoglycans have been shown to inhibit mineral formation *in vitro*, it would be dangerous to extrapolate these findings directly to living tissue. Yet we can say that if they do inhibit in their native state, then changes must occur to negate this inhibitory effect. We believe that high focal concentrations of proteoglycans are formed when calcification is initiated and are required for its initiation. This high focal concentration is normally seen in the lower hypertrophic zone. There must be a mechanism for a concentration of proteoglycan in these sites which are not normally detectable prior to calcification, except when artifically induced. This may result from the limited cleavage of hyaluronic acid and of proteoglycan monomers, which would permit the collapse of proteoglycan aggregates: it could also involve the redistribution of calcium, as indicated by our experimental work.

Once the initial focal deposition of mineral occurs (which may or may not require the presence of chondrocalcin), mineral formation proceeds: mineral growth may be assisted by the presence and continued release of chondrocalcin from hypertrophic cells. Removal of some of the initiating and the surrounding proteoglycan by degradative means may occur to permit continued mineral growth. The local presence of calcium and phosphate ions to support this rapid and localized calcification probably may be assisted by the local mobilization of the unusually high concentrations of calcium present in the mitochondria of the hypertrophic cells (Martin and Matthew, 1970; Arsenis, 1972; Brighton and Hunt, 1976, 1978; Posner, 1978). At the same time inorganic phosphate is most probably mobilized by the high concentration of alkaline phosphatase which is present in calcifying cartilage (Robison, 1923; Väänänen, 1980) and which is primarily concentrated in the matrix in the matrix vesicles (Ali *et al.*, 1970; Matsuzawa and Anderson, 1971; Väänänen and Korhonen, 1979; Hsu *et al.*, 1985). All of these events are normally under strict physiological control. Vitamin D is undoubtedly one of the essential regulatory molecules involved in these processes, in view of the dramatic inhibitory effects of vitamin D deficiency on the growth plate and the rapid restoration of calcification when vitamin D is restored.

Future work will tell us how accurate we are in our assessment of these critical events.

REFERENCES

Ali, S. Y. (1983). Calcification of cartilage. *In* "Cartilage" (B. H. Hall, ed.), pp. 343–378. Academic Press, New York.
Ali, S. Y., Sajdera, S. W., and Anderson, H. C. (1970). *Proc. Natl. Acad. Sci. U.S.A.* **76,** 1513–1520.
Ali, S. Y., Gray, C., Wisby, R., and Phillips, M. (1977). *J. Microsc. (Oxford)* **111,** 65–76.

Anderson, H. C. (1967). *J. Cell Biol.* **35**, 81–92.

Anderson, H. C. (1969). *J. Cell Biol.* **41**, 59–72.

Anderson, H. C. (1985). Matrix vesicle calcification: Review and update. In "Bone and Mineral Research" (W. A. Peck, ed.), Vol. 3, pp. 109–147. Elsevier, Amsterdam.

Arsenault, A. L., and Ottensmeyer, F. P. (1983). *Proc. Natl. Acad. Sci. U.S.A.* **80**, 1322–1326.

Arsenis, C. (1972). *Biochem. Biophys. Res. Comm.* **46**, 1928–1935.

Axelsson, I., Berman, I., and Pita, J. C. (1983). *J. Biol. Chem.* **258**, 8915–8921.

Barckhaus, R. H., Krefting, E.-R., Althoff, J., Quint, P., and Höhling, H. J. (1981). *Cell Tissue Res.* **217**, 661–666.

Bélanger, L. F. (1954). *Can. J. Physiol. Biochem.* **32**, 161–169.

Blumenthal, N., Posner, A. S., Silverman, L., and Rosenberg, L. C. (1979). *Calcif. Tissue Int.* **27**, 75–82.

Bonucci, E. (1967). *J. Ultrastruct. Res.* **20**, 33–39.

Bonucci, E. (1970). *Z. Zellforsch. Mikrosk. Anat.* **103**, 192–217.

Bowness, J. (1968). *Clin. Orthop.* **59**, 233–247.

Bowness, J., and Lee, K. (1967). *Biochem. J.* **103**, 382–390.

Boyde, A., and Shapiro, I. M. (1980). *Histochemistry* **69**, 85–94.

Brighton, C. T., and Hunt, R. M. (1976). *Fed. Proc. Fed. Am. Soc. Exp. Biol.* **35**, 143–147.

Brighton, C. T., and Hunt, R. M. (1978). *Metab. Bone Dis. Relat. Res.* **1**, 199–204.

Buckwalter, J. A. (1983). *Clin. Orthop.* **172**, 207–232.

Buckwalter, J. A., and Rosenberg, L. C. (1982). *J. Biol. Chem.* **257**, 9830–9839.

Buckwalter, J. A., and Rosenberg, L. (1983). *Collagen Rel. Res.* **3**, 489–504.

Buckwalter, J. A., and Rosenberg, L. (1986). *J. Orthop. Res.*, **4**, 1–9.

Buckwalter, J. A., Mower, D., Ungar, R., Schaeffer, J., and Ginsberg, B. (1986). *J. Bone Jt. Surg.* **68A**, 243–255.

Burgeson, R. E., and Hollister, D. W. (1979). *Biochem. Biophys. Res. Commun.* **87**, 1124–1131.

Cecil, R. N. A., and Anderson, H. C. (1978). *Metab. Bone Dis. and Relat. Res.* **1**, 89–95.

Chen, C.-C., Boskey, A. L., and Rosenberg, L. C. (1984). *Calcif. Tissue Int.* **36**, 285–290.

Choi, H. V., Tang, L-H., Johnson, T. L., Pal, S., Rosenberg, L. C., Reiner, A., and Poole, A. R. (1983). *J. Biol. Chem.* **238**, 655–661.

Cuervo, L., Pita, J., and Howell, D. (1973). *Calcif. Tissue Res.* **13**, 1–10.

Davis, W. L., Jones, R. G., Knight, J. P., and Hagler, H. K. (1982). *J. Histochem. Cytochem.* **30**, 221–234.

Dean, D. D., Muniz, O. E., Berman, I., Pita, J. C., Carreno, M. R., Woessner, J. F., and Howell, D. S. (1985). *J. Clin. Invest.* **76**, 716–722.

Di Salvo, J., and Schubert, M. (1967). *J. Biol. Chem.* **242**, 705–710.

Duance, V. C., Shimokomaki, M., and Bailey, A. J. (1982). *Bioscience Rep.* **2**, 223–227.

Franzén, A., Heinegård, D., Reiland, S., and Olsson, S.-E. (1982). *J. Bone Jt. Surg.* **64A**, 558–566.

Gibson, G. J., Schor, S. L., and Grant, M. E. (1981). *Biochem. Soc. Trans.* **9**, 550–551.

Gibson, G. J., Schor, S. L., and Grant, M. E. (1982). *J. Cell Biol.* **93**, 767–774.

Grant, W. T., Sussman, M. D., and Balian, G. (1985). *J. Biol. Chem.* **260**, 3798–3803.

Greulich, R. C., and Leblond, C. P. (1953). *Anat. Rec.* **115**, 559–585.

Hascall, G. K. (1980). *J. Ultrastruct. Res.* **70**, 369–375.

Hinek, A., Reiner, A., and Poole, A. R. (1985a). *J. Cell Biol.*, in press.

Hinek, A., and Poole, A. R. (1985b). Submitted for publication.

Hirschman, A., and Dziewiatkowski, D. D. (1966). *Science (Washington, D.C.)* **154**, 393–395.

Hjertquist, S. O. (1961). *Biol. Sperimentale* **1**, 126–140.

Howell, D. A., and Carlson, L. (1968). *Exp. Cell Res.* **51**, 185–196.

Howell, D., and Pita, J. (1976). *Clin. Orthop.* **118**, 208–229.

Howell, D. S., and Pita, J. C. (1977). *Upsala J. Med. Sci.* **82**, 97–98.

Howell, D., Pita, J., Marquez, J., and Gatter, R. (1969). *J. Clin. Invest.* **48**, 630–641.

Hsu, H. H. T., Munoz, P. A., Barr, J., Oppliger, I., Morris, D. C., Väänänen, H. K., Tarkenton, N., and Anderson, H. C. (1985). *J. Biol. Chem.* **260**, 1826–1831.

Hunziker, E. B., Herrmann, W., Schenk, R. K., Marti, T., Müller, M., and Moor, H. (1981). Structural integration of matrix vesicles in calcifying cartilage after cryofixation and freeze substitution. *In* "Matrix Vesicles," pp. 25–31. Wichtig Editore, Milan.

Jaffe, H. L. (1975). "Metabolic, Degenerative and Inflammatory Diseases of Bones and Joints." Lea & Febiger, Philadelphia.

Kempson, G. E. (1980). The mechanical properties of articular cartilage. *In* "The Joints and Synovial Fluid" (L. Sokoloff, ed.), Vol. II, pp. 177–238. Academic Press, New York.

Landis, W. J., Hauschka, B. T., Rogerson, C. A., and Glimcher, M. J. (1977). *J. Ultrastruct. Res.* **59**, 185–206.

Leung, M. K. K., Fessler, L. I., Greenberg, D. B., and Fessler, J. H. (1979). *J. Biol. Chem.* **254**, 224–232.

Lindenbaum, A., and Kuettner, K. E. (1967). *Calcif. Tissue Res.* **1**, 153–165.

Lohmander, S., and Hjerpe, A. (1975). *Biochim. Biophys. Acta* **404**, 93–109.

Martin, J. H., and Matthews, J. L. (1970). *Clin. Orthop.* **68**, 273–278.

Matsuzawa, T., and Anderson, H. C. (1971). *J. Histochem. Cytochem.* **19**, 810–818.

Matukas, V., and Krikos, G. (1968). *J. Cell Biol.* **39**, 43–48.

Mitchell, N., Shepard, N., and Harrod, J. (1982). *J. Bone Jt. Surg.* **64A**, 32–38.

Morris, D. C., Väänänen, H. K., and Anderson, H. C. (1983). *Metab. Bone Dis. Relat. Res.* **5**, 131–137.

Németh-Csóka, M., and Sarközi, A. (1982). *Acta. Biol. Acad. Sci. Hung.* **33**, 407–417.

Noro, A., Kimata, K., Oike, Y., Shinomura, T., Maeda, N., Yano, S., Takahashi, N., and Suzuki, S. (1983). *J. Biol. Chem.* **258**, 9223–9331.

Pal, S., Tang, L.-H., Choi, H., Habermann, E., Rosenberg, L., Roughley, P., and Poole, A. R. (1981). *Coll. Rel. Res.* **1**, 151–176.

Pita, J. C., Muller, F. J., Morales, S. M., and Alarcon, E. J. (1979). *J. Biol. Chem.* **254**, 10313–10320.

Poole, A. R., Pidoux, I., and Rosenberg, L. (1982a). *J. Cell Biol.* **92**, 249–260.

Poole, A. R., Reddi, A. H., and Rosenberg, L. C. (1982b). *Dev. Biol.* **89**, 532–539.

Poole, A. R., Pidoux, I., Reiner, A., and Rosenberg, L. (1982c). *J. Cell Biol.* **93**, 921–937.

Poole, A. R., Pidoux, I., Reiner, A., Choi, H., and Rosenberg, L. C. (1984). *J. Cell Biol.* **98**, 54–65.

Poole, A. R., Webber, C., Pidoux, I., Choi, H., and Rosenberg, L. C. (1986). *J. Histochem. Cytochem.* **34**, 619–625.

Posner, A. S. (1978). *Ann. N.Y. Acad. Sci.* **307**, 248–249.

Reese, C. A., and Mayne, R. (1981). *Biochemistry* **20**, 5443–5448.

Reinholt, F. P., Engfeldt, B., Hjerpe, A., and Jansson, K. (1982). *J. Ultrastruct. Res.* **80**, 270–279.

Reinholt, F. P., Engfeldt, B., Heinegård, D., and Hjerpe, A. (1985). *Collagen Rel. Res.* **5**, 55–64.

Ricard-Blum, S., Hartmann, D. J., Herbage, D., Payen-Meyran, C., and Ville, G. (1982). *FEBS Lett.* **146**, 343–347.

Robison, R. (1923). *Biochem. J.* **17**, 286–293.

Rosenberg, L. C., Choi, H. U., Tang, L-H., Johnson, T. L., Pal, S., Webber, C., Reiner, A., and Poole, A. R. (1985). *J. Biol. Chem.* **260**, 6304–6313.

Rosenbloom, J., Endo, R., and Harsch, M. (1976). *J. Biol. Chem.* **251**, 2070–2076.

Roughley, P. J., and White, R. J. (1980). *J. Biol. Chem.* **255**, 217–224.

Schenk, R. K., Spiro, D., and Wiener, J. (1967). *J. Cell Biol.* **34**, 275–291.

Scherft, J. P., and Moskalewski, S. (1984). *Metab. Bone Dis. Relat. Res.* **5**, 195–203.

Schmid, T. M., and Conrad, H. E. (1982). *J. Biol. Chem.* **257**, 12444–12450.

Schmid, T. M., and Linsenmayer, T. F. (1985). *J. Cell Biol.* **100**, 598–605.

Schofield, J. D., Uitto, J., and Prockop, D. J. (1974). *Biochemistry* **13**, 1801–1806.

Schraer, H., and Gay, C. V. (1977). *Calcif. Tissue Res.* **23**, 185–188.

Shepard, N., and Mitchell, N. (1976). *J. Histochem. Cytochem.* **24**, 621–629.

Shepard, N., and Mitchell, N. (1977). *J. Histochem. Cytochem.* **25**, 1163–1168.

Shepard, N., and Mitchell, N. (1985). *J. Bone Jt. Surg.* **67A**, 455–464.

Takagi, M., Parmlet, R. T., and Days, F. R. (1983). *J. Histochem. Cytochem.* **31**, 1089–1100.

Thyberg, J. (1974). *J. Ultrastruct. Res.* **46**, 206–218.

Thyberg, J., Lohmander, S., and Friberg, U. (1973). *J. Ultrastruct. Res.* **45**, 407–427.

Tyler, J. A. (1985). *Biochem. J.* **225**, 493–507.

Uitto, J., and Prockop, D. J. (1974). *Biochemistry* **13**, 4586–4591.

Väänänen, H. K. (1980). *Histochemistry* **65**, 143–148.

Väänänen, H. K., and Korhonen, L. K. (1979). *Calcif. Tissue Int.* **28**, 65–72.

van der Rest, M., Mayne, R., Ninomiya, Y., Seidah, N. G., Chretien, M., and Olsen, B. R. (1985). *J. Biol. Chem.* **260**, 220–225.

van der Rest, M., Olsen, B. R., Rosenberg, L. C., and Poole, A. R. (1986). *Biochem. J.* **237**, 923–925.

Vaughan, L., Winterhalter, K. H., and Bruckner, P. (1985). *J. Biol. Chem.* **260**, 4758–4763.

Vittur, F., Pugliarello, M. C., and de Bernard, B. (1971). *Experientia.* **27**, 126–127.

von der Mark, K., van Menxel, M., and Wiedemann, H. (1982). *Eur. J. Biochem.* **124**, 57–62.

Weatherell, J. A., and Weidman, S. M. (1963). *Biochem. J.* **89**, 265–267.

Webber, C., Roughley, P. J., and Poole, A. R. (1985). Submitted for publication.

Weinstein, H., Sach, C., and Schubert, M. (1963). *Science (Washington, D.C.)* **142**, 1073–1075.

Woodward, C., and Davidson, E. (1968). *Proc. Natl. Acad. Sci. U.S.A.* **60**, 201–205.

Wuthier, R. L. (1969). *Calcif. Tissue Res.* **4**, 20–38.

Biochemical Basis of Age-Related Changes in Proteoglycans

Eugene J-M.A. Thonar and Klaus E. Kuettner

Departments of Biochemistry, Medicine, and Orthopedic Surgery, Rush-Presbyterian–
St. Luke's Medical Center, Chicago, Illinois 60612

I. Introduction

Proteoglycans are major components of the extracellular matrix of cartilage. These very large macromolecules which are responsible for the ability of cartilage to undergo reversible deformation show marked age-related changes in structure and composition (Table I). The biochemical basis of these age-related changes remains to a large extent unclear. Are the differences in the composition of proteoglycans derived from immature and adult cartilage the reflection of alterations at the level of biosynthesis, or are they the simple result of more extensive extracellular modifications of the same biosynthetic product in the older tissues? Do the age-related differences arise as a result of changes in the protein or the carbohydrate moieties? Such questions have not been easy to answer because at all ages these complex macromolecules exhibit a large degree of polydispersity with respect to size and number of the glycosaminoglycans which are covalently attached to the core protein (Thonar and Sweet, 1981). It is important to remember that there is an infinitely greater chance of finding in the matrix two proteoglycan molecules that differ in some way from each other than of finding proteoglycan twins identical in every respect. The many drawings or models of cartilage proteoglycans represent simply an average of the molecules studied. Interpretations of chemical analyses are further complicated by recent observations that proteoglycans are heterogeneous, that is, different populations and/or subpopulations exhibit significant differences in either the structure of their core protein and/or their glycosylation patterns (Stanescu *et al.*, 1973; Sweet *et al.*, 1978; Stanescu, 1980; Heinegard and Paulsson,

211

TABLE I

AGE-RELATED CHANGES IN HIGH BUOYANT DENSITY CARTILAGE PROTEOGLYCANS

Parameter	Decrease	No change
Ability to aggregate	Pig laryngeal (Tsiganos and Muir, 1973); human articular (Perricone et al., 1977)	Dog articular (Inerot et al., 1978; human articular (Bayliss and Ali, 1978; Roughley and White, 1980)
Monomer size	Human articular (Roughley and White, 1980); bovine articular (Sweet et al., 1979a; Thonar and Sweet, 1981; Garg and Swann, 1981); rat costal (Oohira and Nagami, 1980)	
Chondroitin sulfate chain size	Human articular (Hjertquist and Wasteson, 1972; Roughley and White, 1980); bovine articular (Sweet et al., 1979a; Thonar and Sweet, 1981; Garg and Swann, 1981)	Dog articular (Inerot et al., 1978)
Core protein size	Dog articular (Inerot et al., 1978); bovine articular (Sweet et al., 1979a; Thonar and Sweet, 1981)	
Chondroitin sulfate/ keratan sulfate[a]	Dog articular (Inerot et al., 1978); human articular (Bayliss and Ali, 1978; Roughley and White, 1980); bovine articular (Sweet et al., 1979a; Thonar and Sweet, 1981; Garg and Swann, 1981)	
Chondroitin 4-sulfate/ chondroitin 6-sulfate[a]	Human articular (Roughley and White, 1980); bovine articular (Sweet et al., 1979a; Thonar and Sweet, 1981; Garg and Swann, 1981)	
O-Linked oligosaccharide/keratan sulfate[a]	Bovine articular (Thonar and Sweet, 1979; Sweet et al., 1979a; Thonar and Sweet, 1981)	

[a] The ratios of chondroitin sulfate to keratan sulfate and chondroitin 4-sulfate to chondroitin 6-sulfate reflect weight/weight ratios, while the ratio of O-linked oligosaccharide to keratan sulfate is on a chain/chain basis.

1984; Heinegard *et al.*, 1985). In some cases, some of these populations can be readily separated from each other. In other cases, because of the polydisperse nature of the proteoglycans within each population, some molecules from different populations may actually have similar physicochemical properties and as a result are impossible to separate and distinguish from one another by conventional biochemical techniques (Heinegard *et al.*, 1982).

In this review, we will attempt to present current ideas about the modifications which cartilage proteoglycans undergo with age. We are quite certain that, with the advent of monoclonal antibodies and recombinant DNA technology, the present concepts will undergo rigorous reexamination in the next few years.

II. Aging and Maturation

A. Definitions

Cartilage proteoglycans are complex macromolecules which enable the tissue to undergo rapid reversible deformation (Hascall and Kimura, 1982). Because they are polydisperse, it has been and remains often extremely difficult to determine their structure and composition. This is compounded by the fact that proteoglycans isolated from different cartilages (Mathews, 1975), from different topographical regions within a cartilage (Sweet *et al.*, 1977a; Thonar *et al.*, 1978), or from cartilages at different ages (see Table I), exhibit both quantitative and qualitative differences. Recent studies have made it quite clear that cartilage proteoglycans exhibit what has been termed age-related changes. For example, it has been argued that senescent chondrocytes synthesize a proteoglycan that is less able to structure water and that is functionally inadequate (Caplan, 1984). Caplan (1984) further argues that this results in tissue failure and fragmentation such as occurs in osteoarthrosis in older people. While this hypothesis is but one of several which contend that qualitative changes in proteoglycan structure and composition play a causative role in osteoarthrosis, it helps illustrate the importance of studies on the changes which cartilage proteoglycans undergo with age and maturation.

Aging designates the developmental sequence of changes that begins before birth and continues throughout the life span (Moment, 1982). Studies on the biological process of aging, as for example in the studies on senescent chondrocytes, are usually centered on determining factors which result in functional impairment (Hadley, 1982). It would be erroneous to equate all age-related or maturation-related changes to the processes of aging as studied by gerontologists. Some of the important changes in proteoglycan structure are most prominent early in life (Table II) (Thonar *et al.*, 1975; Thonar and Sweet, 1981). In some cases, they may simply reflect a new functional requirement, as exemplified by the rapid transition from a non-load-bearing articular cartilage in early fetal life to a load-bearing tissue in late fetal life or early postnatal life (Thonar and Sweet, 1981). Because many of the age-related changes cannot, at least for the present, be considered as playing a causative role in the development of an impairment associated with aging of normal cartilage, we have chosen to refer to changes or differences as being related to age rather than to aging. In time and with an improved understanding of the biochemical basis of individual age-related changes, it is likely that answers to the questions we are trying to solve will provide the basis for an attempt to put into perspective the many complex changes that are associated with the aging process.

B.   *Extracellular Matrix*

Connective tissues are found throughout the body. While by definition they connect other tissues or fill in the space between them, most connective tissues have very specialized functions. Thus, while cartilage and bone provide for movement, support, and protection, the former alone is able to undergo rapid reversible deformation (Caplan, 1984). Cartilages located in different parts of the body may exhibit significant variations in the quantity and quality of macromolecules present in the extracellular matrix (Mathews, 1975). Studies aimed at elucidating the composition of the cartilage matrix are often complicated by the fact that significant topographical differences may exist (Sweet *et al.*, 1977a). For example, water content, which is a close reflection of the amount and arrangement of the collagenous network, varies markedly with depth from the articular surfaces (Thonar *et al.*, 1978; Maroudas, 1980) (Table II). Recent evidence has also shown that the water content of cartilage that is maximally loaded is lower than that of cartilage that does not usually take part in the articulating process (Thonar *et al.*, 1978) (Table II). These topographical and functional variations have complicated interpretation of the observation that there is in most cartilages an age-related decrease in water content. There is evidence that the amount of collagen cross-links increases with age (Cannon, 1983), and it may be argued that this additional restriction on the extent to which proteoglycan can swell is a

TABLE II

AGE-RELATED CHANGES IN WATER CONTENT AND GLYCOSAMINOGLYCANS IN DIFFERENT TOPOGRAPHICAL REGIONS OF BOVINE ARTICULAR CARTILAGE[a,b]

	Water content				Hyaluronic acid % total hexosamine				Chondroitin sulfate:keratan sulfate ratio			
	Articular		Growing		Articular		Growing		Articular		Growing	
	m	M	m	M	m	M	m	M	m	M	m	M
Fetal term 1	90.6	89.9	90.4	89.5	0.94	0.95	0.93	0.94			14.8	13.3
Fetal term 2	89.8	88.4	88.8	87.4	1.16	1.17	1.15	1.20			15.0	12.1
Fetal term 3	89.6	88.2	86.3	85.1	1.15	1.11	1.11	1.11	18.5	14.2	14.6	12.8
Calf	87.1	79.4	82.1	69.9	1.83	2.12	1.89	2.32	19.4	6.1	12.1	8.8
Steer	69.2	67.6			3.60	4.95			2.32	2.01		

[a] After Thonar et al. (1978).

[b] Thin shavings (~1 mm thick) of cartilage were removed from the minimum (m) and maximum (M) weight-bearing areas of the articulating or growing zones of articular cartilage of fetal calves (first, second, and third trimesters of fetal life), of 3-month-old calves, and of 3–10 year old steers. The articulating zone includes the free surface, while the growing zone was sampled 1–2 mm above the calcified layer. The chondroitin sulfate/keratan sulfate ratio is based on the molar ratio of hexosamines present in the two glycosaminoglycans.

major contributor to the exhibited age-related decrease in water content. Additional age-related changes in collagen have been observed. These include increased crystalline organization, increased tensile strength, increased resistance to enzymatic digestion (Cannon, 1983), and decreases in the rate of synthesis and degradation (Prockop et al., 1979).

Cartilages are subjected to continuous stress and deformation as well as to numerous insults that require repair and adjustment to their architecture. Age-related increases in load and stress applied upon the articular cartilage surface have been implicated as the cause of some of the the age-related changes in the composition of the extracellular matrix (Sweet et al., 1978). While many of the changes reflect alterations at the level of synthesis of these matrix macromolecules, it has been argued that some of the changes are the expression of modifications which occur extracellularly and may be a function of time spent in the matrix (Inerot et al., 1978). While there is much disagreement about the rate of turnover of collagen and proteoglycans (Mankin and Lippielo, 1969; Maroudas, 1975; Lohmander, 1976), it is clear some of the molecules have to remain functional for many months (and probably years) and it is possible that some of the exhibited differences in the structure of a matrix macromolecule at two ages are the reflection of the fact that in the older tissue, the molecules have been subjected on average for a longer time to continuous stress and insults as described above. Such time-related modifications may affect the physiological functions of the tissue adversely and may thus lead to the development of pathological processes. For example, Roughley et al. (1986) have obtained evidence that, with age, the human articular cartilage matrix becomes enriched in hyaluronic acid-binding regions which represent degradation products of aggregating proteoglycans. These molecules which remain attached to hyaluronic acid are relatively devoid of glycosaminoglycans, and as a consequence may be less able to fulfil their functional role within the cartilage. This age-related enrichment in hyaluronic acid-binding regions would be most pronounced in individuals who have higher rates of cartilage proteoglycan metabolism (Thonar et al., 1987), possibly predisposing them to degenerative changes in cartilages which are abnormally loaded, i.e., following trauma or injury to a joint.

C.   Cartilage Proteoglycans

The observation that intact proteoglycans could be extracted in large amounts by nondisruptive methods (Sajdera and Hascall, 1969; Hascall and Sajdera, 1969) radically changed the approach to studies on proteoglycan structure and composition. Age-related studies had until then concentrated on describing, often in very great detail, changes in the proportion of the various glycosaminoglycans present. These earlier findings (see review by Matthews, 1975) raised questions which in many cases have remained unanswered for a number of reasons. First, it is becoming increasingly clear that the nature of the age-related changes varies significantly from species to species. For example, during the transition from fetal to postnatal life, the ratio of chondroitin 4-sulfate to chondroitin 6-sulfate increases in bovine cartilage but decreases in the rabbit (Matthews and Glagov, 1966). Venn and Mason (1983) recently made the striking observation that mouse adult intervertebral disc does not contain keratan sulfate, a cartilage glycosaminoglycan which in most species undergoes progressive age-related changes. The contention that mouse cartilages do not contain keratan sulfate is supported by our observation that none of several monoclonal antibodies directed against keratan sulfate recognizes any antigenic determinant in mouse auricular and articular cartilages.

Second, the age-related changes in proteoglycan structure are not uniform throughout the body. These changes may not occur at the same rate in different cartilages or in different topographical regions of the same cartilaginous tissue (Sweet et al., 1977a; Thonar et al., 1978). Of particular significance is the observation that during the transition from late fetal life to early postnatal life, the proteoglycans present in those regions of the articular surface which are maximally loaded exhibited rapid structural and compositional changes that are typical of those found in all cartilages at later stages (Thonar et al., 1978) (Table II). This local prematuration does not occur in the minimally loaded regions of the same articular surfaces. It is likely these modifications, which appear to occur as a response to the intense loading and stresses, modify the physicochemical properties of the matrix rendering it functionally more suitable for the rapid reversible deformation it must undergo when loaded cyclically. It is significant that at later stages of the maturation process, the minimally loaded areas will undergo the same changes. This suggests that the response of immature chondrocytes to pressure is similar, if not identical, to the more progressive changes expressed by all chondrocytes during the process of maturation and development. Importantly, it is worth noting that this response to loading is not a property unique to chondrocytes. Thus, areas of tendon which are subjected to compressive forces become enriched in cartilage-like keratan sulfate-containing proteoglycans, whereas areas which are subjected to tensile forces do not (Vogel and Thonar, 1987).

Third, cartilage proteoglycans exist as multiple populations which in most cases represent different biosynthetic products. Whereby the early findings described changes in the content or quality of glycosaminoglycans isolated by methods which usually degraded the core protein backbone, the modern approach requires that the molecules be isolated and purified as intact proteoglycans. The great majority of the modern studies on the age-related changes in the structure and composition of cartilage proteoglycans focused on the molecules that were of high buoyant density because they were the most abundant and easy to separate from the nonproteoglycan molecules. While age-related differences in the structure of high buoyant density proteoglycans are well documented, for they appear to occur in most cartilages in most species, interpretation of these findings remains difficult. In most cases there was no attempt, because of the difficulty involved, to differentiate between different populations of high buoyant density proteoglycans. Because each population appears to be polydisperse (different number and sizes of chondroitin sulfate and keratan sulfate chains attached to an identical core protein), complete separation of two populations with different average sizes is not possible by conventional methods of separation which make use of size, charge, or density (Heinegard *et al.*, 1982). On the other hand, it is possible to achieve essentially complete separation of different subpopulations of aggregating proteoglycans by electrophoresis in acrylamide–agarose composite gels of low porosity. This method is technically difficult and, at present, can only be used to fractionate a few micrograms of proteoglycans. The rapid emergence of sensitive immunological assays, some of which make use of monoclonal antibodies specific for different protein or carbohydrate epitopes on the proteoglycan molecules, should provide the tools that are required for the study of such small amounts of purified proteoglycan (Baker *et al.*, 1982; Thonar *et al.*, 1985). In addition, the acrylamide–agarose gels can be used to advantage in separating   small   amounts   of   proteoglycans   synthesized   by chondrocytes cultured in the presence of radioactive precursors. These methods (acrylamide–agarose, immunological assays, culture techniques) provide an additional advantage in that they can be used to study very small pieces of cartilage, thereby circumventing the problems associated with topographical variation as described above.

III. PROTEOGLYCAN POPULATIONS: AGE-RELATED CHANGES

The composition of the cartilage matrix varies, sometimes markedly, from species to species and from tissue to tissue. It is thus not surprising that the number and populations of proteoglycans present may show some variation from cartilage to cartilage. It is clear, however, that the major changes which proteoglycans undergo with age have been observed in most animal species and most cartilages studied. The age-related changes which we will describe were obtained, unless otherwise indicated, on bovine articular cartilage. Our choice of animal was to a large extent based on the fact that fresh tissues are readily available from local slaughterhouses. Our decision to select articular cartilage as the tissue of choice was to a large extent motivated by the realization that the requirements for resilience and function may be somewhat different and more pronounced than in other tissues such as nasal, costal, or laryngeal cartilage.

A. Low Buoyant Density Proteoglycans

Cartilage proteoglycans of low buoyant density are present in significant amounts in cartilages (Stanescu and Sweet, 1981; Heinegard *et al.*, 1981; Heinegard and Paulsson, 1984) (Fig. 1b). Rosenberg *et al.*

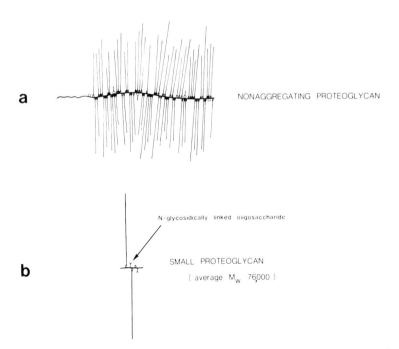

FIG. 1. Nonaggregating cartilage proteoglycans of high (a) and low (b) buoyant density. Reproduced with permission from Heinegard and Paulsson (1984).

(1985) have shown that these proteoglycans consist of a core protein of $M_r = 38,000$ to which one or two dermatan sulfate chains with $M_r = 15,000–24,000$ and oligosaccharides are covalently attached. They have suggested that they are present as two populations (DS-PGI and DS-PGII) which can be separated from each other by taking advantage of the ability of the DS-PGI to self-associate. The ratio of protein to glycosaminoglycan is much higher than in the larger cartilage proteoglycans, and as a consequence, these smaller proteoglycans have much lower buoyant density. They can be recovered in the upper portion of a cesium chloride density gradient well clear of the larger proteoglycans, which are found in the bottom portion of the gradient. They are nonaggregating in nature and may be related to the proteoglycans with similar core proteins which have been found in many connective tissues, including bone, cornea, skin, sclera, and aorta (Heinegard and Paulsson, 1984). While the number of low buoyant density proteoglycans may vary with age or from cartilage to cartilage (Stanescu and Stanescu, 1983; Rosenberg *et al.*, 1985), it is clear that in some cartilages there are actually several times as many low buoyant density as high buoyant density molecules (Heinegard *et al.*, 1985). The presence of such large numbers of nonaggregating small low buoyant density proteoglycans strengthens contentions that they are important for cartilage function.

Biosynthetic experiments (see Section IV below) have clearly shown that these proteoglycans are not derived from the progressive degradation of the larger proteoglycans. Older steer chondrocytes cultured as monolayers synthesize three to five times as many of these small nonaggregating proteoglycans as calf chondrocytes grown under identical conditions (Thonar *et al.*, 1986). Further, the molar ratio of low buoyant to high buoyant density proteoglycans synthesized by these chondrocytes *in vitro* is significantly higher in the steer. Whether different populations of cartilage proteoglycans are synthesized by the same or different cells remains unclear.

B. High Buoyant Density Proteoglycans

Most of the recent reports on age-related changes in proteoglycan structure and composition have restricted themselves to the study of those proteoglycans that can readily be obtained in fairly pure form following isopycnic density gradient ultracentrifugation of a 4 *M* guanidine hydrochloride extract of cartilage. Proteoglycans in the fractions of high density represent the majority (>70%) of the hexosamine- and uronic acid-containing material in the tissue and for the most part are able to form aggregates by interacting noncovalently with hyaluronate and link proteins (Sweet et al., 1979a; Thonar and Sweet, 1981). When the ultracentrifugation is performed under associative conditions, the high buoyant density fraction (A1) contains all three components of the aggregate (Hascall and Kimura, 1982). While it enables one to study the aggregates that are re-formed when the 4 M guanidine hydrochloride extract is brought back to associative conditions, further purification steps are required to separate the proteoglycan monomer from the hyaluronate and the link proteins. This is achieved by subjecting the A1 fraction to a second isopycnic density gradient ultracentrifugation, this time in the presence of 4 M guanidine hydrochloride (Hascall and Kimura, 1982). The high buoyant density fraction (A1D1) obtained in this way contains purified proteoglycan monomers free of the less buoyant hyaluronate and link proteins. Since the A1 fraction does not yield pure aggregates, for it contains significant amounts of nonaggregated monomers, many scientists prefer to subject the 4 M guanidine hydrochloride extract to a one-step purification procedure by performing a single ultracentrifugation in the presence of 4 M guanidine hydrochloride (Hascall and Kimura, 1982). The high buoyant density fraction, which here is called D1, in this case contains essentially the same populations of proteoglycan monomers as the A1D1 fraction. While proteoglycans rich in keratan sulfate appear to have mildly lower buoyant densities than those rich in chondroitin sulfate (Lohmander, 1975; Sweet et al., 1975a) and, as a consequence, some keratan sulfate-rich proteoglycans could have been lost preferentially by those who chose to select the bottom fourth rather than the bottom third of the gradient, this is unlikely to have been a significant problem. In general, the age-related changes in the chemical composition of high buoyant density proteoglycans appear to be similar in most species (Table I). Differences in opinion, when they arose, were often based on interpretation of these data, more particularly how the data were used to reconstruct a three-dimensional picture of the proteoglycan monomer.

The rationale behind some of the research that will come forth in the near future is likely to be based on the data and concepts generated in the course of the many studies on age-related changes in the structure and composition of the high buoyant density proteoglycans. The two best known age-related changes are a decrease in the hydrodynamic size of the proteoglycan monomer and an increase in keratan sulfate content (Inerot et al., 1978; Sweet et al., 1979a; Roughley and White, 1980; Thonar and Sweet, 1981). Other changes have also been observed and are given in Tables I and III.

TABLE III

AGE-RELATED CHANGES IN BOVINE ARTICULAR PROTEOGLYCANS OF HIGH BUOYANT DENSITY[a,b]

	Monomer size (peak K_d on Sepharose 2B)	CS chain size (MW)	KS chain size (MW)	Chondroitin 4-sulfate/ chondroitin 6-sulfate	CS chains/ monomer	KS chains/ monomer	O-Oligosaccharide/ monomer
Fetal term 1	0.26	12,400	3,300	5.5	83	26	16
Fetal term 2	0.27	11,200	3,300	6.0	96	28	16
Fetal term 3	0.28	10,600	3,300	5.8	136	40	24
Calf	0.34	10,400	3,300	8.2	129	48	31
Steer	0.42	8,400	3,700	0.7	80[c]	80[c]	21[c]

[a] Adapted after Sweet et al. (1979a) and Thonar and Sweet (1981).

[b] Monomer size is expressed as the partition coefficient (K_d) of the D1 proteoglycan peak on Sepharose 2B. Chondroitin sulfate and keratan sulfate were prepared as free chains by alkaline borohydride treatment of D1 proteoglycans and were separated from each other by sieve chromatography on BioGel P30. Chondroitin sulfate size was derived from the elution profile of intact free chains on Sephadex G200, while keratan sulfate size was calculated from chemical analysis of purified intact free chains. The ratio of chondroitin 4-sulfate to chondroitin 6-sulfate reflects a weight/weight basis. The numbers of glycosaminoglycan or oligosaccharide chains per monomer were derived indirectly from published Sweet et al. (1979a), Thonar and Sweet (1981), and unpublished biochemical data.

[c] The numbers in steer were derived by assuming that the ratio of chondroitin sulfate-rich to chondroitin sulfate-poor proteoglycan was 1.17—see Fig. 4.

1. DECREASED MONOMER SIZE

The decrease in hydrodynamic size begins in early fetal life and continues progressively throughout maturation and adult life (Table III). Thonar and Sweet (1981) have shown that the decrease in size occurs concomitantly with a significant decrease in the size of the chondroitin sulfate chains, which they argued would be reflected by a significant decrease in the size of the monomers. Similar age-related decreases in chondroitin sulfate chain size have been observed in other species (Table I). Since chondroitin sulfate chains are thought to radiate in more than one plane perpendicular to the core protein, a small change in chondroitin sulfate chain length will result in a major change in the diameter of the cylindrical conformation of the hydrated proteoglycan (Fig. 2). Thonar and Sweet (1981) argued that there appears to be in addition an age-related decrease, at least in postnatal life, in the size of the core protein. Inerot et al. (1978) have found that in canine and human postnatal articular cartilages the age-related

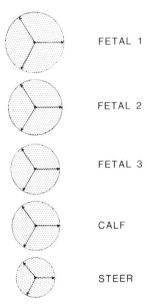

FETAL 1

FETAL 2

FETAL 3

CALF

STEER

FIG. 2. Representation of the age-related decrease in the diameter of the high buoyant density cartilage proteoglycan. The arrows radiating from the center represent chondroitin sulfate average chain lengths as determined by Thonar and Sweet (1981). The decrease in chain size during maturation and development (fetal term 1 → steer) brings about a decrease of 54% in the hydrodynamic size of the proteoglycan monomer (assuming the length of the polysaccharide attachment region remains constant).

decrease in size of the monomer is not accompanied by an appreciable decrease in the size of the chondroitin sulfate chains; they have further postulated that the decrease in size is the direct result of a decrease in the size or length of the core protein.

2.   INCREASE IN KERATAN SULFATE CONTENT

The keratan sulfate content of cartilage proteoglycans has been considered one of the most reliable indicators of the age of the tissue. The finding of Venn and Mason (1983), who showed that this glycosaminoglycan is not found in the mouse intervertebral disc, a tissue which in other species is particularly rich in keratan sulfate (Adams and Muir, 1976; Adams et al., 1977; Stevens et al., 1979), suggests it may not be an essential ingredient for adequate cartilage function during adult life when enormous loads and stresses are placed upon the resilient cartilage structures. This contention is also supported by the recent observation that the incidence of degenerative joint changes does not appear to be higher than normal in humans with macular corneal dystrophy whose cartilage proteoglycans do not contain normally sulfated keratan sulfate (Thonar et al., 1985; unpublished observations). Enrichment with keratan sulfate may occur at different rates in different species. It has been shown that this enrichment is the result for the most part of increased substitution of the core protein with keratan sulfate chains (Table III). As a consequence, the ratio of chondroitin sulfate chains to keratan sulfate chains decreases (Sweet et al., 1979a; Thonar and Sweet, 1981). While chain size does not appear to change appreciably with age in bovine articular cartilage, there appears to be some age-related increase in human articular cartilage (Roughley and White, 1980). Whereas the work of Heinegard and Axelsson (1977) led to the discovery of a keratan sulfate-rich region which in bovine nasal cartilage proteoglycans contained the majority of the keratan sulfate, it is now clear and is often understated that in many cases a significant proportion of the keratan sulfate is located in other areas of the polysaccharide attachment region. In bovine articular cartilages, only about half of the keratan sulfate chains are located in the keratan sulfate-rich region irrespective of the age (Sweet et al., 1979a; Thonar and Sweet, 1981). In other words, the age-related enrichment occurs throughout the major portion of the core protein rather than in the keratan sulfate-rich region specifically.

Cartilage proteoglycans contain O-linked oligosaccharides which bear great similarities to the linkage region of the keratan sulfate molecule (Thonar and Sweet, 1979; Lohmander et al., 1980). In bovine articular cartilage proteoglycans, the age-related enrichment in keratan sulfate is not accompanied by an equivalent decrease in the total number of sialic acid-containing O-linked oligosaccharides present (Thonar and Sweet, 1979; 1981). While this suggests the O-linked oligosaccharides and keratan sulfate chains bear no direct biosynthetic relationship, the contention that the total number of keratan sulfate and O-linked oligosaccharide chains attached to the core protein may be constant at all ages cannot be ruled out. Additional studies to test this hypothesis are likely to be stimulated by the recent observation that the rat chondrosarcoma high buoyant density proteoglycan contains, in addition to the sialic acid-containing O-linked oligosaccharides, as many as 60 galactosamine residues O-linked individually to the core protein (Thonar *et al.*, 1983).

C. Aggregating vs Nonaggregating Proteoglycans

A number of recent reports have shown that cesium sulfate or sucrose rate zonal centrifugation could be used to separate, on the basis of size, nonaggregating high buoyant density proteoglycans from proteoglycan aggregates which contain the aggregating proteoglycan monomers (Hoffman, 1979; Kimata *et al.*, 1982; Hascall and Kimura, 1982).

1. NONAGGREGATING PROTEOGLYCANS

Heinegard and his co-workers (1982) have shown that there are small amounts of nonaggregating proteoglycans which on the basis of chemical and immunological analyses appear to represent for the most part a population that is not derived from extracellular degradation of the aggregating proteoglycans (Fig. 1a). In addition, there may be proteoglycan molecules which have been altered in the matrix such that they are no longer able to interact with hyaluronic acid. In fetal and postnatal bovine articular cartilage, at least 85% of the high buoyant density proteoglycans can be recovered in aggregates sedimenting at the bottom of cesium sulfate gradients. A comparison of calf and steer showed that there was no significant difference in the relative proportion of aggregating and nonaggregating high buoyant density proteoglycans. These nonaggregating proteoglycans have lower keratan sulfate contents than their aggregating counterparts (keratan sulfate antigen*: calf aggregating, 2.7%; calf nonaggregating, 1.2%; steer aggregating, 13.2%; steer nonaggregating, 4.1%). Their pattern

* Keratan sulfate content was determined by an ELISA with inhibition step using a monoclonal antibody specific for keratan sulfate.

of migration in acrylamide–agarose gels suggests they are hetero-geneous in nature and as a consequence they may prove difficult to study further. The age-related differences in the composition of the high buoyant density proteoglycans, such as we presented and dis-cussed above, are not the result of differences in the nonaggregating proteoglycans at different ages. It is becoming increasingly clear that essentially all of the age-related differences which have been pre-viously observed in bovine articular cartilage high buoyant density proteoglycans are still present when the studies are restricted to ag-gregating proteoglycans purified free of nonaggregating proteo-glycans.

2. Aggregating Proteoglycans

While it may be fairly simple to qualify the low buoyant density nonaggregating proteoglycan and high buoyant density aggregating proteoglycan as distinct populations, as they have quite different core proteins (Stanescu and Sweet, 1981; Heinegard and Paulsson, 1984), it has been far more difficult to obtain clear evidence that aggregating proteoglycans exist as more than one population or subpopulation. Based on the earlier work of others and their own observations, Heinegard and his co-workers (1981, 1982, 1984) postulate that in adult bovine cartilages there are two major subpopulations of ag-gregating proteoglycans; these differ in hydrodynamic size and ker-atan sulfate content. Since these differences represent the two major age-related changes in structure and composition of cartilage pro-teoglycans, one may then ask whether age-related differences are not the simple reflection of changes in the proportion of these two sub-populations at different ages. There has been evidence that acryl-amide–agarose composite gels are able to differentiate between differ-ent subpopulations of high buoyant density proteoglycans in many species (Stanescu et al., 1973; Sweet et al., 1978; Stanescu, 1980; Heinegard et al., 1985; Buckwalter et al., 1985). Recent work has now shown that calf bovine aggregating cartilage proteoglycans migrate predominantly as a single band, whereas the corresponding proteogly-cans in steer migrate as two bands (Thonar et al., 1984; Heinegard et al., 1985) (Fig. 3). It has been argued that the slower moving band represents a chondroitin sulfate-rich subpopulation, whereas the fast-er moving band represents keratan sulfate-rich proteoglycans (Heinegard and Paulsson, 1984). The subpopulation which Heinegard and Paulsson (1984) referred to as being keratan sulfate-rich is also poor in chondroitin sulfate. While this subpopulation has a relatively

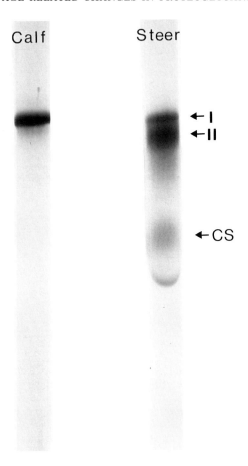

FIG. 3. Electrophoresis of purified aggregating proteoglycans in composite acryl-amide–agarose gels. Calf and steer aggregating proteoglycans were purified using ce-sium sulfate velocity gradients, reduced with SDS and mercaptoethanol, and electro-phoresed for 1 hr in composite gels consisting of 1.2% acrylamide and 0.7% agarose. Band I represents the chondroitin sulfate-rich proteoglycans and band II the chondroitin sulfate-poor proteoglycans. The mobility of free chondroitin sulfate chains (CS) which were incorporated in the steer sample is also shown.

high keratan sulfate to chondroitin sulfate molar ratio, it contains apparently only half as many keratan sulfate chains per core protein as the other subpopulation of aggregating proteoglycans (Heinegard *et al.*, 1985). Consequently, we will refer to it as the chondroitin sulfate-poor rather than keratan sulfate-rich subpopulation (see also Fig. 4).

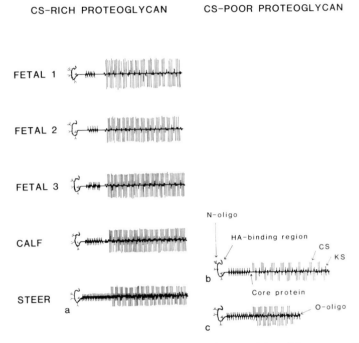

FIG. 4. Age-related changes in the structure and composition of the aggregating proteoglycans of bovine articular cartilage. Fetal and immature calf cartilages contain chondroitin sulfate-rich proteoglycans almost exclusively. Steer cartilage contains chondroitin sulfate-rich proteoglycans (a) and chondroitin sulfate-poor proteoglycans (b,c). Recent studies suggest some chondroitin sulfate-poor proteoglycans (b) have core proteins that are similar in size to the core proteins of the chondroitin sulfate-rich proteoglycans while others (c) may have somewhat shorter core proteins. CS, Chondroitin sulfate; KS, keratan sulfate.

D. *Chondroitin Sulfate-Rich vs Chondroitin Sulfate-Poor Proteoglycans*

1. CHONDROITIN SULFATE-RICH PROTEOGLYCAN

The present representation of the chondroitin sulfate-rich proteoglycan (Heinegard and Paulsson, 1984; Heinegard *et al.*, 1985) must take into consideration a number of observations which make it quite clear that there are significant changes in the composition of chondroitin sulfate-rich proteoglycans during fetal life. At all stages of fetal life, high buoyant density proteoglycans have a low content of keratan sulfate and migrate as a single band in acrylamide–agarose gels (Stanescu et al., 1973). Those results also suggested that the fetal and early postnatal forms of the chondroitin sulfate-rich proteoglycan have slightly different mobilities and thus may not be equated (Stanescu et al., 1973). It is possible the slower mobility of the single band seen in fetal life is the reflection of the presence of longer chondroitin sulfate and of resultant larger hydrodynamic size (see Table III). Additional differences in glycosylation of chondroitin sulfate-rich proteoglycans at different ages must also be considered. First, there is evidence that the proteoglycans that populate the matrix in early fetal life are less substituted with chondroitin sulfate chains (Table III). In other words, there appear to be significant differences in the sizes and numbers of chondroitin sulfate chains attached to the core protein of the chondroitin sulfate-rich proteoglycan at different ages. Second, in the present model (Heinegard and Paulsson, 1984), the chondroitin sulfate-rich proteoglycan is depicted as having no keratan sulfate-rich region. This does not take into consideration previous demonstrations that approximately half of the keratan sulfate of bovine articular cartilage high buoyant density proteoglycans is present in a keratan sulfate-rich region, irrespective of the age (Thonar and Sweet, 1981) and relative proportions of the two proteoglycan subpopulations. Third, S. Bjornsson, K. E. Kuettner, and E. J.-M. A. Thonar (unpublished observations) have found that the composition of the proteoglycans which migrate as band I in acrylamide–agarose gels was quite different for calf and steer. Calf band I proteoglycans contained longer chondroitin sulfate chains and higher molar ratios of chondroitin 4-sulfate/chondroitin 6-sulfate and chondroitin sulfate chains/keratan sulfate chains than the steer band I proteoglycans. On the other hand, the size of the core protein was identical at the two ages. These results clearly suggest that the core protein of the chondroitin sulfate-rich proteoglycans is differentially glycosylated at different ages. In Fig. 4, we offer a model of the different forms of chondroitin sulfate-rich proteoglycans which populate the matrix at different stages of fetal and early postnatal life when the keratan sulfate-rich proteoglycan is either absent or present in very small amounts. This subpopulation of chondroitin sulfate-rich proteoglycans undergoes major changes with age and maturation. As a consequence, one must conclude that the age-related changes in the structure and composition of the aggregating proteoglycans of bovine articular cartilage are not exclusively based on the emergence with time of a subpopulation of chondroitin sulfate-poor proteoglycans.

2. CHONDROITIN SULFATE-POOR PROTEOGLYCAN

The single most important and salient feature of the model proposed by Heinegard and Paulsson (1984) for the keratan sulfate-rich proteoglycan is that it is depicted as having a shorter core protein. In the ensuing paragraphs, we will try to put into perspective a number of findings that either support or contradict the contention that the core protein of this proteoglycan, which we refer to as being chondroitin sulfate-poor (see Fig. 4), is shorter than that of the chondroitin sulfate-rich species.

a. Different Core Protein Size. Differences in the ratio of protein to chondroitin sulfate or keratan sulfate originally led Heinegard (1977) to postulate that high buoyant density proteoglycans contained

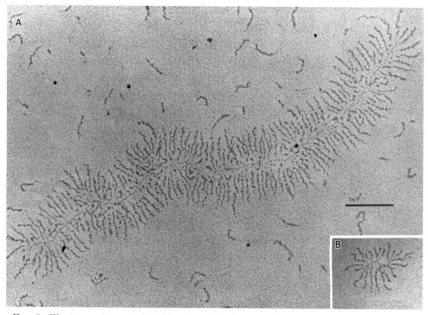

FIG. 5. Electron micrographs of bovine articular cartilage proteoglycans purified by equilibrium density gradient centrifugation performed using associative conditions. A large calf proteoglycan aggregate (A) consisting of a central hyaluronic acid filament and multiple aggregated monomers. Most calf proteoglycan aggregates were about one-half this size. Nonaggregated calf proteoglycan monomers can also be seen. The frequency distribution of monomer lengths is shown in Fig. 6. Steer proteoglycan aggregates, an example of which is shown in the inset (B), are generally smaller than calf proteoglycan aggregates. Bar = 500 nm; final magnification ×26,000. Reproduced with permission from Buckwalter *et al.* (1985).

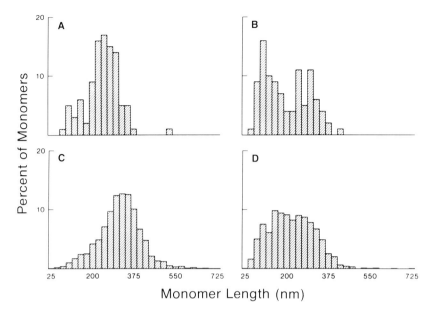

Monomer Length (nm)

Fig. 6. Histograms showing frequency distributions of monomer lengths. (A) Nonaggregated calf monomers. The distribution appears unimodal but slightly asymmetric. A small population of monomers <200 nm long creates the asymmetry. (B) Nonaggregated steer monomers. The apparent bimodal distribution of monomer lengths suggests that a distinct population of monomers <200 nm long exists in steer articular cartilage. (C) Aggregated calf monomers. The distribution is almost symmetrical. (D) Aggregated steer monomers. The distribution contains a large population of monomers <200 nm in length. Reproduced with permission from Buckwalter et al. (1985).

core proteins of varying lengths. In addition, when spread as mono-layers and viewed in the electron microscope (Fig. 5), high buoyant density proteoglycans exhibit extreme polydispersity in the length of their long axis representing their core protein (Fig. 6) (Buckwalter et al., 1985). This polydispersity is present in both aggregated and nonaggregated proteoglycans and, therefore, it cannot be argued that it is a reflection of the fact that some of the proteoglycans have lost (to a smaller or greater extent) portions of the end containing the hyaluronic acid-binding region. In addition to the polydispersity which is present at all ages and irrespective of whether there are one or two subpopulations detected by electrophoresis in acrylamide–agarose gels, there appears to be a statistically significant ($p < 0.005$) age-related decrease in the size of the core protein as measured from the electron microscope spreads. In calf there appears to be a single sub-

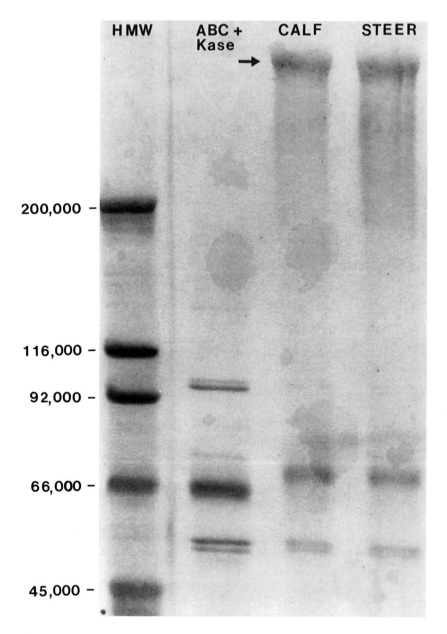

FIG. 7. Electrophoresis of core proteins of purified calf and steer aggregating proteoglycans. Calf and steer aggregating proteoglycans were digested with chondroitinase ABC and keratanase in the presence of protease inhibitors. The resulting core proteins were heated at 100°C for 5 min in 1% SDS–5% mercaptoethanol and subjected to SDS–

population with a range of sizes, while in steer two subpopulations can be discerned (Fig. 6). While at present there is no direct evidence to suggest that the steer subpopulation with apparently shorter core proteins, as seen in the electron microscope, represents the faster moving band prominent in electrophoretic runs of steer samples (Fig. 3), it is tempting to do so. At this stage one cannot underestimate the possibility that differences in glycosylation patterns, such as are known to be present, contribute to a substantial degree to the development of the observed polydispersity and in some cases heterogeneity as a result of differences in spreading upon the grid.

Additional suggestions that the core proteins are of different sizes stem from the observations that the amino acid composition of the core proteins of calf and steer high buoyant density proteoglycans is different (Sweet *et al.*, 1979a; Thonar and Sweet, 1981) and that one polyclonal antibody appeared to recognize one or more protein antigenic determinants in one subpopulation but not the other (Wieslander and Heinegard, 1981). While these findings indicate differences are present, they may simply reflect differences in the amino acid composition or sequence of two core proteins of similar or identical size.

b. Similar Core Protein Size. Evidence to suggest that a significant proportion of the chondroitin sulfate-rich and chondroitin sulfate-poor aggregating proteoglycans present in bovine steer cartilage have core proteins of similar size comes from a number of observations by S. Bjornsson, K. E. Kuettner, and E. J.-M. A. Thonar. First, core protein obtained following digestion of aggregating proteoglycan with chondroitinase ABC and keratanase was subjected to chromatography on Sepharose CL 4B and electrophoresis in acrylamide gels. There were no major differences in the elution profile of calf and steer core proteins on Sepharose CL 4B, suggesting that differences in sizes of calf and steer proteoglycans, such as are observed by chromatography on Sepharose 2B or electrophoresis, are largely the reflection of differences in glycosylation patterns at the two ages.

In a separate experiment these core proteins were subjected to electrophoresis in acrylamide gels (Fig. 7). The core proteins of calf or steer proteoglycans migrated as a sharp band with the same apparent

---

PAGE using 3–6% gradient gels. The position of migration of the core proteins is shown by an arrow. The majority of the calf and steer aggregating proteoglycans have a core protein with apparently identical molecular weight (~330,000). High molecular weight standards (HMW) and the enzymes (ABC + Kase) were run as controls. The molecular weights are indicated on the left-hand side. Bovine serum albumin present in the core protein preparations migrated with reduced mobility (Bjornsson *et al.*, unpublished data).

mobility, suggesting they are of similar if not identical size. A small proportion of the core proteins migrated further as a broad band. We have used electrophoresis in acrylamide–agarose gels to separate chondroitin sulfate-rich from chondroitin sulfate-poor aggregating proteoglycans from steer. The purified subpopulations were then digested with chondroitinase ABC and keratanase and subjected to electrophoresis in acrylamide gels. The results showed that the smaller core proteins which migrated as a broad band were derived exclusively from chondroitin sulfate-poor proteoglycans. Importantly, a significant proportion of the core proteins of these chondroitin sulfate-poor proteoglycans migrated as a sharp band with the same mobility as the core proteins of the chondroitin sulfate-rich proteoglycans. These findings indicate that, unlike chondroitin sulfate-rich proteoglycans, chondroitin sulfate-poor aggregating proteoglycans exhibit considerable heterogeneity with respect to the size of their core proteins (Fig. 4). The relative proportions of the larger and smaller core proteins have not been established as yet. What is clear is that the great majority (>80%) of the core proteins of calf and steer proteoglycans are of identical size, thereby suggesting that the age-related differences in the composition of calf and steer proteoglycans reflect to a significant extent differential glycosylation upon a similar-size protein backbone. We have obtained evidence that the ratio of chondroitin sulfate chains to keratan sulfate chains in those proteoglycans with apparently similar-sized core proteins is at least twice higher in calf than in steer. Additional evidence to suggest that the major age-related differences in the composition and structure of high buoyant density proteoglycans are not simple reflections of different core protein sizes comes from analysis of peptide maps of cyanogen bromide-derived fragments of the chondroitinase ABC + keratanase-treated aggregating proteoglycans (S. Bjornsson, K. E. Kuettner, and E. J.-M. A. Thonar, unpublished observations). The maps did not differ significantly at the two ages (Fig. 8). While it is possible one of the peptides is repeated in the sequence, the results are consistent with the interpretation that there are no major differences in the size and composition of the core protein from calf and steer aggregating proteoglycans.

The results of our studies, as well as the work of others, lead us to try to explain the biochemical basis of the changes which bovine articular aggregating proteoglycans undergo during fetal development and postnatal life (Fig. 4). Since the great majority (>85%) of the high buoyant density proteoglycans are of the aggregating type, our interpretation can be extended in more general terms as descriptive of changes in high buoyant density proteoglycans. Our conclusions are summarized as follows:

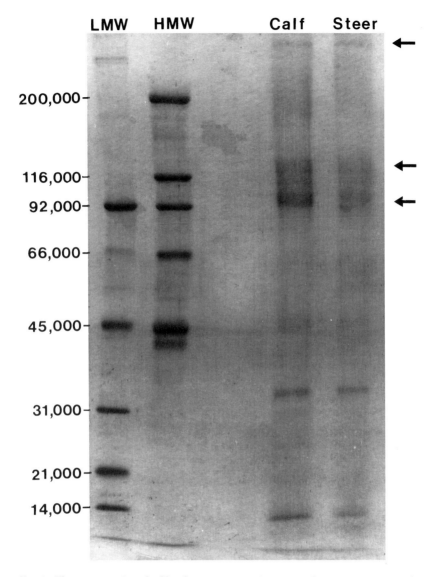

FIG. 8. The core proteins of calf and steer aggregating proteoglycans were prepared as described in the legend to Fig. 7. They were recovered by precipitation with 10 volumes of ethanol and dissolved in 70% formic acid–2% cyanogen bromide. Incubation was allowed to proceed under nitrogen at 25°C for 20 hr. The digestion products were lyophilized, reconstituted, and treated with 1% SDS–5% mercaptoethanol at 100°C for 5 min and subjected to SDS–PAGE using 6–16% gradient gels. Three prominent cyanogen bromide cleavage products were detected (see arrows). There were no noticeable differences between calf and steer. Low molecular weight (LMW) and high molecular weight (HMW) standards were run as controls.

1. The great majority of the age-related changes are the result of changes at the level of biosynthesis and are not caused by differential extracellular degradation of the same biosynthetic product (see Section IV below).

2. There is a progressive age-related decrease in the size of the chondroitin sulfate chains attached to the core protein (Figs. 2 and 4). This is a most significant contributor to the age-related decrease in proteoglycan size.

3. The number of keratan sulfate chains attached to the core protein of the chondroitin sulfate-rich subpopulation (Fig. 3, band I) increases progressively with age and maturation. This age-related change in glycosylation of the subpopulation that is present in both immature and adult cartilages contributes to a great extent to what has been termed an age-related enrichment in keratan sulfate content in aggregating proteoglycans (Fig. 4).

4. The appearance after birth of a new subpopulation of apparently smaller aggregating proteoglycans poor in chondroitin sulfate (Fig. 3, band II) contributes to some, but not all, of the exhibited age-related relative enrichment in keratan sulfate and decrease in hydrodynamic size (Fig. 4).

5. Both individual subpopulations of aggregating proteoglycans undergo age-related changes with respect to the number and size of chondroitin sulfate chains and the number of keratan sulfate chains covalently attached to the core protein.

Age-related changes may express themselves differently in other species and in other cartilages. Consequently, while the end result may be similar for most cartilages—that is, smaller proteoglycans poor in glycosaminoglycans populate the adult matrix—when and how those many changes take place may vary. It will be interesting to find out what age-related changes occur in mouse proteoglycans which appear to lack keratan sulfate.

IV. Biosynthesis

Bovine articular chondrocytes plated at high density and cultured in Ham's F12 medium–10% fetal calf serum as monolayers for several weeks will remain phenotypically stable in that they will continue to synthesize high molecular weight proteoglycans and type II collagen as the major matrix-associated molecules (Kuettner *et al.*, 1982a,b). We have extended these findings by showing that articular chondrocytes derived from calf and steer bovine articular cartilage will, when cultured in an identical fashion, continue to synthesize proteoglycans that are typical of those found in the calf or steer cartilage matrix, respectively (Thonar *et al.*, 1986). The great majority of the high buoyant density proteoglycans synthesized by those chondrocytes on day 5 of culture were able to form aggregates. The proteoglycan monomers were of larger hydrodynamic size in calf than in steer (Table IV). As in the tissue of origin, this age-related change was accompanied by a concomitant decrease in the size of the chondroitin sulfate chains. In addition, the proteoglycans exhibited significant differences in their glycosylation pattern (Table IV). Keratan sulfate represented ~14% of the glycosaminoglycans in steer but only 5.5% in calf. These data indicate that when denuded of their extracellular matrix and cultured under identical conditions, calf and steer chondrocytes will continue to synthesize, for at least 5 days in culture, proteoglycans that are similar to those found in the respective calf and steer matrix from which the cells were isolated. This observation is of great importance, for it suggests that there exists within chondrocytes a program that dictates the quality of the proteoglycan synthesized at different ages. The apparent inability of molecules in the fetal calf serum present in the medium to affect adversely the expression of the native steer phenotype suggests this program is not under the direct control of the extracellular environment.

The difference between calf and steer proteoglycans did not become accentuated during a 5-hr chase period following a 10-min pulse with $^{35}SO_4$, and it may thus be argued that the differences truly reflect

TABLE IV

CHARACTERIZATION OF HIGH BUOYANT DENSITY (^{35}S) D1 PROTEOGLYCANS SYNTHESIZED BY CALF AND STEER ARTICULAR CHONDROCYTES CULTURED AS MONOLAYERS[a]

	Calf	Steer
Monomer size (peak K_d on Sepharose 2B)	0.27	0.35
Chondroitin sulfate chain size (peak K_d on Sepharose G200)	0.35	0.45
Keratan sulfate chain size (peak K_d on BioGel P30)	0.38	0.38
[^{35}S]Chondroitin sulfate/[^{35}S]keratan sulfate	20.8	7.6

[a] Sizes are expressed in each case as the average partition coefficient of the peak (peak K_d), with higher values reflecting decreased hydrodynamic size (Thonar *et al.*, 1986). The ratio of [^{35}S]chondroitin sulfate to [^{35}S]keratan sulfate was derived following purification of each glycosaminoglycan as intact free chains and determination of ^{35}S label incorporated.

differential expression at the biosynthetic level at the two ages. As in the tissue of origin, only one subpopulation of aggregating proteoglycan was present in calf while two were found in steer (Fig. 9). Chondrocyte cultures should thus prove extremely useful in further probing the biochemical basis of the age-related differences.

In separate experiments, calf and steer chondrocytes were pulsed in the presence of β-D-xylosides in order to learn more about the mechanisms which determine chain length. The results showed that steer chondrocytes had a greater capacity to synthesize chondroitin sulfate chains than calf chondrocytes (Fig. 10). At both ages, chain size decreased with increasing concentration of β-D-xylosides and concurrent

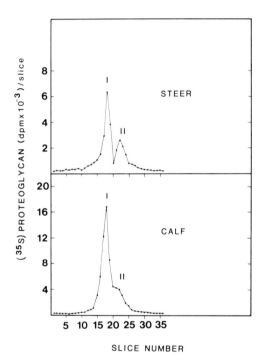

SLICE NUMBER

Fig. 9. Acrylamide–agarose electrophoresis of ^{35}S-labeled aggregating proteoglycans. ^{35}S-Labeled proteoglycans synthesized by immature calf or adult steer bovine chondrocytes (Thonar et al., 1986) were purified as aggregating proteoglycans using cesium sulfate velocity gradients. The proteoglycans were dissociated by heating in 1% SDS–5% mercaptoethanol and electrophoresed in composite gels (1.2% acrylamide–0.7% agarose). Slices were cut (1 mm thick) and radioactivity measured by scintillation counting. At both ages, the majority of the ^{35}S label was present in the chondroitin sulfate-rich proteoglycan (band I). The proportion of ^{35}S label present in the chondroitin sulfate-poor proteoglycan was significantly higher in steer (35%) than in calf (20%).

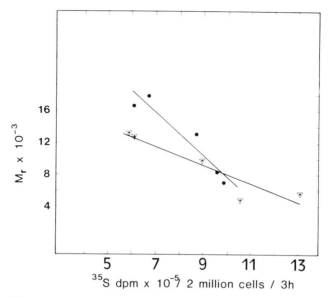

FIG. 10. Effect of rate of synthesis on the size of chondroitin sulfate chains synthesized by calf (●) and steer (△) chondrocytes. Chondrocytes were pulsed with ³⁵SO₄ in the presence of β-D-xylosides. The rate of chondroitin sulfate synthesis was found to increase with increasing xyloside concentration (0–2 mM). Chondroitin sulfate chain sizes were calculated from the partition coefficients on a Sephadex G200 column. The results show that age-related differences in chondroitin sulfate chain length are not the simple results of differences in rates of synthesis at the two ages.

increase in the rate of synthesis. At each xyloside concentration used, chondroitin sulfate chains were longer in calf than in steer. These results provide additional support for the existence of an inherent program which directs chondroitin sulfate chain size and, by inference, proteoglycan hydrodynamic size at different ages. The results also suggest that age-related differences in the length of the chondroitin sulfate chains are not the simple result of differences in the quality of the core protein which acts as the usual acceptor, as it was clear that significant differences in the size of the chains were still present when essentially all the chains were added upon identical β-D-xyloside acceptors.

It has been stated in the past that chondrocytes would "age" in culture as evidenced by a decrease in hydrodynamic size of the proteoglycan monomer (Pacifici et al., 1981; Fellini et al., 1981; Caplan, 1984). We compared in our system proteoglycans synthesized on day 5 and day 21 and observed a similar decrease in size of both calf and steer proteoglycans. It was noteworthy that day 21 calf proteoglycans were of larger hydrodynamic size than the corresponding day 21 steer proteoglycans. More importantly, the ratio of chondroitin sulfate to keratan sulfate in steer proteoglycans was significantly higher on day 21, indicating that the long-term culture conditions affected the synthesis of keratan sulfate and chondroitin sulfate differentially. The depression in keratan sulfate synthesis was the result in part of a decrease in keratan sulfate chain size. These observations with respect to the synthesis of keratan sulfate suggest that the decrease in chondroitin sulfate chain size observed with time in culture may not represent an age-related process. The data are consistent with the interpretation that the decrease in hydrodynamic size of calf and steer proteoglycans with increasing culture time is the result for the most part of long-term culture effects which produce a progressive decrease in the length of all glycosaminoglycan chains.

In summary, we would like to contend that the quality of the proteoglycans synthesized at different ages is, *in vivo*, under the regulation of an intracellular program that is not readily reversible but which may under extremely adverse conditions be altered. Whether, where, and when such alterations occur in pathological states is well worth investigating.

V. PROTEOGLYCANS IN DISEASE

The structure and composition of cartilage proteoglycans undergoes significant changes in a number of pathological conditions. In osteoarthrosis, there is evidence that some of the adult chondrocytes synthesize fetal-type or immature proteoglycans which may, in some way, be unsuitable for the repair process. In cartilage tumors, there may be a relationship between keratan sulfate chain length and content and the degree of malignancy.

A. Osteoarthrosis

Osteoarthrosis is characterized by articular cartilage destruction, subchondral bone changes which include cyst formation and sclerosis, and the appearance of new cartilage and osteophytes around the periphery of the articular surface (Sweet *et al.*, 1977b). A number of observations made more than 10 years ago suggested that in osteoarthrotic cartilage the sizes of the glycosaminoglycan chains (Mankin and Lippielo, 1971), the amount of keratan sulfate (Bollet and Nance, 1966), and the amount of chondroitin 4-sulfate (Mankin and

Lippielo, 1971) all decreased. These changes have been interpreted to reflect an increase in the proportion of immature proteoglycans in the cartilage matrix (Sweet *et al.*, 1977b). In order to test this hypothesis further, Sweet *et al.* (1977b) analyzed and characterized the high buoyant density proteoglycans present in different topographical regions of the osteoarthrotic human femoral head (Fig. 11). The conclusions which were drawn from the data are shown in Table V. While some of the changes may have resulted from increased enzymatic degradation of the proteoglycans, a number of changes were consistent with an increase in the proportion of immature proteoglycans synthesized. This increase in the synthesis of immature proteoglycans by the adult chondrocytes appeared more prominent in the severely fibrillated areas and the repaired osteophytic cartilage than in the apparently intact portions of the articular cartilage. It is not known at this time if the changes result from an alteration in the synthesis of all chondrocytes or from the emergence of a dormant immature cell population.

The type of osteoarthrosis we are considering is thought to result from failure of normal cartilage under abnormal load (Sweet *et al.*, 1977b). The initial step might well be collagen disruption and/or fracture. This is accompanied, even at an early stage, by cartilage repair. It is postulated that the balance between destruction and repair deter-

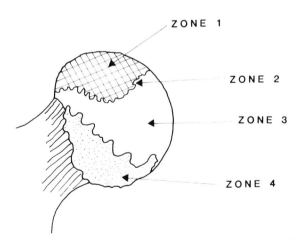

FIG. 11. Diagrammatic representation of the different zones of articular cartilage of the femoral head in tilt deformity osteoarthrosis. Zone 1, superior loaded surfaces denuded of cartilage. Zone 2, area of variable extent consisting of obviously fibrillated but not discolored cartilage. Zone 3, ring of unfibrillated but discolored cartilage. Zone 4, marginal osteophytes covered by white cartilage. Reproduced with permission from Sweet *et al.* (1977b).

TABLE V

ANALYSIS OF THE MATRIX OF HUMAN OSTEOARTHROTIC AND NORMAL HUMAN ADULT
ARTICULAR CARTILAGES[a,b]

	Osteoarthrotic cartilage			Normal cartilage
	Zone 2	Zone 3	Zone 4	
Hexosamine (μg/100 mg dry wt)	1700	6880	8800	5810
Uronic acid (μg/100 mg dry wt)	1450	2910	7270	3920
Chondroitin sulfate (μg/100 mg dry wt)	1334	2779	6952	2858
Keratan sulfate (μg/100 mg dry wt)	289	3468	1549	1779
Hyaluronic acid (μg/100 mg dry wt)	48	150	199	802
Chondroitin sulfate/keratan sulfate	4.62	0.80	4.49	1.61

[a] Adapted after Sweet et al. (1977b).

[b] The amounts of chondroitin sulfate, keratan sulfate, and hyaluronic acid present are given as micrograms of hexosamine present in each glycosaminoglyclan per 100 mg of dry tissue. The three zones of osteoarthrotic cartilage are described in Fig. 11. The ratio of chondroitin sulfate to keratan sulfate is expressed as a ratio of hexosamine/hexosamine.

mines whether the changes are progressive or not. In zone 2 destruction outruns repair; in zone 3 these processes are more or less in equilibrium (Fig. 11). If zone 4 represents a state of repair without any corresponding destruction, then the pattern of proteoglycans found in osteophytic cartilage must represent the newly synthesized "repair" proteoglycans.

A relative depletion of hyaluronic acid was noted in all zones of osteoarthrotic cartilage (Table V). This low hyaluronic acid content is typical of immature cartilage. Since it was prominent not only in zones 2 and 4 but in the apparently intact cartilage as well, it may be argued that the metabolism of all the cartilages on the articular surface is altered and that there may be a generalized increase in the synthesis and reorganization of a more immature matrix. There is no indication to suggest these changes play an etiological role. On the other hand, it is likely the switch may contribute to the pathogenesis of the progressive lesions and apparent failure at repair. Sweet et al. (1977b) suggested the newly synthesized matrix may be inappropriate in the adult, be unable to bear load adequately, and, moreover, be more susceptible to proteoglycan degradation. The immature matrix of zone 4 appeared to be intact, but then it is loaded minimally or not at all.

The severe loss of proteoglycans in some osteoarthrotic tissues is not accompanied by as significant a loss of collagen (Muir, 1980). We believe it is possible that the immature proteoglycans synthesized in the attempt at repair may be too large to be effectively incorporated within the tightly knit, highly cross-linked (Cannon, 1983) collagen network in the matrix in zones 2 and 3. In zone 4, on the other hand, the synthesis of large amounts of both collagen and proteoglycans occurs concurrently and it is likely these elements are organized into a newly formed matrix that grows rapidly in mass without undue problems. In other words, we would like to contend that the failure of the repair process in zones 2 and 3, the load-bearing regions, may be attributable to the reversal to the synthesis of fetal-like proteoglycans that may be too large to be incorporated into the existing collagen network in replacement of proteoglycans that have been lost. This hypothesis is but one of many which have tried to implicate abnormalities in proteoglycan metabolism in the development or relentless progression of the osteoarthrotic disease processes.

B. Cartilage Tumors

The extracellular matrix of human chondrosarcoma is rich in typical aggregating cartilage proteoglycans (Thonar et al., 1979). While early reports suggested that the proteoglycans did not contain significant amounts of keratan sulfate (Meyer et al., 1956; Anderson et al., 1963), it is now evident that the matrix is, in many cases, populated by proteoglycans which are rich in keratan sulfate (Sweet et al., 1975b; Pal et al., 1978; Thonar et al., 1979). In addition, in one such tumor the aggregating proteoglycans were found to exist as a single subpopulation containing two forms of keratan sulfate (Thonar et al., 1979). A larger species (~23 monosaccharides on average) was found in the keratan sulfate-rich region exclusively, while a smaller form (~13 monosaccharides on average) was present in both the keratan sulfate-rich and chondroitin sulfate-rich regions. Pal et al. (1978) showed that keratan sulfate chain size and content appear to correlate well with the degree of malignancy as established by histopathological examination. The ratio of cell to matrix is higher in the more malignant tumors, and histologically the tissue is not unlike rapidly metabolizing normal fetal cartilage.

Additional studies of the changes which proteoglycans undergo at different stages of tumor development may help clarify our understanding of changes which take place in normal cartilages. Other tumors of cartilage are likely to prove as interesting to study. Cartilage tumor cells, as osteoarthrotic cartilage cells, appear to have the ability to revert to the synthesis of fetal-like proteoglycans that are poor in keratan sulfate. For instance, the proteoglycans present in the matrix of a primary chordoma were fairly rich in keratan sulfate (keratan sulfate, 24%; chondroitin sulfate, 76%) (Sweet *et al.*, 1979b). Subsequent analysis of a secondary metastasis in the same patient revealed that the keratan sulfate content was significantly reduced (keratan sulfate, 12%; chondroitin sulfate, 88%). Chordoma cells are thought to be derived from notochordal rests. Assuming the cells present in the metastasis and the primary tumor were all derived from the same original cell, it would suggest that a cell has the potential to synthesize proteoglycans of different shapes and sizes.

ACKNOWLEDGMENTS

We are indebted to Dr. S. Bjornsson, who made available some of his unpublished results, and to our many collaborators who contributed to the many studies described in this review. We thank Ms. Verhonda Hearon for preparing the manuscript. The review was prepared with grant support from the National Institutes of Health (AG 04736 and AM 09132) and the Illinois chapter of the Arthritis Foundation. Dr. Eugene Thonar is the recipient of support for work from the William Noble Lane Medical Research Organization.

REFERENCES

Adams, P., and Muir, H. (1976). *Ann. Rheum. Dis.* **35**, 289–296.
Adams, P., Eyre, D. R., and Muir, H. (1977). *Rheumatology Rehabilitation* **16**, 22–29.
Anderson, C. E., Ludowieg, J., Eyring, E. J., and Horowitz, B. (1963). *J. Bone Jt. Surg.* **45A**, 753–764.
Baker, J. R., Caterson, B., and Christner, J. E. (1982). In "Methods in Enzymology" (V. Ginsburg, ed.), Vol. 83, Part D., pp. 216–235. Academic Press, New York.
Bayliss, M. T., and Ali, S. Y. (1978). *Biochem. J.* **176**, 683–693.
Bollet, A. J., and Nance, J. L. (1966). *J. Clin. Invest.* **45**, 1170–1177.
Buckwalter, J. A., Kuettner, K. E., and Thonar, E. J.-M. A. (1985). *J. Orthop. Res.* **3**, 251–257.
Cannon, D. J. (1983). In "Altered Proteins and Aging" (R. C. Adelman and G. S. Roth, eds.), pp. 161–168. CRC, Boca Raton, Florida.
Caplan, A. I. (1984). *Sci. Am.* **251**, 84–94.
Fellini, S. A., Pacifici, M., and Holtzer, H. (1981). *J. Biol. Chem.* **256**, 1038–1043.
Garg, H. G., and Swann, D. A. (1981). *Biochem. J.* **193**, 459–468.
Hadley, E. C. (1982). In "Testing the Theories of Aging" (R. C. Adelman and G. S. Roth, eds.), pp. 115–136. CRC, Boca Raton, Florida.
Hascall, V. C., and Kimura, J. H. (1982). In "Methods in Enzymology" (L. W. Cunningham and D. W. Frederiksen, eds.), Vol. 82, Part A, pp. 769–800. Academic Press, New York.
Hascall, V. C., and Sajdera, S. W. (1969). *J. Biol. Chem.* **244**, 2384–2396.
Heinegard, D. (1977). *J. Biol. Chem.* **252**, 1980–1989.
Heinegard, D., and Axelsson, I. (1977). *J. Biol. Chem.* **252**, 1971–1979.
Heinegard, D., and Paulsson, M. (1984). In "Extracellular Matrix Biochemistry" (K. A. Piez and A. H. Reddi, eds.), pp. 277–328. Elsevier, Amsterdam.

Heinegard, D., Paulsson, M., Inerot, S., and Carlstrom, C. (1981). *Biochem. J.* **197**, 355–366.

Heinegard, D., Paulsson, M., and Wieslander, J. (1982). *Semin. Arthritis Rheum.* **11**, 31–33.

Heinegard, D., Wieslander, J., Scheehan, J., Paulsson, M., and Sommarin, Y. (1985). *Biochem. J.* **225**, 95–106.

Hjertquist, S.-O., and Wasteson, A. (1972). *Calcif. Tissue Res.* **10**, 31–37.

Hoffman, P. (1979). *J. Biol. Chem.* **254**, 11854–11860.

Inerot, S., Heinegard, D., Audell, L., and Olssen, S.-E. (1978). *Biochem. J.* **169**, 143–146.

Kimata, K., Kimura, J. H., Thonar, E. J.-M. A., Barrach, H.-J., Rennard, S. I., and Hascall, V. C. (1982). *J. Biol. Chem.* **257**, 3819–3826.

Kuettner, K. E., Pauli, B. U., Gall, G., Memoli, V. A., and Schenk, R. K. (1982a). *J. Cell Biol.* **93**, 743–750.

Kuettner, K. E., Memoli, V. A., Pauli, B., Wrobel, N. C., Thonar, E. J.-M. A., and Daniel, J. C. (1982b). *J. Cell Biol.* **93**, 751–757.

Lohmander, S. (1975). *Eur. J. Biochem.* **57**, 549–559.

Lohmander, S. (1976). Ph.D. thesis, Karolinska Instituet, Stockholm.

Lohmander, S., DeLuca, S., Nilsson, B., Hascall, V. C., Caputo, C. B., Kimura, J. H., and Heinegard, D. (1980). *J. Biol. Chem.* **255**, 6084–6091.

Mankin, H. J., and Lippielo, L. (1969). *J. Bone Jt. Surg.* **51A**, 1591–1600.

Mankin, H. J., and Lippielo, L. (1971). *J. Clin. Invest.* **50**, 1712–1719.

Maroudas, A. (1975). *Biorheology* **12** , 233–248.

Maroudas, A. (1980). *In* "The Joints and Synovial Fluid" (L. Sokoloff, ed.), Vol. II, pp. 239–291. Academic Press, New York.

Mathews, M. B. (1975). "Connective Tissue Macromolecular Structure and Evolution." Springer-Verlag, Berlin and New York.

Mathews, M. B., and Glagov, S. (1966). *In* "Biochimie et Physiologie du Tissu Conjonctif" (P. Comte, ed.), pp. 21–26. Societe Ormeco et Imprimerie du Sud-Est a Lyon, Lyon.

Meyer, K., Davidson, E., Linker, A., and Hoffman, P. (1956). *Biochim. Biophys. Acta* **21**, 506–518.

Moment, G. B. (1982). *In* "Testing the Theories of Aging" (R. C. Adelman and G. S. Roth, eds.), pp. 1–23. CRC, Boca Raton, Florida.

Muir, I. H. M. (1980). *In* "The Joints and Synovial Fluid II" (L. Sokoloff, ed.), Vol. II, pp. 27–94. Academic Press, New York.

Oohira, A., and Nagami, H. (1980). *J. Biol. Chem.* **255**, 1346–1350.

Pacifici, M., Fellini, S. A., Holtzer, H., and DeLuca, S. (1981). *J. Biol. Chem.* **256**, 1029–1037.

Pal, S., Strider, W., Margolis, R., Gallo, G., Lee-Huang, S., and Rosenberg, L. (1978). *J. Biol. Chem.* **253**, 1279–1289.

Perricone, E., Palmoski, M. J., and Brandt, K. E. (1977). *Arthritis Rheum.* **20**, 1372–1380.

Prockop, D. J., Kivirikko, K. I., Tuderman, L., and Guzman, N. A. (1979). *New Engl. J. Med.* **301**, 75–85.

Rosenberg, L. C., Choi, H. U., Tang, L.-H., Johnson, T. L., Pal, S., Webber, C., Reiner, A., and Poole, A. R. (1985). *J. Biol. Chem.* **260**, 6304–6313.

Roughley, P. J., and White, R. J. (1980). *J. Biol. Chem.* **255**, 217–224.

Roughley, P. J., Poole, A. R., Campbell, I. K., and Mort, J. S. (1986). *Ortho. Trans.* **11**, 209.

Sajdera, S. W., and Hascall, V. C. (1969). *J. Biol. Chem.* **244**, 77–87.

Stanescu, V. (1980). In "Biochimie des Tissus Conjonctifs Normaux et Pathologiques" (A. M. Robert and L. Robert, eds.), pp. 151–159. CNRS, Paris.

Stanescu, V., and Stanescu, R. (1983). Biochim. Biophys. Acta **757**, 377–381.

Stanescu, V., and Sweet, M. B. E. (1981). Biochim. Biophys. Acta **673**, 101–113.

Stanescu, V., Maroteaux, P., and Sobczak, E. (1973). Biomed. Express **19**, 460–463.

Stevens, R. L., Ewins, R. J. F., Revell, P. A., and Muir, H. (1979). Biochem. J. **179**, 561–572.

Sweet, N. B. E., Thonar, E. J.-M. A., and Immelman, A. R. (1975a). S. Afr. J. Sci. **71**, 347–348.

Sweet, M. B. E., Thonar, E. J.-M. A., and Immelman, A. R. (1975b). Biochim. Biophys. Acta **437**, 71–86.

Sweet, M. B. E., Thonar, E. J.-M. A., and Immelman, A. R. (1977a). Biochim. Biophys. Acta **500**, 173–186.

Sweet, M. B. E., Thonar, E. J.-M. A., Immelman, A. R., and Solomon, L. (1977b). Ann. Rheum. Dis. **36**, 387–398.

Sweet, M. B. E., Thonar, E. J.-M. A., and Immelman, A. R. (1978). Arch. Biochem. Biophys. **189**, 28–36.

Sweet, M. B. E., Thonar, E. J.-M. A., and Marsh, J. (1979a). Arch. Biochem. Biophys. **198**, 439–448.

Sweet, M. B. E., Thonar, E. J.-M. A., Berson, S. D., Skikne, M. I., Immelman, A. R., and Kerr, W. A. (1979b). Cancer **44**, 652–660.

Thonar, E. J.-M. A., and Sweet, M. B. E. (1979). Biochim. Biophys. Acta **584**, 353–357.

Thonar, E. J.-M. A., and Sweet, M. B. E. (1981). Arch. Biochem. Biophys. **208**, 535–547.

Thonar, E. J.-M. A., Sweet, M. B. E., and Immelman, A. R. (1975). S. Afr. J. Sci. **71**, 347–348.

Thonar, E. J.-M. A., Sweet, M. B. E., Immelman, A. R., and Lyons, G. (1978). Calcif. Tissue Res. **26**, 19–21.

Thonar, E. J.-M. A., Sweet, M. B. E., Immelman, A. R., and Lyons, G. (1979). Arch. Biochem. Biophys. **194**, 179–189.

Thonar, E. J.-M. A., Lohmander, L. S., Kimura, J. H., Fellini, S. A., Yanagashita, M., and Hascall, V. C. (1983). J. Biol. Chem. **258**, 11564–11570.

Thonar, E. J.-M. A., Kimura, J. H., Gershon, G., Matijevitch, B., and Kuettner, K. E. (1984). Ortho. Trans. **8**, 76.

Thonar, E. J.-M. A., Lenz, M. E., Klintworth, G. K., Caterson, B., Pachman, L., Glickman, P., Katz, R., Huff, J., and Kuettner, K. E. (1985). Arthritis Rheum. **28**, 1367–1376.

Thonar, E. J.-M. A., Buckwalter, J. A., and Kuettner, K. E. (1986). J. Biol. Chem. **261**, 2467–2474.

Thonar, E. J.-M. A., Schnitzer, T., and Kuettner, K. E. (1987). J. Rheum., in press.

Tsiganos, C. P., and Muir, H. (1973). In "Connective Tissue and Aging" (H. G. Vogel, ed.), pp. 132–137. Excerpta Medica, Amsterdam.

Venn, G., and Mason, R. M. (1983). Biochem. J. **215**, 217–225.

Vogel, K. G., and Thonar, E. J.-M. A. (1987). Ortho. Trans., **12**, 46.

Wieslander, J., and Heinegard, D. (1981). Biochem. J. **199**, 81–87.

Extracellular Matrix Components of the Synapse

Regis B. Kelly

*Department of Biochemistry and Biophysics, University of California, San Francisco,
California 94413-0448*

Steven S. Carlson

Department of Physiology, University of Washington, Seattle, Washington

and

Pico Caroni

Hirnforschungs-Institut, Universität Zurich, Zurich, Switzerland

I. Introduction

Since neurons are ectodermal in origin, it is not surprising that their fate is intimately determined by interactions with extracellular matrix (ECM). Neurite extension (Edgar *et al.*, 1984; Lander *et al.*, 1985), neuron adhesion (Cole *et al.*, 1985; Vlodavsky *et al.*, 1982), neuron–neuron interaction (Schubert and La Corbiere, 1985), and neuron migration (Boucaut *et al.*, 1984) are all regulated in some way by components of the ECM. ECM components also play a crucial role in synaptic formation or maintenance. The purpose of the present chapter is to review what is known about ECM components of the synaptic cleft and how they might be involved in synaptic function.

II. ECM Components Are Involved in Synaptic Regeneration

The now classic experiments of Dr. McMahan and his colleagues on the frog neuromuscular junction showed that the exquisite precision with which a regenerating nerve rediscovers its old synaptic region lies not in the muscle itself but in the basal lamina that surrounds the muscle. Molecules in the ECM that reside at the earlier synaptic junctional region have the capacity to anchor the incoming nerve terminal at exactly the correct target (Marshall *et al.*, 1977; Sanes *et al.*, 1978). Indeed the basal lamina is capable of instructing the nerve terminal where to have active zones, morphologically recognizable regions that are the sites of exocytosis (Glicksman and Sanes, 1983). This work has focused attention on unique components of the ECM that must play a role in synaptic regeneration and perhaps even in synapse formation and synapse maintenance. What properties are expected of such specific ECM components? They should not be muscle type specific. Although regenerating axons can identify an old synaptic site with great specificity, they are poor at finding the muscle they originally innervated (for review, see Purves and Lichtman, 1985). Even vagal nerves that normally innervate heart muscle can recognize an empty synaptic site on striated muscle (Landmesser, 1971, 1972). Second, the guiding molecules should be concentrated at the synaptic regions. This eliminates the common ECM components, collagen IV, fibronectin, or laminin, since these are distributed over the muscle surface (Sanes, 1982; Chiu and Sanes, 1984). Although molecules restricted to the synaptic regions are candidates for nerve-targeting molecules, the tiny amount of muscle membrane involved in synapse formation makes isolation of such molecules a prodigiously difficult task. Success, where it has come, has resulted from more indirect approaches.

ECM components of the synaptic junction can also play a role in organizing the receptor distribution of the muscle cell. When myoblasts proliferate inside the basal lamine and fuse to give a muscle fiber, the acetylcholine receptors (AChR) cluster under the site of the old nerve ending (Burden *et al.*, 1979; McMahan and Slater, 1984). Presumably a molecule or molecules in the old synaptic site region can link itself to AChR, causing them to cluster at the original site. The receptors must be linked therefore to ECM components which are in turn linked to the nerve terminals. Such linkage, direct or indirect, forms a molecular bridge linking pre- and postsynaptic elements. Elucidation of the molecules involved in such a bridge will hopefully lead to insight into specific synapse formation. The existence of such molecules is not pure speculation; they can be readily seen in conventional thin sections and are especially prominent in etched preparations (Hirokawa and Heuser, 1982).

Although the present evidence supports a role of ECM only in nerve regeneration, a developmental role is also not unlikely. During development the growth cone of the nerve contacts the muscle fiber membrane in the absence of obvious extracellular material. As development continues, however, synaptic vesicles appear in the nerve terminal, and basal lamina appears between pre- and postsynaptic membranes before it is detectable elsewhere on the muscle fiber (Kullberg et al., 1977). The insertion of ECM material causes the distance between the pre- and postsynaptic membranes to increase to the separation found in mature synapses. As development proceeds the basal lamina spreads until it covers the whole muscle cell (Sanes et al., 1984). Additional evidence is presented later that synaptic development as well as synaptic regeneration is regulated via the ECM.

The association between exocytotic release sites and extracellular matrix might not be restricted to the neuromuscular junction. In pituitary cells the site of hormone secretion is restricted to those regions of the plasma membrane that abut ECM regions (Vila-Porcile and Oliver, 1980).

Both the nature of the components unique to the synaptic region, and their site of origin are presently unknown. Either the muscle or the nerve could in principle secrete them into the synaptic cleft. A neuronal origin for the unique molecules has an obvious appeal, since the nerve terminal already has a mechanism for secreting locally, namely, the exocytotic secretion of neurotransmitters. Neurons are already known to secrete components of basal lamina (Alitalo et al., 1980; Hampson et al., 1983; Burgess and Kelly, 1984; Culp et al., 1980). If ECM molecules are indeed released by nerve terminals, they must be transported down the neuronal axon. The next section considers the evidence for fast axonal transport of such molecules in the central nervous system.

III. AXONAL TRANSPORT OF SULFATED MATERIALS

The unusual morphology of the neuron permits easy identification of those newly synthesized proteins destined for nerve terminals. When radioactive precursors are injected near the cell body, those newly synthesized macromolecules destined for the nerve terminals are selectively transported down the axon. Since the axons are often a considerable distance from the cell body, dissection of axons free of cell bodies is simple. Newly synthesized proteins in the axon can be shown by two-dimensional gel analysis to be a subpopulation of proteins synthesized in the cell body (e.g., Kelly et al., 1980). It has been known for some time that sulfate-labeled glucosaminoglycans are transported down the axon at very fast rates (200–400 mm/day) (Elam et al., 1970; Elam, 1982). The label was recovered in both heparin sulfate and chondroitin sulfate. Turnover studies revealed the surprising result that the sulfated macromolecules appeared to turn over with a significantly shorter half-time ($\sim$1 day) than either fucose-labeled or leucine-labeled nerve terminal proteins (for review, see Elam, 1979). Indeed, sulfate turned over more rapidly than glucosamine in glycosaminoglycan (GAG; Margolis and Margolis, 1973).

There is little evidence at present that the proteoglycans sent to the nerve terminal are incorporated into ECM. Histochemical stainings for sugars and lectin binding showed accumulation of sugar groups in the synaptic cleft (for review, see Mahler, 1979), but the nature of the sugar groups is not known. Left unanswered in these studies was whether the source of the carbohydrate in the synaptic cleft was pre- or postsynaptic. No evidence that ECM regulates synaptogenesis in the central nervous system has yet been presented.

IV. Proteoglycans in Secretory Vesicles

One possible vehicle for the sulfate-labeled material that undergoes fast axonal transport is synaptic vesicles. It has been appreciated for some time that the presence of sulfated proteoglycans is one of the few characteristics shared by almost all secretory vesicles, including those in neurons, in exocrine, endocrine, neutrophil, and mast cells and in polymorphonuclear leukocytes (Giannattasio *et al.*, 1979). Their presence in vesicles has been attributed to a packaging function, a trophic function, transport to the cell surface, or chance.

Secretory cell lines in culture also have proteoglycans in their secretory vesicles. The pituitary tumor cell line AtT-20 stores a chondroitin sulfate proteoglycan in its secretory vesicles and secretes it coordinately with the secretion of the pituitary hormone ACTH (Moore *et al.*, 1983; Burgess and Kelly, 1984). It is the only sulfated molecule detected in the secretory vesicles whose sulfation is unaffected by tunicamycin (H.-P. Moore and R. B. Kelly, unpublished observations). The addition of GAG chains to the protein core of the chondroitin sulfate proteoglycan can be inhibited in AtT-20 cells by xyloside analogs. Alternatively, a cell line clone can be isolated that lacks the sulfated proteoglycan. In neither case was sorting, storage or release of the ACTH affected (Burgess and Kelly, 1984). Since an intracellular role for secretory vesicle proteoglycans seemed unlikely in this cell type, an extracellular role became more attractive.

Proteoglycans are also present in cholinergic synaptic vesicles from electric organ. Vesiculin, a charged molecule in vesicles initially incorrectly identified as a protein, was subsequently reidentified as a glycosaminoglycan (Stadler and Whittaker, 1978). Vesiculin probably does not exist as such but is an inadvertent degradation product of a membrane-associated heparan sulfate proteoglycan. Two groups have reported the presence of such a proteoglycan in *Torpedo* and *Narcine* synaptic vesicles (Stadler and Dowe, 1982; Carlson and Kelly, 1983). This highly antigenic molecule elutes with an apparent size of ~100,000 kDa relative to protein (Carlson and Kelly, 1983). It is almost certainly a membrane protein, since it is not extracted from the vesicle membrane by alkali stripping and can be constituted into liposomes (Carlson, *et al.*, 1986).

The initial interest in a membrane-associated vesicle protein with an antigenic determinant on the luminal membrane surface centered on a need for a marker of exocytosis. Chromaffin cells expose dopamine β-hydroxylase transiently during exocytosis (Wildmann *et al.*, 1981; Lingg *et al.*, 1983; Dowd *et al.*, 1983; Phillips *et al.*, 1983; Patzak *et al.*, 1984), but the use of this marker is restricted to a very small number of cell types. However, synaptosomes from resting electric organ bound monoclonal antibodies to the proteoglycan as effectively as those from stimulated ones (Miljanich *et al.*, 1982). Nor could an increase be seen when synaptosomes from electric organ were stimulated to undergo exocytosis *in vitro* (E. S. Schweitzer and R. B. Kelly, unpublished). The results could be explained by postulating a large pool of the antigenic sites on the outside of resting nerve terminals. This explanation, combined with the failure to find an intracellular role for the vesicular proteoglycan in AtT-20 cells (Burgess and Kelly, 1984), suggested that synaptic vesicles might be transporting proteoglycans to the outside of the cell. An appealing conjecture for the purpose of such a transfer was to deposit unique molecules in the ECM at the site of exocytosis. Subsequent experiments, discussed below, support the idea that the nerve terminal contributes material to the ECM, but are confusing about the role synaptic vesicles might play in the delivery.

V. A Synaptic Junctional Proteoglycan Related to the Vesicle Form

The data on the synaptic vesicle proteoglycan suggested that immunoelectron microscopy should be used to look for an extracellular pool of cross-reacting material. To allow antibody penetration, electric organ was homogenized gently to generate large membrane sheets. Only sheets derived from the innervated face of the electric cells were covered with intact nerve terminals (Buckley *et al.*, 1983). Primary and secondary antibodies readily bind to the outside of these nerve terminals in such preparations, which can then be used to determine the location of binding both by immunoelectron microscopy and sub-cellular fractionation.

Immunoelectron microscopy using a horseradish peroxidase second antibody showed that the majority of the binding of an antibody to the vesicle proteoglycan SV1 antigen was to the material lying between the pre- and postsynaptic membranes. This was not an artifact of fixation, since a second monoclonal antibody to a nerve terminal component, which was not a vesicle proteoglycan, had ready access to other regions of the nerve terminal membrane (Buckley *et al.*, 1983). These observations have been extended to antibodies to other synaptic vesicle antigens (Fig. 1). The SV4 antibody, which also recognizes the intra-vesicular proteoglycan, similarly binds to the synaptic cleft region, while SV2, a cytoplasmic antibody, only binds to the occasional lysed terminal.

In recent attempts to quantitate the antigen distribution, we have used a second antibody coupled to gold beads to detect antibody binding. With the improved quantitation it is clear that binding is only to the innervated face of the cell, and is absent from regions of basal lamina that lie between nerve terminals and from the ECM at the noninnervated face. We conclude that the antigenic site on the intra-vesicular proteoglycan is also present on the outside of the nerve terminal (Carlson *et al.*, 1986).

We have evidence that the SV4 antigen found outside the nerve terminal is associated with ECM. If anti-vesicle proteoglycan antibodies (anti-SV4) are bound to the outside of intact nerve terminals and the second antibody is radioactive, it is possible to follow the antigen through subcellular fractionation experiments. The two antigenic determinants, SV1 and SV4, found on the vesicle proteoglycan are now found to behave differently. SV4, but not SV1, is recovered in the ECM fraction (Caroni *et al.*, 1985). If unlabeled ECM is prepared, the specific activity of the SV4 antigen is close to that found in synaptic vesicles (Fig. 2). This is not an artifact, since another veiscle antigen, SV2, is absent from the ECM. Further evidence that the SV4 and SV1 antigens are quite different was obtained by studying axonal transport. The data described until now could be explained if SV4 antigens are deposited by the postsynaptic cell in the cleft, from which they are endocytosed by recycling synaptic vesicles. This explanation was shown to be invalid by demonstrating that the SV4 antigen is transported rapidly down the axon, away from its site of synthesis in the cell body (Caroni *et al.*, 1985). In contrast, the SV1 antigen is not

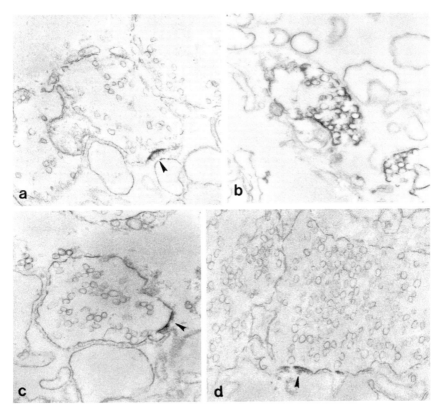

Fig. 1. Localization of antibody-binding sites on intact and disrupted nerve terminals with immunoperoxidase techniques. After incubation of electric organ nerve terminals with monoclonal antibodies to SV1 (a), SV3 (d), and SV4 (c), peroxidase reaction product (arrowheads) was associated with the outside of the presynaptic membrane. In contrast, after incubation with antibody to SV2 (b), the reaction product appeared to be localized to the outside of synaptic vesicles in nerve terminals that were apparently lysed during preparation of the tissue. ×22,025.

transported down the axon but is generated at the nerve terminal. We conclude that the SV4 antigen which associates with the ECM is made in the cell body and transported down the nerve terminal, while the SV1 antigen on the vesicle proteoglycan arises by some modification of the vesicle proteoglycan that occurs at the nerve terminal.

Modifications of proteoglycans at the cell surface are not unknown. This could involve proteolysis of the core protein, or endoglycosidase action at the carbohydrate side chains (Robinson *et al.*, 1978; Thumb-

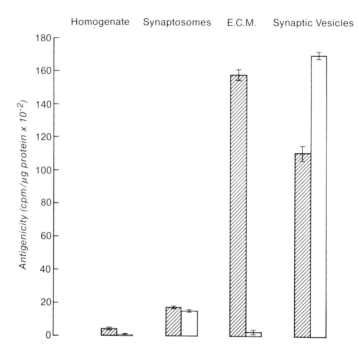

FIG. 2. The antigenic sites per milligram protein in different fractions isolated from electric ray electric organ. The number of antigenic sites were estimated by dissolving the samples in 1% sodium dodecyl sulfate (0.2 M NaCl–10 mM HEPES, pH 7.0) and spotting them on nitrocellulose filters. Monoclonal antibodies against the SV4 antigen (7A7) and against the SV2 antigen (10H3) were bound to the adsorbed protein and the amount bound quantified using a radioactive second antibody (Caroni *et al.*, 1985). For an antigen only in synaptic vesicles (SV2, open bars) the antigenic sites per milligram protein increases several hundred fold on isolating vesicles from the homogenate. The vesicle antigens are enriched in the synaptosomal fraction, as expected for a preparation of pinched-off nerve terminals. No SV2 antigen can be found in the purified ECM. In contrast, the SV4 antigen (7A7, hatched bars) is only enriched 10- to 20-fold in going from homogenate to vesicles, since the SV4 antigen is not restricted to vesicles. The region of the electric organ where the SV4 antigen is also found at high specific activity is the ECM.

erg *et al.*, 1982; Castellot *et al.*, 1982; Lark and Culp, 1984), or both. It is very likely that the molecule bearing the SV4 antigen is transported down the axon, externalized, after which it becomes associated with the extracellular matrix. A plausible hypothesis is that after modification the molecule loses its ability to bind to the ECM and the resulting molecule is internalized and marked for degradation. It is noteworthy that electron microscopy of the neuronal cell bodies shows SV1, the nerve terminal-acquired antigen in multivesicular bodies, presumably en route to lysosomal degradation.

We conclude, then, that a synaptic vesicle proteoglycan shares SV4 antigenic determinant with an ECM component of the synaptic junction, but not the SV1 determinant.

VI. Purification and Characterization of the ECM Proteoglycan

The vesicle proteoglycan has already been characterized (Carlson and Kelly, 1983). To compare it to the cross-reacting ECM material, we have used the SV4 antigen to purify and characterize the ECM molecule (Carlson, 1986). It requires severely denaturing conditions (4 M guanidine HCl–2% Nonidet P-40 or 1% SDS) to solubilize it from the ECM. After solubilization, it is very large, with an uncorrected sedimentation rate in a sucrose–SDS density gradient of ~11S. It also elutes ahead of blue dextran from a Sephacryl S-1000 column run in denaturing concentrations of guanidine hydrochloride. It is highly negatively charged, eluting from a DEAE column run at pH 5.0 at a salt concentration of 0.7 M, conditions compatible with charged sulfate groups. The molecule carrying the SV4 antigen binds to a laminin column from which it is eluted by heparan sulfate GAG chains. It has a large pronase-resistant domain or domains each about one-tenth the size of the whole molecule on Sephacryl S-1000. These domains may be the regions at which the negatively charged carbohydrate chains attach to protein core, since they retain the capacity to bind to DEAE at pH 5 and to be eluted at 0.7 M salt. A pronase-resistant domain enriched in O-linked sugars has been found on bovine and human low-density lipoprotein receptors (Russell *et al.*, 1984; Yamamoto *et al.*, 1984). Finally, the molecule carrying the SV4 antigen can be reconstituted into liposomes and so presumably has a transmembrane domain.

Because of the heterogeneous size and glycosylation of large glyco-proteins, it is difficult to be certain that purification yields a single molecular species until the protein core can be isolated. In this case we know that the ECM glycoprotein is a minor component, since it does not elute with the bulk of ECM protein on sizing or ion-exchange columns. Material purified by immunoadsorption using a monoclonal antibody to the SV4 antigen has properties indistinguishable from that purified by conventional means. With the caution that what we are describing might still be a family of glycoproteins sharing an unusual epitope, it is convenient for the moment to consider it a single molecule. We know that the molecule is on the nerve terminals and has a hydrophobic domain that allows it to be anchored to the membrane. Finally, we know it purifies with the basal lamina fractions in the presence of detergent and has the capacity to bind laminin, an ECM component. Because of its physical properties, the class of molecules that carry the SV4 antigenic determinant can be considered as terminal anchorage proteins (TAPs). We have no evidence as yet whether or not anchorage is their primary role *in vivo*.

Although the TAP shares antigenic determinants with a synaptic vesicle proteoglycan, and has physical properties associated with proteoglycans, it is still premature to call it a proteoglycan. Some of the classic tests have been ambiguous, which may be because the carbohydrate chains are clustered on a pronase-resistant core. A large negatively charged molecule with somewhat similar properties has been described by Chernoff *et al.* (1983) in the growing tips of neuroblastoma.

VII. The Antigenic Determinant
of Electric Organ TAP

To characterize further the electric organ TAP bearing the SV4 epitope, monoclonal antibodies have been raised to the isolated molecule that have allowed us to distinguish other antigenic domains. We designate these epitopes SV4a, SV4b, and so on, with the original epitope being SV4a. Since pronase digestion reduces the ability of the residual oligosaccharides to bind to nitrocellulose paper, the conventional assays of antigenicity cannot be used to measure whether the antigens are pronase sensitive. Antigenicity in the pronase-digested material can still be measured, however, in a competitive assay, or after cross-linking it chemically to a microtiter dish. Since a considerable fraction of the pronase-digested antigenic material binds to DEAE–Sephacryl, is not included on a G-100 Sephadex column and has its antigenicity destroyed by periodate, it is likely to be associated with a glycosaminoglycan (Caroni, unpublished observation).

The association with carbohydrate of at least three epitopes, that are present on the synaptic vesicle proteoglycan and the terminal anchorage protein, was a surprise. Carbohydrates more commonly give rise to shared antigens, not unique ones. Second, it is not usual for carbohydrates to be on a small subclass of proteins. Asparagine-linked complex oligosacchardies, for example, are present in many types of membrane and secreted proteins. Unique carbohydrate antigens are not unprecedented, however, especially in neural tissue. For example, some but not all of the adhesion molecule N-CAM have a carbohydrate epitope that is also found on some of a quite separate adhesion molecule, L1, that is involved in cerebellar migration. The epitope is found in neuronal, glial, but not fibroblastic cells. It is also present on lymphocytes, including natural killer (NK) cells (Kruse et al., 1984). It is possible that all types of adhesion molecules might be capable of expressing this epitope (the L2/HNK-1 epitope). In support of this hypothesis, a new molecule, the J1 antigen, isolated because it carries the L2/HNK-1 epitope, turns out to be a third class of adhesion molecules, a class involved in neuron–astrocyte adhesion (Kruse et al., 1985). A second class of carbohydrate epitopes restricted to its neuronal distribution are the stage-specific embryonal antigens SSEA-3 and SSEA-4. These are present in a subclass of dorsal root ganglion cells of the rat. The subclass of neurons sharing the epitope would appear to be functionally significant, since they innervate a defined lamina of the spinal cord (Dodd et al., 1984). The carbohydrate epitopes SSEA-3 and SSEA-4 are found on both glycoproteins and glycolipids. A fucose-containing glycolipid has also been shown to be expressed during development of the mammalian bribrain (Yamamoto et al., 1985). Indeed, the use of monoclonal antibodies to define developmental stages has led us to appreciate that many such antigens have oligosaccharide properties (Feizi, 1985). Finally, specific lectin binding at the neuromuscular junction (Sanes and Cheney, 1982) makes it likely that special oligosaccharide moieties are enriched there.

The unique oligosaccharide determinants on the synaptic vesicle proteoglycan and the electric organ terminal anchorage protein are, therefore, not without precedent in the biological literature. However, in none of these cases has it been established whether the oligosaccharide moiety participates in cell–cell interaction, although that is an obviously attractive possibility (Kruse et al., 1984, 1985).

VIII. Insertion of the Anchorage Protein

As mentioned earlier, the presence of the same antigenic determinants on the vesicle proteoglycan and a terminal anchorage protein suggested that the synaptic vesicle might be involved in transporting the anchorage protein. Unfortunately, the relationship of the vesicle proteoglycan to the synaptic junctional anchorage protein is not clear at present. The different properties of the two molecules are compared in Table I. We do not know with complete confidence whether the differences are real. They could conceivably result from the different

TABLE I

COMPARISON OF VESICLE PROTEOGLYCAN AND TERMINAL ANCHORAGE PROTEIN

Properties	Vesicle proteoglycan	Terminal anchorage protein
SV4a, SV4b, SV4c, SV4d antigen	Present	Present
SV1	Present	Absent
Elution from DEAE–Sephacel	0.4 M salt	0.7 M salt
Associates with liposomes	Yes	Yes
Sedimentation velocity	2.2S	11S

purification procedures used to isolate ECM and synaptic vesicles. If the differences are not real but arise during isolation, then synaptic vesicles might be capable of delivering the proteoglycan to the synaptic junction ECM when there is a space available for it (Fig. 3a). If there are no free sites on the basal lamina, the molecule would be recycled with the vesicle membrane and eventually returned to the cell body. Alternatively, if the differences between the two molecules are real, then delivery of the anchorage protein could involve a transport vesicle other than the synaptic vesicle (Fig. 3b). After modification which reduces its size, or removes charged groups and perhaps exposes the SV1 antigen, the modified proteoglycan can be internalized by a vesicle during membrane recycling. A third possibility is that the two proteoglycans show no similarity in their polypeptide backbones but only share a glycosylation site (Fig. 3c), whose function, if any, is not yet known.

We have made preliminary attempts to distinguish between these models. The models in Fig. 3b and c predict that both forms of molecule will be transported down the axon, while the Fig. 3a model predicts only one form, presumably the larger. Preliminary evidence shows the presence of both forms at axonal ligatures and in the cell bodies of neurons innervating the electric organ. Definitive evidence, however, will require isolation of the protein core of both glycoproteins.

IX. OTHER COMPONENTS OF THE ECM AT THE SYNAPSE

It is helpful to look at the findings on the nerve terminal proteoglycan in the context of other molecules associated with extracellular matrix at the synapse. A partial list of these is given in Table II. The molecule whose properties most parallel the nerve terminal proteoglycan may be the heparan sulfate proteoglycan found in the

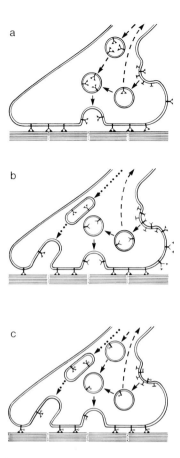

Fig. 3. Three models of how the synaptic vesicle proteoglycan and the extracellular terminal anchorage protein (TAP) might be related. (a) The TAP is transported down the axon in an organelle that is the precursor of the synaptic vesicle. When that vesicle undergoes exocytosis the vesicle proteoglycan attaches to any vacant site on the basal lamina. If a site is unavailable the proteoglycan is internalized during the coated vesicle-mediated endocytosis of vesicle membranes. The endocytosed vesicle may either be refilled with transmitter and go through further exocytotic cycles, or it may travel retrogradely in the axon to be degraded in the cell body. The apparent differences between the vesicle proteoglycan and the anchorage protein (Table I) could be explained if, for example, intravesicular degradation occurred during vesicle recycling. (b) A second possibility is that synaptic vesicles never contain the intact TAP. An unspecified transport vesicle inserts the anchorage protein at the nerve terminal, where it binds to the basal lamina. While in the cleft it acquires the SV1 antigen, which can be detected extracellularly in the synaptic junction (Fig. 1). After partial destruction is becomes reduced in size, and no longer attaches to the basal lamina. It can now be internalized in coated vesicles during vesicle membrane recycling and either exocytosed again or returned by axonal transport to the cell body. (c) The vesicle proteoglycan may be neither identical to (a) or derived from (b) the anchorage protein but be an unrelated protein that happens to bear the same antigenic oligosaccharide. An explanation why not all vesicle proteins, and not all ECM molecules have that oligosaccharide is not obvious.

TABLE II

MOLECULES ASSOCIATED WITH THE SYNAPTIC JUNCTION

Molecule	References
Heparin sulfate proteoglycan	Anderson and Fambrough (1983); Bayne *et al.* (1984); Caroni *et al.* (1985)
ECM antigens: JS1, JS2, JS3	Sanes and Hall (1979); Silberstein *et al.* (1982)
Laminin	Daniels *et al.* (1984); Bayne *et al.* (1984)
Collagen	Vogel *et al.* (1983); Bayne *et al.* (1984)
Clustering factor	Fallon *et al.* (1985); Wallace *et al.* (1985)
Acetylcholine esterase	Hall and Kelly (1971); Grassi *et al.* (1983)
N-CAM	Reiger *et al.* (1985); Couvault and Sanes (1985)
Lectin-binding site	Sanes and Cheney (1982)

neuromuscular junction of the frog, *Xenopus laevis*, and made by isolated muscle cells of frog in culture (Anderson and Fambrough, 1983). When *Xenopus* muscles were cultured without neurons, plaques of proteoglycan were seen by immunocytochemistry, showing that the molecule is postsynaptic in origin. Furthermore, when plaques of AChR were seen they were always associated with plaques of proteoglycan.

A similar result was obtained with chicken muscle (Bayne *et al.*, 1984). A monoclonal antibody to a heparin sulfate proteoglycan recognized synapses on sections of muscle cells, although it had weak reactivity with other ECM regions of the muscle. Furthermore, when chick muscle cells were grown in culture without neurons, all AChR clusters colocalized with clusters of the heparan sulfate proteoglycan. Indeed the fine structure of the AChR distribution almost exactly matched that of the proteoglycan at the light microscope level. This congruence led the authors to suggest that the two molecules were linked, and that the linkage was important in controlling the AChR distribution on the muscle.

The colocalization of AChR with the heparan sulfate proteoglycan in embryonic muscle cells is highly reminiscent of the association of synapse-specific antigens (JS1, JS2, and JS3) with AChR observed with cultured muscle cells (Silberstein *et al.*, 1982). Unfortunately the nature of these three antigens is poorly defined. Coclustering of laminin and receptors has been reported (Daniels *et al.*, 1984; Bayne *et al.*, 1984), as has coclustering of another common ECM component, collagen (Vogel *et al.*, 1983; Bayne *et al.*, 1984). It would appear then that muscle cells prior to innervation can have clusters of AChR that are already associated with the conventional elements of the ECM: proteoglycan, laminin, and collagen. This association does not seem to be fortuitous. Treatment of muscle cells with agents that stimulate collagen biosynthesis favors the aggregation of receptors (Kalcheim *et al.*, 1982). Furthermore, laminin itself can cause AChR clustering (Vogel *et al.*, 1983). One of many possible models to account for this interaction is to propose that AChR is inserted into the plasma membrane in association with the heparan sulfate proteoglycan. By interaction with the proteoglycan molecules the collagen and laminin molecules can then bring about clustering. The role of the nerve would therefore be to bind to one of these clusters or to raft several of them together into an aggregate the size of a synapse.

This model is obviously too simple. A factor has been isolated from the basal lamina of *Torpedo* electric organ that facilitates AChR clustering at remarkably low concentrations, and is not laminin (Fallon *et al.*, 1985). The clustering factor is loosely associated with the basal lamina, since it can be removed by a pH 5.0 wash. Monoclonal antibodies to the clustering factor block its effectiveness. These antibodies also bind to the synaptic cleft region of the *Torpedo* neuromuscular junction, and not to extrajunctional regions. Although other clustering factors have been described, no other has been shown to have a synaptic location. As yet the specific role of the clustering factor is not known. It might, for example, be a lectin that links complexes of proteoglycan and AChR much more efficiently than laminin or collagen can link them. Detailed description of the properties of this clustering factor are awaited with enthusiasm.

Other molecules are also found in the synaptic cleft. Acetylcholine esterase is perhaps the best known. It is thought that the collagen tail of the asymmetric form facilitates its anchorage in the synaptic cleft (Massouli and Bon, 1982; Grassi *et al.*, 1983), but cannot explain its specific location there. Although the specific junctional binding site for cleft acetylcholine esterase is not known, the AChR clustering factor can also induce esterase clustering (Wallace *et al.*, 1985). The ubiquitous N-CAM has also been found to be enriched at the neuromuscular junction. This molecule, originally isolated by its ability to promote neuron–neuron association is now found to be made by glial (Noble *et al.*, 1985; Keilhauer *et al.*, 1985) and muscle cells (Rieger *et al.*, 1985; Couvault and Sanes, 1985), and may in fact also promote neuron–astrocyte and astroycyte–astrocyte interactions (Keilhauer *et al.*, 1985). N-CAM is found to become widely distributed over the muscle surface after denervation and so is unlikely to play a specific role in synaptic regeneration at the original sites. Finally, as mentioned earlier, a unique lectin-binding site has been observed in the neu-romusuclar junction (Sanes and Chaney, 1982).

The data are too sparse at present to provide a satisfying picture of the role of ECM in synaptic regeneration in the peripheral nervous system. Initial formation of peripheral synapses and the involvement, if any, of central nervous system synapses with ECM molecules is even less clear. What is emerging, however, is tantalizing to those intrigued by ECM. Unidentified components of the ECM are clearly specifying synaptic regeneration. What about synaptic development? The postsynaptic clusters of AChR are tightly associated with a pro-teoglycan prior to synapse formation (Bayne *et al.*, 1984). The nerve terminal also has the capacity to insert proteoglycan into its plasma membrane when it is mature, but perhaps also during development. It is known that growth cones are rich in ECM material (Culp *et al.*, 1980), and that synaptic vesicles can fuse with nerve terminal plasma mem-brane prior to contact with the target (Young and Poo, 1983; Hume *et al.*, 1983). ECM components within the synaptic vesicle could thus be exposed, although such exposure is apparently not essential (Hender-son *et al.*, 1984). Membrane-attached protoeglycans on both the nerve terminal and the AChR cluster could interact with the collagens and laminins of the basal lamina to glue the pre- and postsynaptic sides of the synapse to the basal lamina. But is the basal lamina only a non-specific glue? The regeneration results suggest that it may be a matrix for anchoring specific recognition molecules. Obviously one easy way of assuring that specific recognition molecules are restricted to the junc-tional region is to secrete them from the nerve terminal. After all, the whole structure of the nerve is geared toward exocytosis of signaling molecules exclusively toward receptors on its target cell.

ACKNOWLEDGMENTS

We would like to thank Dr. Kathleen Buckley for permission to publish her electron micrographs. This work from the authors' lab described in this paper was supported by an Muscular Dystrophy Association award to Dr. Kelly and NIH grant NS 09878 and NS 16073 to Dr. Kelly. P.C. was supported by an EMBO fellowship (ALTF-57-1982). The authors are also grateful to colleagues in the proteoglycan field who have helped ac-quaint us with its mysteries. We also wish to thank Leslie Spector for her care and attention in assembling this manuscript.

REFERENCES

Alitalo, K., Karkinen, M., Vaheri, I., Rohde, H., and Timpl, R. (1980). *Nature (London)* **287,** 465–466.
Anderson, M. G. and Fambrough, D. M. (1983). *J. Cell Biol.* **97,** 1396–1411.
Bayne, E. K., Anderson, M. J., and Fambrough, D. M. (1984). *J. Cell Biol.* **99,** 1486–1501.

Boucaut, J. C., Darribere, T., Poole, T. J., Aoyama, H., Yamada, K. M., and Thiery, J. P. (1984). *J. Cell Biol.* **99**, 1822–1830.

Buckley, K. M., Schweitzer, E. S., Miljanich, G. M., O'Grady, L. C., Kushner, P., Reichardt, L. F., and Kelly, R. B. (1983). *Proc. Natl. Acad. Sci. U.S.A.* **80**, 7342–7347.

Burden, S. J., Sargent, P. B., and McMahan, U. J. (1979). *J. Cell Biol.* **82**, 412–425.

Burgess, T. L., and Kelly, R. B. (1984). *J. Cell Biol.* **99**, 2229–2230.

Carlson, S. S., and Kelly, R. B. (1983). *J. Biol. Chem.* **258**, 11082–11091.

Carlson, S. S., Caroni, P., and Kelly, R. B. (1986). *J. Cell Biol.* **103**, 509–520.

Caroni, P., Carlson, S. S., Schweitzer, E., and Kelly, R. B. (1985). *Nature (London)* **314**, 441–443.

Castellot, J. J., Fayreau, C. V., Karnovsky, M. J., and Rosenberg, R. D. (1982). *J. Biol. Chem.* **257**, 11256–11260.

Chernoff, E. A. G., Maresh, G. A., and Culp, L. A. (1983). *J. Cell Biol.* **96**, 661–668.

Chiu, A. Y., and Sanes, J. R. (1984). *Dev. Biol.* **103**, 456–467.

Cole, G. J., Schubert, D., and Glaser, L. (1985). *J. Cell Biol.* **100**, 1192–1199.

Couvault, J., and Sanes, J. R. (1985). *Proc. Natl. Acad. Sci. U.S.A.* **82**, 4544–4548.

Culp, L. A., Ansbacher, R., and Domen, C. (1980). *Biochemistry* **19**, 5899–5907.

Daniels, M. P., Vigny, M., Sonderegger, P., Bauer, H. C., and Vogel, E. (1984). *Int. J. Dev. Neurosci.* **2**, 87–99.

Dodd, J., Solter, D., and Jessel, T. M. (1984). *Nature (London)* **311**, 469–472.

Dowd, D. J., Edwards, C., Englert, D., Mazurkiewicz, J. E., and Ye, H. Z. (1983). *Neuroscience* **10**, 1025–1033.

Edgar, D., Timpl, R., and Thoenen, H. (1984). *EMBO J.* **3**, 1463–1468.

Elam, J. S. (1979). *In* "Complex Carbohydrates of Nervous Tissue" (R. U. Margolis and R. K. Margolis, eds.), pp. 235–267. Plenum, New York.

Elam, J. S. (1982). *J. Neurochem.* **39**, 1220–1230.

Elam, J. S., Goldberg, J. M., Radin, N. S., and Agranoff, B. W. (1970). *Science* **170**, 458–460.

Fallon, J. R., Nitkin, R. M., Reist, N. E., Wallace, B. G., and McMahan, W. J. (1985). *Nature (London)* **315**, 571–573.

Feizi, T. (1985). *Nature (London)* **314**, 53–57.

Giannattasio, G., Zanini, A., and Meldolesi, J. (1979). *In* "Complex Carbohydrates of Nervous Tissue" (R. M. Margolis and R. K. Margolis, eds.), pp. 327–345. Plenum, New York.

Glicksman, M. A., and Sanes, J. R. (1983). *J. Neurocytol.* **12**, 661–671.

Grassi, J., Massoulie, J., and Timpl, R. (1983). *Eur. J. Biochem.* **133**, 31–38.

Hall, Z. W., and Kelly, R. B. (1971). *Nature* **232**, 62–63.

Hampson, I. N., Kumar, S., and Gallagher, J. T. (1983). *Biochim. Biophys. Acta* **763**, 183–190.

Henderson, L. P., Smith, M. A., and Spitzer, N. C. (1984). *J. Neurosci.* **4**, 3140–3150.

Hirokawa, N., and Heuser, J. E. (1982). *J. Neurocytol.* **11**, 487.

Hume, R. I., Role, L. W., and Fischbach, G. D. (1983). *Nature (London)* **305**, 632–634.

Kalcheim, C., Duksin, D., and Vogel, Z. (1982). *J. Biol. Chem.* **257**, 12722–12727.

Keilhauer, G., Faissner, A., and Schachner, M. (1985). *Nature (London)* **316**, 728–730.

Kelly, A. S., Wagner, J. A., and Kelly, R. B. (1980). *Brain Res.* **185**, 192–197.

Kruse, J., Mailhammer, R., Wernecke, H., Faissner, A., Sommer, I., Goridis, C., and Schachner, M. (1984). *Nature (London)* **311**, 153–155.

Kruse, J., Keilhauer, G., Faissner, A., Timpl, R., and Schachner, M. (1985). *Nature (London)* **316**, 146–148.

Kullberg, R. W., Lentz, T. L., and Cohen, M. W. (1977). *Dev. Biol.* **60**, 101–129.

Lander, A. D., Fujii, D. K., and Reichardt, L. F. (1985). *Proc. Natl. Acad. Sci. U.S.A.* **82**, 1882–1890.

Landmesser, L. (1971). *J. Physiol. (London)* **213**, 707–725.

Landmesser, L. (1972). *J. Physiol. (London)* **220**, 243–256.

Lark, M. W., and Culp, L. A. (1984). *J. Biol. Chem.* **259**, 212–217.

Lingg, G., Fischer-Colbrie, R., Schmidt, W., and Winkler, H. (1983). *Nature (London)* **301**, 610–611.

McMahan, U. J., and Slater, C. R. (1984). *J. Cell Biol.* **98**, 1453–1473.

Mahler, H. J. (1979). *In* "Complex Carbohydrates of Nervous Tissue" (R. U. Margolis and R. K. Margolis, eds.), pp. 165–184. Plenum, New York.

Margolis, R. K., and Margolis, R. U. (1973). *Biochim. Biophys. Acta* **304**, 413–420.

Marshall, L. M., Sanes, J. R., and McMahan, U. J. (1977). *Proc. Natl. Acad. Sci. U.S.A.* **74**, 3073–3077.

Massoulie, J., and Bon, S. (1982). *Annu. Rev. Neurosci.* **5**, 57–106.

Matthew, W. D., Tsavaler, L., and Reichardt, L. F. (1981). *J. Cell Biol.* **91**, 257–269.

Miljanich, G. P., Brasier, A. R., and Kelly, R. B. (1982). *J. Cell Biol.* **94**, 88–96.

Moore, H.-P., Gumbiner, B., and Kelly, R. B. (1983). *J. Cell Biol.* **97**, 810–817.

Noble, M., Albrechtsen, M., Moller, C., Lykes, J., Bock, E., Goridis, C., Watanabe, M., and Rutishauser, U. (1985). *Nature (London)* **316**, 725–728.

Patzak, A., Bock, G., Fischer-Colbrie, R., Schauenstein, K., Schmidt, W., Lingg, G., and Winkler, H. (1984). *J. Cell Biol.* **98**, 1817–1825.

Phillips, J. H., Burridge, Wilson, S. P., and Kirschner, N. (1983). *J. Cell Biol.* **97**, 1906–1917.

Purves, D., and Lichtman, L. W. (1985). "Principles of Neural Development." Sinauer Assoc. Inc., Massachusetts.

Rieger, F., Grumet, M., and Edelman, G. M. (1985). *J. Cell Biol.* **101**, 285–293.

Robinson, H. C., Horner, A. A., Hook, M., Ogren, S., and Lindahl, U. (1978). *J. Biol. Chem.* **253**, 6687–6693.

Russell, D. W., Schneider, W. J., Yamamoto, T., Luskey, K. L., Brown, M. S., and Goldstein, J. L. (1984). *Cell* **37**, 577–585.

Sanes, J. R. (1982). *J. Cell Biol.* **93**, 442–451.

Sanes, J. R., and Cheney, J. M. (1982). *Nature (London)* **300**, 646–648.

Sanes, J. R., and Hall, Z. W. (1979). *J. Cell Biol.* **83**, 357–370.

Sanes, J. R., Marshall, L. M., and McMahan, U. J. (1978). *J. Cell Biol.* **78**, 176–198.

Sanes, J. R., Feldman, D. H., Cheney, J. M., and Lawrence, J. C. (1984). *J. Neurosci.* **4**, 446–473.

Schubert, D., and LaCorbiere, M. (1985). *J. Cell Biol.* **100**, 56–63.

Silberstein, L., Inestrosa, N. C., and Hall, Z. W. (1982). *Nature (London)* **295**, 143–145.

Stadler, H., and Dowe, G. H. C. (1982). *EMBO J.* **1**, 1381–1384.

Stadler, H., and Whittaker, V. P. (1978). *Brain Res.* **153**, 408.

Thumberg, L., Backstrom, G., Wasteson, A., Robinson, H. C., Ogren, S., and Lindahl, U. (1982). *J. Biol. Chem.* **257**, 10278–10283.

Vila-Porcile, E., and Oliver, L. (1980). *In* "Synthesis and Release of Adenohypophyseal Hormones" (M. Jutisz and K. W. McKerns, eds.), pp. 67–103. Plenum, New York.

Vlodavsky, I., Levi, A., Lax, I., Fuks, Z., and Schlessinger, J. (1982). *Dev. Biol.* **93**, 285–301.

Vogel, Z., Christian, C. N., Vigny, M., Bauer, H. C., Sonderegger, P., and Daniels, M. P. (1983). *J. Neurosci.* **3**, 1058–1068.

Wallace, B. J., Nitkin, R. M., Reist, N. E., Fallon, J. R., Moayeri, N. N., and McMahan, U. J. (1985). *Nature (London)* **315**, 574–577.

Wildmann, J., Dewair, M., and Matthaei, H. (1981). *J. Neuroimmunol.* **1**, 353–364.

Yamamoto, T., Davis, C. G., Brown, M. S., Schneider, W. J., Casey, M. L., Goldstein, J. L., and Russell, D. W. (1984). *Cell* **39**, 27–38.
Yamamoto, M., Boyer, A. M., and Schwarting, G. A. (1985). *Proc. Natl. Acad. Sci. U.S.A.* **82**, 3045–3049.
Young, S. H., and Poo, M. (1983). *Nature (London)* **305**, 634–635.

Blood Vessel Proteoglycans

Thomas N. Wight, Michael W. Lark*,
and Michael G. Kinsella

*Department of Pathology, School of Medicine, University of Washington, Seattle,
Washington 98195*

I. Introduction

Although proteoglycans (PG) constitute a minor component of vascular tissue (2–5% by weight) as compared to cartilage (50% by weight),
a number of studies over the years have demonstrated that these macromolecules are of enormous importance in influencing such arterial
properties as viscoelasticity, permeability, lipid metabolism, hemostasis, and thrombosis (Wight, 1980; Camejo, 1982; Berenson *et al.*, 1984).
In addition, more recent studies indicate that these macromolecules
influence, or are influenced by, many cellular events characteristic of
arterial development and disease such as cellular proliferation, migration, and adhesion (Wight, 1985). A number of methodological developments within the last few years have led to a better understanding
of the structure, location, and function of these complex protein polysaccharides in arterial wall biology. These methods, initially applied to
cartilage, have now been adapted for soft tissues such as blood vessels,
and include the use of chaotropic solvents (Sajdera and Hascall, 1969)
and protease inhibitors (Oegema *et al.*, 1975) for efficient extraction,
molecular-sieve and ion-exchange chromatography for separation and
purification (e.g., Chang *et al.*, 1983), and monoclonal antibodies and
recombinant DNA technology (see Caterson *et al.*, this volume;
Sandell, this volume) for structure and localization analyses. These developments have led to the concept that PG in tissue such as blood vessel
share a common basic structure with those present in cartilage but
differ structurally in a number of subtle ways. The significance of
these structural similarities and differences may indeed hold the key
to our complete understanding of the functional significance of these
macromolecules within blood vessels and soft connective tissue in gen-

* Present address: Department of Experimental Pathology, Merck and Co., Inc., P.O.
Box 2000, Rahway, New Jersey 07065.

267

eral. The purpose of this review will be to highlight the more recent studies (i.e., since 1980) which define the structure and distribution of arterial PG, the cells that synthesize them, and their functional role within the blood vessel wall. Major developments in the field of arterial PG and glycosaminoglycans (GAGs) prior to 1980 have been previously reviewed (Wight, 1980).

II. BIOCHEMISTRY OF ARTERIAL PROTEOGLYCANS

Following the initial studies of Eisenstein *et al.* (1975) and Oegema *et al.*, (1979), who used methods developed for extracting PG from cartilage to extract PG from blood vessels, several groups have analyzed arterial PG. Salisbury and Wagner (1981) effectively separated two populations of human aortic PG based on hydrodynamic size: a large chondroitin sulfate proteoglycan (CSPG) which is capable of forming high molecular weight aggregates with hyaluronic acid (HA), and a smaller dermatan sulfate proteoglycan (DSPG). Heparan sulfate protoeglycans (HSPG) were shown to be present but not in sufficient quantity to characterize fully.

CSPG appears to be a major PG present in blood vessels. In addition to human tissues, CSPGs have also been characterized from pig (Breton *et al.*, 1981), pigeon (Rowe and Wagner, 1985; Wagner *et al.*, 1983a), and bovine (Seethanathan *et al.*, 1980; Schmidt *et al.*, 1982a; Kapoor *et al.*, 1981; Murray, 1983; Radhakrishnamurthy *et al.*, 1986) aortas. Although structural differences in this PG class isolated from different species have been reported, most studies indicate that the large aortic CSPG consists of a core protein of M_r ~200,000 to which are attached 10–15 chondroitin sulfate chains each of M_r ~40,000. The finding that this PG class is capable of forming high molecular weight link-stabilized aggregates with hyaluronic acid indicates some similarity to the CSPG present in cartilage. However, Heinegård *et al.* (1985) recently demonstrated that the large aortic CSPG, while containing the HA-binding region, had different peptide maps and showed only partial homology when compared to the aggregating CSPG in cartilage, suggesting differences in primary structure in the core proteins of these two CSPG. A single link protein (M_r 49,000) has also been identified in aortic tissue (Vijayagopal *et al.*, 1985a), indicating that this tissue, like cartilage, possesses accessory proteins which function to stabilize the PG aggregate.

A class of PG, smaller in hydrodynamic size and containing dermatan sulfate, is also present in blood vessels. This class of PG appears to differ from the CSPG in (1) the size of the protein core, (2) the number of GAG and oligosaccharide chains, and (3) the ability to aggregate with HA. Rowe and Wagner (1985) isolated a DSPG from pigeon aorta and determined the molecular weight of the protein core to be ~50,000 after chondroitinase ABC digestion. Since there may be small oligosaccharides attached to the protein core following chondroitinase ABC digestion, the molecular weight of 50,000 may be a slight overestimate. Interestingly, this protein core is similar in size to the protein cores of DSPG isolated from a variety of other tissues such as skin (Damle *et al.*, 1982; Pearson and Gibson, 1982; Glossl *et al.*, 1984), tendon (Vogel and Heinegard, 1985), cartilage (Rosenberg *et al.*, 1985), cornea (Hassell *et al.*, 1984), and sclera (Cöster and Fransson, 1981). Heinegard *et al.* (1985) have demonstrated that antisera against a small PG from cartilage (i.e., M_r ~76,000) which contains chondroitin sulfate chains cross-reacts with a small PG from aorta which contains chondroitin sulfate–dermatan sulfate chains but does not effectively cross-react with the small DSPG in bone, cornea, sclera, and tendon. Furthermore, peptide maps of the core proteins of these PGs indicate similarities between the small aortic and cartilage PG but differences when compared to the DSPG isolated from bone, skin, sclera, and tendon. A small DSPG is believed to be located at the d band of the collagen fibril (Scott and Orford, 1981) in tendon, and some studies indicate that a DSPG isolated from tendon is capable of inhibiting type I and II collagen fibrillogenesis *in vitro* (Vogel *et al.*, 1984). Whether the small DSPG present in blood vessels exhibits a similar activity is not yet clear, but a periodic association of a small PG matrix granule has been associated with blood vessel collagen (Wight, 1980) as well as in collagen gels populated by arterial smooth muscle cells (Lark and Wight, 1986). A similar enrichment of DSPG has also been seen within collagen gels from cultures of lung (Vogel *et al.*, 1981) and skin (Gallagher *et al.*, 1983) fibroblasts and endothelial cells (Winterbourne *et al.*, 1983). Kapoor *et al.* (1981) noted the presence of at least two populations of aortic DSPGs present in the bovine aorta. One population was enriched in iduronic acid (~75% of the total uronic acid residues), while the other contained small amounts of iduronic acid but was enriched in glucuronic acid residues. Similar species of DSPG have been observed to be synthesized by arterial smooth muscle cells *in vitro* (Lark and Wight, 1986). It is still unclear whether these two DSPG species are separate gene products or were generated by posttranslational modification of a single gene family of PG. The aortic iduronic acid-rich DSPGs tend to self-aggregate (Kapoor *et al.*, 1986), similar to DSPGs isolated from other tissues (Fransson *et al.*, 1979).

Much less is known about the HSPG present in arterial tissue. Salisbury and Wagner (1981) noted the presence of heparan sulfate in tissue extracts of human aorta but were unable to isolate monomer populations containing heparan sulfate due to the small amount of HSPG present and their resistance to extraction. Vijayagopal *et al.* (1983) demonstrated that only 12% of the total HSPG from bovine aorta could be extracted with chaotropic solvents such as 4 *M* guanidine HCl, while considerably more heparan sulfate (~70% of the total) could be extracted if the tissue was first treated with collagenase and elastase. Physicochemical studies done on the extractable aortic HSPG revealed them to be of high buoyant density but considerably smaller than the CSPG present in aortic tissue. The difficulty in extracting HSPG from blood vessels using agents or conditions which minimize degradation has seriously limited the amount of new information concerning the chemistry of this group of arterial PG. Considerably more information is available on the HSPG synthesized by arterial cells *in vitro* (see Section V).

III. Morphology of Arterial Proteoglycans

Recent histochemical studies confirmed earlier reports that PGs appear to be most concentrated in the intimal regions of blood vessels and progressively decrease across the arterial wall toward the adventitia (Bartholomew and Anderson, 1983a) (Fig. 1). Several groups have begun to use antibodies directed against different PG to map their distribution within blood vessel walls. Antibodies have been generated which recognize (1) specific disaccharide stubs present on the protein core of cartilage PG after chondroitinase ABC digestion (Christner *et al.*, 1980; also see Caterson *et al.*, this volume), (2) the protein core of cartilage PG (Mangkornkanok-Mark *et al.*, 1981), or (3) the entire PG monomer from aortic tissue (Lark *et al.*, 1985; Bartholomew and Anderson, 1983b). These studies indicate that antisera prepared against cartilage CSPG core protein and GAG will cross-react with PG present in blood vessels. Using antibodies which were able to distinguish between chondroitin 4-sulfate, chondroitin 6-sulfate, and dermatan sulfate, Cleary and Baker (1987) demonstrated that chondroitin 6-sulfate was the major PG found in the porcine aorta and appeared to be concentrated in the inner third of the vessel wall. On the other hand, DSPG was localized to the outer portion of the thoracic aorta but to the inner portion of the abdominal aorta. These studies indicate that CSPG and DSPG have different distributions depending on the region of the vasculature analyzed. Using antibodies directed against a HSPG isolated from a basement membrane-producing tumor (Hassell

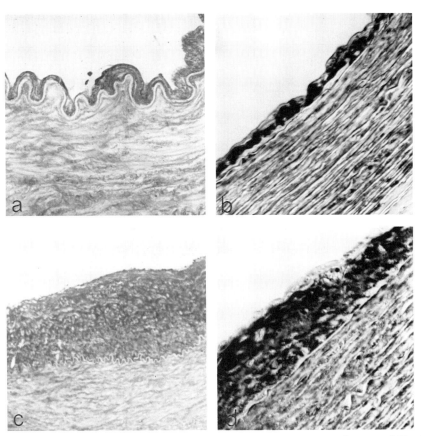

Fig. 1. (a) Light micrograph illustrating that the narrow intima of a normal blood vessel stains more intensely with alcian blue than the underlying medial layer. The two layers are separated by a "wavy" band of elastic tissue. ×240. Reproduced with permission from Wight and Ross (1975). *J. Cell Biol.* **67**, 660–674. (b) Section immunostained with a monoclonal antibody against aortic CSPG using the indirect peroxidase method (Lark *et al.*, 1985). Note that the normal narrow intima is more intensely stained than the underlying medial layer. ×265. (c) Section taken from an artery undergoing intimal hyperplasia (early atherosclerosis) and stained with alcian blue. Intense alcianophilia is seen in the thickened intima. ×250. Reproduced with permission from Wight and Ross (1975). *J. Cell Biol.* **67**, 660–675. (d) Section taken from an artery comparable to that shown in (c) and immunostained with a monoclonal antibody against aortic CSPG, as described for (b). Intense immunostaining is seen in the thickened intima.

et al., 1980), Clowes *et al.* (1984) demonstrated cross-reactivity with aortic tissue and found HSPG to be enriched in the subendothelial region of the arterial wall and capillaries. In addition, intense immunostaining for CSPG has been observed in vessels that have undergone intimal hyperplasia in response to either experimental injury or cholesterol feeding (Lark *et al.*, 1985) (Fig. 1). These studies further support the concept that the layer of the blood vessel wall involved in developing atherosclerotic lesions, the intima, is enriched in PG.

Electron-microscopic observations have also contributed to our understanding of the distribution of PG within the blood vessel wall. A limited number of studies *in vivo* indicate that some PG are located on the endothelial cell surface. Simionescu *et al.* (1981) demonstrated that a significant number of anionic sites present on capillary endothelial cells could be removed by perfusion with chondroitinase ABC and/or heparinase, indicating the probable presence of cell surface CSPG and/or HSPG. Other histochemical evidence has been presented to indicate that HSPG is present on the surface of capillary endothelium and is depleted at the tips of growing capillaries (Ausprunk *et al.*, 1981). Proteoglycans are also associated with the subendothelial basement membrane and basal laminae surrounding smooth muscle cells. Small ruthenium red-positive granules are associated with basement membranes within blood vessels (Wight, 1980; Huang *et al.*, 1984; Richardson *et al.*, 1980, 1981), and these granules have been shown to be susceptible to nitrous acid degradation, indicating the presence of heparan sulfate (Huang *et al.*, 1984) (Fig. 2). Using radiolabeled antisera generated against a core protein preparation of HSPG isolated from a basement membrane tumor, Clowes *et al.* (1984) demonstrated labeling of those sites containing basement membrane or basal laminae in the arterial wall. Similar results have been demonstrated in continuous capillaries in the mouse (Charonis *et al.*, 1983). Cheminitz and Collatz-Christensen (1983) used low-temperature embedding of arterial tissue to demonstrate the interstitial PG could be preserved as rodlike structures (20–50 nm in width and ~200 nm long), while Coltoff-Schiller and Goldfischer (1981) used toluidine blue and osmium ferrocyanide to demonstrate that PG present in the rat aorta is preserved as filamentous structures and not as granules. Similar filamentous structures identified as PG have been identified in arterial smooth muscle cell cultures using a variety of cationic dyes (Chen and Wight, 1984). It appears that the granules observed in the ruthenium red preparations result from the collapse of individual PG monomeric units which exist as extended bottlebrush structures in their native state (Hascall, 1980). Two different sizes of ruthenium

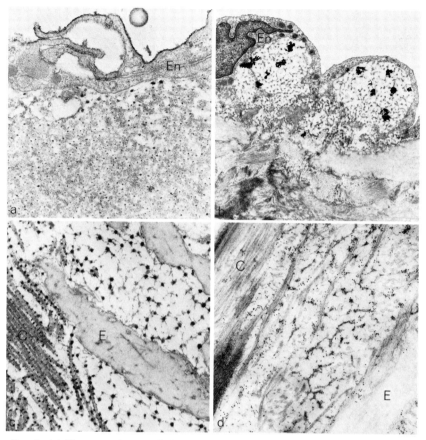

FIG. 2. (a) Electron micrograph demonstrating small round ruthenium red-positive granules within the basement membrane below endothelial cells (En). ×28,000. Reproduced with permission from Wight and Ross (1975). *J. Cell Biol.* **67**, 660–675. (b) Electron-micrographic autoradiogram prepared by incubating aortic sections with ^{125}I-labeled anti-HSPG antisera generously supplied by Dr. John Hassell (National Institutes of Health). Radiographic grains are confined only to those regions that contain basement membrane. En, Endothelium. ×11,000. Reproduced with permission from Clowes *et al.* (1984). *Histochemistry* **80**, 379–384. (c) Electron micrograph from a portion of the intercellular matrix of a blood vessel processed in the presence of ruthenium red. Numerous large granules fill the interstitial space between elastic fibers (E), while the smaller granules are located along the collagen fibrils (C). ×51,000. Reproduced with permission from Wight (1980). *Prog. Hemostasis Thromb.* **5**, 1–39. (d) Aortic section prepared by incubation with a monoclonal antibody to aortic CSPG and a secondary antibody conjugated to colloidal gold (Lark *et al.*, 1985). Immunogold labeling was present along short fibrillar aggregations within the interstitial space. Collagen (C) and elastic fiber (E). ×51,000.

red-positive granules have been identified in the arterial wall (see review, Wight, 1980). The smaller class (10–20 nm in diameter) is always present associated with basement membrane and basal laminae as well as attached to the collagen fibrils at their major period band. The larger class of ruthenium red-positive granule (20–50 nm in diameter) is seen dispersed throughout the interstitial matrix and not intimately associated with other extracellular matrix structures (Fig. 2). These size differences may reflect chemical differences in the PG preserved. For instance, as stated above, the smaller granule present associated with basement membranes have been shown to contain HSPG. Biochemical studies of the basement membrane HSPGs reveal it to be a hydrodynamically small molecule of about 130,000 to 150,000 Da (Kanwar *et al.*, 1981; Fujiwara *et al.*, 1984; Kobayashi *et al.*, 1983). On the other hand, the aortic CSPG is a hydrodynamically large molecule (M_r ~1.4 × 10^6: Oegema *et al.*, 1979; Radhakrishnamurthy *et al.*, 1986), which could be represented by the large interstitial ruthenium red-positive granules. Recent studies using a monoclonal antibody against aortic CSPG reveals that this PG is confined to regions of the arterial interstitium which contains the large ruthenium red granule (Lark *et al.*, 1985). Similarly, the small granules observed associated with collagen fibrils could represent the small DSPG known to be present in blood vessels. Studies have shown that aortic DSPG is a hydrodynamically smaller molecule than aortic CSPG (Salisbury and Wagner, 1981; Rowe and Wagner, 1985). The possibility that the size of the ruthenium red-positive granules reflects the chemical size of individual PG monomers has been emphasized by Iozzo *et al.* (1982), who have shown a direct relationship between the number and size of the ruthenium red-positive granules and the quantity and molecular size of the PG isolated from the interstitial matrix of normal and malignant colon. Similar techniques have been applied to blood vessels as well (Huang *et al.*, 1984; Richardson *et al.*, 1980, 1981). These workers have used quantitative morphological techniques to demonstrate that the number of ruthenium red-positive granules increases in the interstitial space in areas of blood vessels undergoing intimal hyperplasia. Their results correlated well with results from biochemical studies which demonstrated that these regions preferentially accumulated PG (Wight *et al.*, 1983).

IV. PROTEOGLYCANS IN ATHEROSCLEROSIS

Proteoglycans have been implicated in the pathogenesis of atherosclerosis for many years. A number of studies have demonstrated changes in the composition and content of arterial GAGs in both spontaneous and experimentally induced atherosclerosis in a variety of species (see reviews, Wight, 1980; Camejo, 1982; Berenson et al., 1984). Generally, these studies demonstrate that PG accumulate early in the intimal layer of blood vessels involved in developing atherosclerosis. The accumulation of extracellular matrix PG is thought to predispose the arterial wall to further complications such as lipid accumulation, calcification, and thrombosis, which are all events known to be affected by PG. However, little is known as to why PG specifically accumulate within this region of blood vessels. Studies using the balloon injury model of experimental atherosclerosis indicate that PG accumulate within regions of injured vessels characterized by endothelial regrowth and not within regions devoid of endothelium (Huang et al., 1984; Richardson et al., 1980, 1981; Wight et al., 1983) (Fig. 3). These studies raise the interesting possibility that during healing, the regenerating endothelium contributes to the PG composition of the arterial wall and/or in some way influences the metabolism of PG present in this region of blood vessel. Salisbury et al. (1985a) have presented evidence to indicate that reendothelialized regions of rabbit aortas incorporated approximately twice as much radioisotopic PG pre-

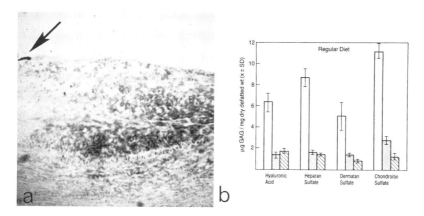

FIG. 3. (a) Section stained with alcian blue and taken at the interface between deendothelialized and reendothelialized intimas of injured rabbit aortas. The deendothelialized zone is marked at the surface with carbon black (arrow). Note the marked alcianophilia in the reendothelialized zone. ×93. Reproduced with permission from Wight (1985). Fed. Proc. Fed. Am. Soc. Exp. Biol. 44, 381–385. (b) Quantity of glycosaminoglycans expressed as micrograms per gram of dry, defatted weight of tissue in reendothelialized (unshaded bars), deendothelialized (shaded bars), and uninjured (hatched bars) rabbit aortas. Reproduced with permission from Wight et al. (1983). Am. J. Pathol. 113, 156–164.

cursors as deendothelialized regions and that the specific radioactivities of all the GAGs in the reendothelialized regions were also approximately twice that found in the deendothelialized portions at early time points following injury (2–3 weeks). These metabolic data suggest that accumulation is in part due to increased synthesis of PG by the resident cells in the reendothelialized areas. Using quantitative autoradiography, Merrilees and Scott (1985) demonstrated increased labeling of PG in reendothelialized regions as compared to deendothelialized areas, while Alavi and Moore (1985) using a modified organ culture system also demonstrated elevated GAG-synthetic activity in segments of reendothelialized vessels as compared to deendothelialized segments. These findings suggest that endothelial cells may be affecting arterial PG content by (1) increasing their own PG-synthetic activity or that by adjacent smooth muscle cells, and (2) acting as a reverse barrier preventing loss or turnover of PG.

Little is known concerning whether the structural properties of intact PG are altered in developing atherosclerotic plaques. In a pilot study, Wagner *et al.* (1983b) noted that CSPG isolated from human fatty-fibrous plaques exhibited less of a tendency to form aggregates with HA, and the aggregates that were present in the fatty-fibrous arterial plaques were smaller than those present in normal aorta. Altered aggregate formation may have some bearing on arterial calcification, a complication of advanced atherosclerotic plaques, since the degree of PG aggregate formation may influence calcification (see Poole and Rosenberg, this volume). Other minor changes in the structure of arterial PG have been noted. For example, Rowe and Wagner (1985) found that DSPG isolated from atherosclerosis-susceptible pigeons contain GAGs which were smaller than similar chains isolated from DSPGs of pigeons which were less susceptible to this disease. Whether these differences influence the functional properties of arterial PG and whether other differences exist must await further investigation.

Two conditions associated with rapid onset of atherosclerosis are diabetes (Ruderman and Haudenschild, 1984) and hypertension (Connolly *et al.*, 1983). In alloxan-diabetic dogs, Sirek and co-workers (1980, 1981) have reported altered aortic GAG content, while alterations in GAGs have been found in a number of other tissues of diabetic animals including kidney, liver, and implanted tumors. For example, Rohrbach *et al.* (1982, 1983) have shown that the EHS basement membrane tumor when transplanted into diabetic mice contains 50–80% less heparan sulfate when compared to tumor tissue transplanted into nondiabetic mice. This decrease was shown to be due to reduced synthesis of HSPG. Similar results were reported for $^{35}SO_4$ incorporation into rat glomerular basement membrane GAGs in experimental diabetes (Cohen and Surma, 1981), while Kjellén et al. (1982) demonstrated that HSPGs isolated from livers of diabetic rats had a reduced sulfate content as compared to normal rats. Brown et al. (1982) also demonstrated reduced $^{35}SO_4$ incorporation into PG in glomeruli and aorta of diabetic rats, indicating that disturbances in aortic PG metabolism occur in diabetes. In concert with alterations in HSPG content and structure within diabetic glomerular basement membrane, the permeability of the microvascular bed is increased (Joyner et al., 1981), suggesting PG involvement in the control of vascular permeability. This suggestion is further supported by the finding that removal of heparan sulfate from glomerular basement membrane markedly alters glomeruli permeability (Kanwar et al., 1980). Altered vascular permeability may be associated with increased uptake and deposition of plasma lipoproteins within arterial tissue, a condition known to exist in diabetes as well as in atherosclerosis.

Hypertension also has been shown to contribute to changes in arterial PG metabolism. Hollander et al. (1968) demonstrated in dogs with surgically produced coarctation, that the incorporation of radioactive sulfate into GAGs was significantly increased in the hypertensive portion of the aorta as compared with the normotensive portion. Crane (1962a,b) reported that in experimental hypertension, incorporation of $^{35}SO_4$ into PG was increased in the hypertensive vascular lesions and in mesenteric arteries, whereas Boueck et al. (1976) demonstrated increased PG synthesis focally at bifurcations of arteries where shear stress was elevated. It is interesting to note in this regard that arterial smooth muscle cells subjected to physical shear stress such as stretching (Leung et al., 1976) or centrifugal force (Merrilees et al., 1977) respond by increasing the synthesis of PG. In recent studies, Reynertson et al. (1986a,b) confirmed earlier work which demonstrated increased accumulation of PG in hypertensive arteries by comparing the content and metabolism of arterial PG within spontaneously hypertensive and normotensive rats. These investigations demonstrated a specific increase in aortic CSPG in the hypertensive strain. These changes could lead to alterations in the viscoelasticity of the vessel wall and changes in vascular resistance. The mechanism responsible for PG increase in hypertensive vessels needs to be determined.

Proteoglycans have been implicated in fundamental events associated with atherosclerosis such as lipid accumulation, calcification (see Poole and Rosenberg, this volume), and thrombosis (see Marcum et al., this volume). A number of studies indicate the PG can interact with lipoproteins in the presence of calcium to form insoluble complexes, and it has been suggested that these complexes are the major reason why lipoproteins accumulate in the arterial wall during formation of atherosclerotic lesions (see review, Camejo, 1982). For example, regions of blood vessels that have been experimentally injured accumulate PG in the neointima only in reendothelialized areas, while deendothelialized regions and uninjured areas have low amounts of intimal PG (Wight *et al.*, 1983). If these animals are placed on lipid-rich diets, only those regions containing high amounts of PG (i.e., reendothelialized portions) accumulate lipoprotein (Falcone *et al.*, 1980). These findings support the hypothesis that PG trap lipoproteins, causing them to accumulate. Camejo and his colleagues (1982, 1983) have been studying the interaction of arterial PG with lipoprotein and have found that arterial CSPG exhibits a marked affinity for low-density lipoproteins (LDL). These investigators found that LDL isolated from hypercholesterolemic individuals or individuals diagnosed as having acute and chronic cardiovascular disease exhibit more of a tendency to complex with arterial PG than LDL from normocholesterolemic or nondiseased patients. These lipoproteins were characterized by a high ratio of cholesterol ester to triglycerides and high apparent isoelectric points, indicating positive charge. Further studies revealed that the "high-reacting LDL," which exhibited increased affinity for PG, contained small amounts of sialic acid. Removal of sialic acid from normal LDL by neuraminidase markedly increased this activity for the complexing PG. These results indicate that at low ionic strength and physiological calcium concentration and pH, the surface charge of LDL is an important modulator of the interaction with arterial PG.

Vijayagopal *et al.* (1981) have also studied the interaction between LDL and very low-density lipoprotein (VLDL) and a PG preparation isolated from bovine aorta. The preparation, which contained both chondroitin sulfate and dermatan sulfate, forms insoluble complexes with both LDL and VLDL in the presence of calcium at physiological pH and ionic strength. Interestingly, modifying the free amino groups of LDL by acetylation almost completely abolishes insoluble-complex formation. It was also noted that in the absence of calcium, no insoluble complexes form. Removal of the protein core after β-elimination or protease treatment of the PG abolishes the insoluble complex-forming ability of the PG. These findings are interpreted to mean that destruction of the protein core releases free GAG chains which reduce the molecular charge density and interfere with insoluble-complex formation. The role of net negative charge in complex formation is emphasized by the observation that desulfation of the PG drastically reduces its ability to interact with LDL (Vijayagopol et al., 1981).

In addition to coulombic forces influencing complex formation, nonpolar mechanisms may be involved. For example, Camejo (1982) found that once aggregates formed, ionic detergents were required for their dissociation, suggesting hydrophobic associations involving the proteins that constitute the LDL and the PG molecule. Christner and Baker (1987) have shown that the core protein prepared from cartilage CSPG by chondroitinase digestion can interact with LDL and this interaction appears to be highly specific for the HA-binding region of the core protein. Thus, it appears that two mechanisms are involved in LDL–PG interaction: one that is coulombic, involving charge groups of the GAG and protein moieties of the molecule, and one that is nonpolar, involving hydrophobic associations between the protein components of the two sets of molecules.

Mourao and Bracomonte (1984) isolated PG from human aorta and obtained a preparation which contained about equal amounts of chondroitin 4- and 6-sulfate and dermatan sulfate. The CSPG and DSPG exhibited similar LDL-binding properties, as judged by their elution from immobilized LDL columns by buffers of increasing salt concentrations. Mourao et al. (1981) also demonstrated that chondroitin 6-sulfate bound avidly to plasma LDL. Bovine aorta HSPG has been isolated and its LDL-binding properties examined (Vijayagopol et al., 1983). It forms only about one-tenth as much precipitate and about one-third as much soluble complex with LDL as compared to the CS–DSPG preparation, and these complexes are much more easily disrupted by high salt concentrations. These results indicate that specific families of PG exhibit different affinities for LDL.

A number of papers refer to the isolation of PG–LDL complexes from normal and atherosclerotic arteries. Such complexes contain LDL, VLDL, and calcium, and morphological evidence indicates that these complexes are highly aggregated. Recent work indicates that multiple pools of LDL–PG complexes are present within blood vessels (Vijayagopal et al., 1983; Srinivasan et al., 1982, 1984). For example, complexes extracted with saline contained chondroitin 6-sulfate and HA as the major and minor GAG and were cholesterol ester enriched. On the other hand, LDL–PG complexes isolated after collagenase treatment contain mostly HA and minor amounts of chondroitin 6-sulfate, while complexes isolated after elastase contained only HA and were cholesterol ester poor. These findings suggest that particular

LDL–PG complexes may occur within different regions of the extracellular matrix within blood vessels and there may be multiple types of PG–LDL interactions.

Additional support for the involvement of the CSPG in LDL–PG complex formation comes from a study by Hoff and Wagner (1986), who demonstrated that following 11 weeks on a hypercholesterolemic diet, porcine aortic GAG concentrations did not differ from that of normocholesterolemic animals, even though lipoproteins were shown to accumulate within the aorta. However, the amount of aortic chondroitin sulfate increased and dermatan sulfate decreased in the hypercholesterolemic animal, indicating that cholesterol feeding may influence types of PG that accumulate but not total quantity. Intense immunostaining using an anti-aortic CSPG antiserum has been observed in atherosclerotic lesions in the rabbit following lipid feeding (Lark *et al.*, 1985). The mechanism responsible for this specific accumulation is unclear. It may be that lipoproteins influence the synthesis of PG by the resident cells of the arterial wall. LDL have been shown to increase the synthesis of PG by cultured bovine smooth muscle cells, but it is unclear whether specific types of PG synthesized are affected (Pilai *et al.*, 1980; Wosu *et al.*, 1983, 1984; Pietila, 1982).

The above studies demonstrate that PG within the extracellular matrix of blood vessels may bind lipoproteins, causing the accumulation of lipoprotein in this tissue. However, lipid and/or lipoprotein are not only found within the extracellular matrix but are also found within cells of the arterial wall during the genesis of the atherosclerotic lesion. Two cell types that are known to accumulate lipid in developing atherosclerotic lesions are smooth muscle cells and macrophages (see review, Ross, 1986). Recent studies suggest that PG may play a role in the intracellular lipid accumulation within atherosclerotic lesions as well. For example, it is known that macrophages contain very few receptors for LDL, and LDL will not accumulate within macrophages when these cells are incubated with high concentrations of LDL. However, if LDL is modified by acetylation, acetoacylation, or malondialdehyde derivation, there is a dramatic increase in the uptake of modified LDL by macrophages (Goldstein *et al.*, 1979; Traber and Kayden, 1979; Schechter *et al.*, 1981; Brown *et al.*, 1979; Mahley *et al.*, 1979; Fogelman *et al.*, 1980). Enhancement of LDL uptake also occurs when macrophages are incubated with medium containing dextran sulfate (Basu *et al.*, 1979). These findings suggest that charge alteration of LDL molecules influence their recognition by macrophages. Thus, overall charge density of lipoproteins could be affected by complexing with negatively charged PG. Falcone *et al.* (1984)

demonstrated that insoluble complexes of LDL, heparin, fibronectin, and collagen were taken up more rapidly in combination and to a greater extent by macrophages than when these cells were incubated with LDL alone. Furthermore, this study demonstrated that catabolism of the endocytosed LDL complex was greatly diminished, causing cholestrol ester to accumulate within these cells. It is interesting to note that sulfated polyanions such as dextran sulfate inhibit the fusion of phagosomes with lysosomes within macrophages by modifying membrane fluidity of the lysosome (Kielian *et al.*, 1982; Kielian and Cohn, 1982). This finding may partially explain why LDL taken up as a heterogeneous complex that contains a polysulfated GAG, is not degraded and accumulates in macrophages, leading to foam cell formation. However, decreased degradation of LDL may not be the only mechanism by which lipid accumulates in macrophages. Salisbury *et al.* (1985b) demonstrated increased cholesterol ester synthesis when macrophages were incubated with aortic PG and plasma LDL. In addition, Vijayagopal *et al.* (1985b) showed that LDL complexed to aortic PG–HA aggregate was taken up by macrophages but degraded more rapidly than LDL complexed to PG monomer. This enhanced degradation was also accompanied by increased cholesterol ester synthesis by the macrophage. It was also demonstrated in this study that acetyl-LDL completely inhibited the degradation of the LDL–PG complex, indicating that the complex may be recognized by the modified LDL receptor of the macrophage.

V. Cell Culture Studies

Cell culture techniques have allowed a closer examination of the cell types responsible for the synthesis of PG. A number of early studies focused on the types of GAGs synthesized by arterial smooth muscle and endothelial cells (see review, Wight, 1980). These studies demonstrated that vascular endothelial cells synthesize and secrete principally heparan sulfate and hyaluronic acid. On the other hand, vascular smooth muscle cells were shown to synthesize and secrete chondroitin sulfate and dermatan sulfate with less heparan sulfate and hyaluronic acid. The relative amounts of each of these isomers synthesized by these two cell types appears to depend on the donor and age of the species, portion of the vasculature used to culture the cells, and the metabolic state of the cells while they are being maintained in cell culture.

Methods used to isolate intact PG from aortic tissue have recently been applied to organ and cell culture, and these studies reveal, in

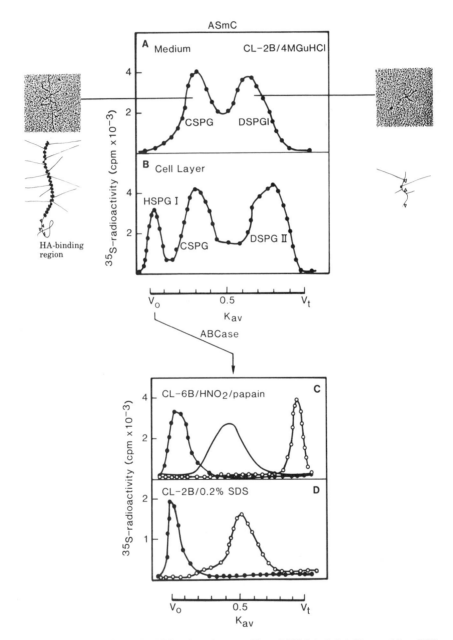

FIG. 4. Top panels (A, B) Molecular-sieve profiles of ^{35}S-labeled 4 *M* guanidine HCl extracts of medium (A) and cell layer (B) from near-confluent cultures of arterial smooth muscle cells (*Macaca nemestrina*). The major peaks that elute on Sepharose

general, that PG synthesized in an *in vitro* environment are similar to those present in intact arteries. Such studies indicate that rat, pigeon, bovine, pig, and monkey arterial smooth muscle cells synthesize and secrete principally CSPG and DSPG, the majority of which is released into the culture medium (~80%) and the remainder deposited within the pericellular matrix (Chang *et al.*, 1983; Wagner *et al.*, 1982; Vijayagopal *et al.*, 1980; Merrilees and Scott, 1982; Wight and Hascall, 1983; Schmidt *et al.*, 1982b; 1984; Arumugham and Anderson, 1984; Horn *et al.*, 1983; Diehl *et al.*, 1983) (Fig. 4). Considerably less HSPG is synthesized by these cells (~10% of the total) and appears to be enriched and concentrated in the cell layer fraction.

Biochemical and ultrastructural studies of CSPG derived from macaque arterial smooth muscle cells indicate that this PG contains about 10 to 15 GAG chains of M_r ~43,000 with a 6-sulfate to 4-sulfate ratio of 2, a large number of O-linked oligosaccharides, and few N-linked oligosaccharides (Chang *et al.*, 1983; Wight and Hascall, 1983; Wight *et al.*, 1985, 1986). This molecule appears to be similar to the CSPG present in human and bovine aorta (Salisbury and Wagner, 1981; Schmidt *et al.*, 1982b; Kapoor *et al.*, 1981; Murray, 1983; Radhakrishnamurthy *et al.*, 1986), although detailed analysis of the oligosaccharides present in the aortic CSPG has not been carried out. Furthermore, it has been demonstrated that this PG synthesized *in vitro* has the ability to interact with HA and form "link"-stabilized aggregates (Chang *et al.*, 1983). Since the arterial smooth muscle cells also synthesize HA, the above studies indicate that these cells are capable of synthesizing all of the components necessary for formation of high molecular weight aggregates. These same cells form native PG aggregate *in vitro* (Wight and Hascall, 1983). Recently, a monoclonal antibody has been generated against bovine arterial PG which recognizes a CSPG synthesized by arterial smooth muscle cells (Lark *et al.*, 1985) (Fig. 5). Chondroitinase AC-II digestion of the immunoprecipitated

---

CL-2B under dissociative conditions have been identified by a variety of techniques including susceptibility to proteases, chondroitinases, nitrous acid (bottom panel), and electron microscopy (top panel). Analysis indicated that these cells synthesize and secrete a large CSPG and a smaller DSPG. In addition, an HSPG has been identified in the cell layer which voids on CL-2B in the absence of detergent (D, ●) but is included if eluted in the presence of detergent (D, ○). This material is susceptible to nitrous acid (C, ○) and papain (C, ———), indicating that it contains heparan sulfate chains of ~40,000 MW. A model for CSPG and DSPG is presented in (a): light lines, GAG chains; ●, O-linked oligosaccharides; △, N-linked oligosaccharides. The long axis of the molecule represents the core protein.

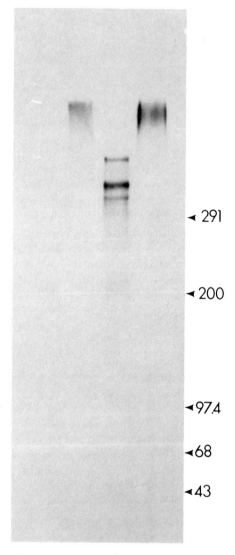

Fig. 5. SDS–PAGE (3–7½%) of immunoprecipitated CSPG from arterial smooth muscle cell cultures (*Macaca nemestrina*). Cells were radiolabeled with [³⁵S]methionine for 48 hr. The medium was collected and the proteoglycans purified by passing the media through a DEAE column. Bound PG was eluted with 8 *M* urea–1 *M* NaCl. Immunoprecipitation was performed using *Staphylococcus aureus* with M709 (Dako mouse monoclonal anti-DRC 1) as a control (lane 1), MAb 941 (lane 2) (Wight and Lark, 1985). The immunoprecipitated material was digested with chondroitinase AC-II while still associated with the *S. aureus* pellet (lane 3). Lane 4 represents CSPG pooled from a Sepharose CL-2B column (see Fig. 4).

methionine-labeled CSPG followed by SDS–PAGE reveals three radiolabeled bands between M_r 400,000 and 500,000 (T. K. Yeo and T. N. Wight, unpublished observation) (Fig. 5). Ultrastructural analysis of this PG reveals that the long axis of the molecule measures approximately 280 to 330 nm, similar to the size of the core protein of the major CSPG present in cartilage. Radhakrishnamurthy et al. (1986) estimated the large aortic CSPG core protein to be ~180,000 MW. The large size of the core preparation of CSPG from cultured arterial smooth muscle cells (i.e., 400,000–500,000 MW) is probably due to the large number of O-linked oligosaccharides which remain attached to the core following chondroitinase AC-II digestion (Chang et al., 1983). Although this form of CSPG is clearly the major form of CSPG secreted by arterial smooth muscle cells, several studies suggest that CSPG is also located on the surface of cultured arterial smooth muscle cells (Nilsson et al., 1983; Fritze et al., 1985), fibroblasts (Hedman et al., 1983), and cultured human malignant melanoma cells (Hellström et al., 1983; Bumol and Reisfeld, 1982; Wilson et al., 1981, 1983; Ross et al., 1983; Bumol et al., 1984; Garrigues et al., 1986).

A second population of smaller PG which contain dermatan sulfate are also synthesized by arterial smooth muscle cells. These cells appear to synthesize at least two classes of DSPG which differ primarily in the percentage of iduronic acid that they contain. One class appears in the culture medium and contains a small amount of iduronic acid ($\leq 20\%$) (Lark and Wight, 1986). The other class is deposited in the cell layer matrix and contains a higher percentage of iduronic acid (~50%) and is predominant in smooth muscle cell cultures maintained on collagen gels (Lark and Wight, 1986). A similar observation was made for fibroblasts (Vogel et al., 1984; Gallagher et al., 1983) and endothelial cells (Winterbourne et al., 1983). Kapoor et al. (1981, 1986) recognized two types of DSPG which vary in their iduronic acid content within intact blood vessels. It remains to be shown whether these two DSPG differ in their ability to interact with other matrix components such as collagen. Both DSPG appear to be of similar size. Electron microscopy of purified DSPG prepared from arterial smooth muscle cells and bovine aorta reveals a molecule with a long axis of ~100 nm with two or three side projections. These measurements would correspond to a core protein M_r ~50,000 if the protein was fully extended, which is close to the M_r of the core proteins of DSPG found in a number of tissues (Damle et al., 1982; Pearson and Gibson, 1982; Glossl et al., 1984; Vogel and Heinegård, 1985; Rosenberg et al., 1985; Hassell et al., 1984; Cöster and Fransson, 1981) including blood vessels (Rowe and Wagner, 1985). In vitro translation experiments with RNA from arterial smooth muscle cells and an antibody against DSPG-II from arterial cartilage indicate that these cells synthesize DSPG core precursor of M_r ~41,000 (L. Sandell and T. N. Wight, 1986, unpublished observation). In addition, antisera prepared against a small DSPG synthesized by skin fibroblasts (Glossl *et al.*, 1984) immunoprecipitates a small DSPG (M_r ~100,000) synthesized by arterial smooth muscle cells (T. K. Yeo and T. N. Wight, unpublished observation). Chondroitinase ABC treatment of this immunoprecipitated methionine-labeled DSPG revealed two protein bands on SDS–PAGE at M_r 47,000 and 45,000, similar to those results reported for skin fibroblasts.

The third class of PG synthesized by arterial smooth muscle cells contains heparan sulfate as the major GAG. This class of PG represents a minor fraction of the total PG synthesized and secreted by monkey arterial smooth muscle cells and is usually found in higher amounts in extracts of the cell layer as compared to the medium. M. W. Lark and T. N. Wight (unpublished observations) have recently characterized the HSPG families synthesized by monkey arterial smooth muscle cells and have been able to define at least three size classes of HSPG and the presence of free-chain heparan sulfate. The majority of the HSPG present in cell layer extracts elutes in the void volume of Sepharose CL-2B column in the presence of 4 M guanidine HCl but shifts to an included position of K_{av} = 0.5 upon detergent treatment (Fig. 4). The GAG chains of this PG have an M_r ~40,000 and contain a significant proportion of both N- and O-sulfate residues. This molecule has an affinity for a hydrophobic binding resin (octyl-Sepharose) and may represent the intercalated plasma membrane form of the HSPG. A similar molecule is not found in the culture medium. Recent studies by Fritze *et al.* (1985) have identified a population of HSPG on the surface of rat arterial smooth muscle cells which give rise to heparan sulfate chains by trypsin treatment. These chains are capable of inhibiting smooth muscle cell growth *in vitro* (see Marcum *et al.*, this volume). These studies indicate that this class of PG has important biological functions and may play critical roles in those events that form the basis of arterial development and disease.

The pattern of PG synthesis by arterial smooth muscle cells fluctuates with the physiological state of the cells. For example, quiescent arterial smooth muscle cells increase their synthesis of PG when stimulated to divide, and this increase occurs principally during the G_1 phase of the cell cycle (Wight *et al.*, 1985, 1986; Wight, 1985) (Fig. 6). Furthermore, the stimulation of PG synthesis in dividing cell populations results principally from an increase in the synthesis and secre-

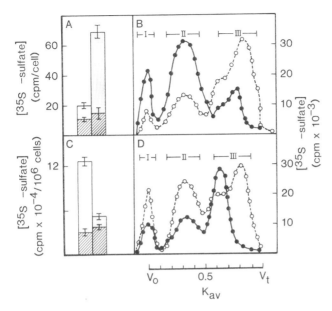

FIG. 6. Proteoglycan synthesis by arterial smooth muscle cells. (A, B) Arterial smooth muscle cells (*Macaca nemestrina*) were made quiescent by maintenance in low serum (0.1%)-containing medium. Growth was stimulated by the addition of 5% serum and the cultures were labeled with $^{35}SO_4$ for 24 hr. Panel (A) demonstrates that newly synthesized PG accumulates to a greater extent in the growth-stimulated cultures (shaded bars) than in the quiescent cultures (unshaded bars). Hatched bars represent percentage of activity present in the cell layer; open bars represents radioactivity present in the medium. (B) Elution profile on Sepharose CL-2B of $^{35}SO_4$-labeled 4 *M* guanidine HCl extracts of cell layers of growing (●) and quiescent (○) cultured smooth muscle cells. The major peak of activity in the quiescent cultures elutes at K_{av} 0.78 and contains dermatan sulfate, while the major peak in the growing cultures elutes at K_{av} 0.31 and contains chondroitin sulfate. (C, D) Arterial smooth muscle cells (*Macaca nemestrina*) were grown on collagen gels (type I) for 7 days and then labeled with $^{35}SO_4$ for 48 hr. Cultures were then extracted for proteoglycans with 4 *M* guanidine HCl. Cultures on collagen (shaded bars) accumulated significantly less radiolabeled sulfate when compared to plastic (unshaded bars). However, a much larger percentage of radiolabeled PG was present in the 4 *M* guanidine HCl extract of the cell layer from cultures maintained on collagen than cultures maintained on plastic. Molecular-sieve chromatography (Sepharose CL-2B) of these cell layer extracts reveals an enrichment of peak III (DSPG) in cultures on collagen (●- -●) as compared to cultures on plastic (○- -○).

tion of CSPG. Other studies (Hollmann *et al.*, 1986) demonstrate that several enzymes involved in the biosynthesis of chondroitin sulfate, such as xylosyl transferase, *N*-acetylgalactosaminyl transferase I, and two sulfotransferases, increase during the proliferative phase in arterial smooth muscle cell cultures. The significance of this modulation during growth stimulation is not understood but illustrates stimulated synthesis of a specific PG during an event characteristic of development and disease. It will be important to determine whether this modulation is directive, permissive, or merely coincident with cellular proliferation.

Another factor that appears to influence the types and amounts of PG synthesized by these cells is the nature of the substratum upon which the cells are grown. When arterial smooth muscle cells are cultured on hydrated collagen gels (type I collagen), they decrease their overall PG accumulation (expressed on a per cell basis) but increase the amount of iduronic acid-rich dermatan sulfate present in the extracellular matrix (Lark and Wight, 1986) (Fig. 6). Preliminary pulse–chase studies indicate that the accumulation of this specific proteoglycan within the cell layer is due partly to decreased turnover and partly to increased synthesis. These results illustrate that the surrounding extracellular environment markedly influences the synthetic phenotype of arterial cells. The consequence of a highly localized concentration of DSPG within the extracellular matrix of blood vessels may have some impact on lipid accumulation, since dermatan sulfate has been shown to bind avidly to LDL (Iverius, 1972), the major lipoprotein that accumulates during atherosclerosis. Other "extracellular" factors shown to influence PG synthesis by cultured arterial smooth muscle cells include oxygen content (Pietila and Jaakkola, 1984), heparin (Harris *et al.*, 1981), prostaglandins (Pietila *et al.*, 1980), lipids (Pillai *et al.*, 1980; Wosu *et al.*, 1983, 1984; Peitila, 1982), and ascorbic acid (Scott-Burden *et al.*, 1983).

Endothelial cells also synthesize and secrete PG. Buonassissi (1973, 1983) was one of the first to recognize that vascular endothelial cells have heparan sulfate associated with their cell surface. Gamse *et al.* (1978) later presented evidence that heparan sulfates isolated from endothelial cells were more heavily N-sulfated than heparan sulfates from arterial smooth muscle cells and were therefore considered to be more heparin-like. Since these initial observations, a number of investigators have characterized the nature of the PG synthesized and secreted by these cells *in vitro* (Kramer *et al.*, 1982; Oohira *et al.*, 1983; Kinsella and Wight, 1986a,b; Buonassissi and Coburn, 1983). Oohira *et al.* (1983) recognized the appearance of at least two species of HSPG synthesized by bovine aortic endothelial cells based on sedimentation in a cesium chloride density gradient. A hydrodynamically large, low buoyant density HSPG ($K_{av} = 0.2$ on Sepharose CL-4B) resembles the large HSPG described by Kramer *et al.* (1982) from a cloned line of bovine aorta endothelial cells, as well as the HSPG synthesized by basement membrane-producing cells (PYS-2) (Oohira *et al.*, 1982). Oohira *et al.* (1983) described a smaller HSPG (K_{av} = 0.45 on Sepharose CL-4B) of high buoyant density, and this class of PG resembled the major PG produced by kidney glomerular cells (Kanwar *et al.*, 1981). Kinsella and Wight (1986) have extended studies of the PG synthesized by cultured aortic endothelial cells and have found at least three distinct size classes of HSPG and three size classes of DSPG (Fig. 7). The two larger species of HSPG appear to be identical to the large low buoyant density HSPG described by Oohira *et al.* (1983), and were found to require detergent for their enhanced recovery (Kinsella and Wight, 1985, 1986b). In addition, the large HSPG was capable of forming disulfide-bonded aggregates and bore GAG chains with M_r ~38,000. After heparatinase treatment, the M_r of the protein core preparations from the large HSPG was estimated to be ~220,000 (Kinsella and Wight, 1986b). In contrast, the smaller (K_{av} = 0.45 on Sepharose CL-4B) HSPG present in endothelial cell cultures contained

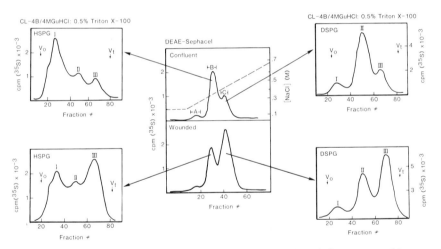

FIG. 7. The central panels represent typical DEAE–Sephacel chromatographic profiles of $^{35}SO_4$-labeled material from the cell layers of confluent (above) or wounded (below) aortic endothelial cell cultures. Peak B, known to contain HSPG, was pooled and rechromatographed on Sepharose CL-4B (left panels). Profiles of similarly pooled and rechromatographed peak C samples, which contain DSPG, are represented in the right panels. Molecular-sieve profiles indicated the presence of three (I, II, and III) subclasses of HSPG and three subclasses of DSPG. The subclasses of apparent small hydrodynamic size were found to accumulate more rapidly in wounded cultures than in confluent cultures. Reproduced with permission from Wight *et al.*, Ciba Symposium #124.

smaller GAG chains (M_r ~22,000), and a smaller core protein (~50,000) similar to the high-density HSPG described by Oohira et al. (1983). In addition, all subclasses of HSPG carried sizable proportions of nonsulfated oligosaccharide in addition to the GAG chains. Winterborne et al. (1983) examined the degree and location of the sulfate residues present in heparan sulfate GAG synthesized and secreted by endothelial cells. Their study demonstrated that the average degree of sulfation was low (~0.6 sulfate groups per disaccharide), similar to heparan sulfate isolated from whole aorta (Höök et al., 1974). However, it was shown that the sulfate groups were not evenly distributed along the heparan sulfate chain but arranged in groups or blocks. Such "heparin-like" regions could be expected to be excised by the platelet heparatinase, which has a specificity for unsulfated regions of the chain (Oosta et al., 1982). These observations may have biological relevance, since recent observations by Castellot et al. (1982) indicate that a platelet heparatinase may be required to release a "heparin-like" inhibitor of smooth muscle cell growth that is synthesized by endothelial cells (see Marcum et al., this volume).

The finding of multiple classes of HSPG and DSPG in aortic endothelial cultures has raised questions concerning whether these PG subclasses arise by synthesis of a number of similar monomers, or whether such apparent diversity is the result of the accumulation of intermediates in the processing and degradation of these macromolecules. Hassell et al. (1985) have also recognized two forms of EHS sarcoma cell-derived HSPG which, like endothelial cell HSPG, were separated on the basis of density gradient centrifugation and hydrodynamic size. Recent studies from this laboratory (Ledbetter et al., 1985) suggest that the larger HSPG is a precursor for the smaller class of HSPG in this tissue. Preliminary evidence from our laboratory (Kinsella and Wight, 1985), based largely on pulse–chase studies, suggests that a similar precursor–product relationship may hold among the HSPG subclasses present in endothelial cell cultures. Therefore, the array of HSPG in these cultures may differ structurally as the result of combined action of proteases and endoglycosidases during metabolic processing. It remains to be determined which of these enzymatic activities are exogenous and which are cell associated.

Proteoglycan metabolism by endothelial cells is also influenced by events related to arterial development and disease. For example, endothelial migration is a fundamental event during new blood vessel growth and also occurs in response to arterial injury as an early event in the development of the atherosclerotic plaque. Kinsella and Wight (1986a) found that confluent aortic endothelial cells, subjected to multiscratch wounding to induce migration, increased their synthesis of PG by ~4-fold during their peak of migratory activity (Fig. 8). Analysis of PGs synthesized by migrating endothelial cells indicated a shift toward the synthesis and deposition of CSPG rather than HSPG which was the major PG class synthesized by nonmigrating cells. Autoradiographic analysis of these cultures revealed that migrating cells exhibited increased radiolabeling with $^{35}SO_4$, and inhibition of DNA synthesis did not interfere with the elevated PG-synthetic profile or migration of the endothelial cells. The significance of this modulation in regulating endothelial migration as well as influencing the PG content of the arterial wall awaits further study.

Recent studies suggest that soluble factors such as interleukins released from blood cells also stimulate PG synthesis by endothelial cells. For example, Montesano et al. (1984, 1985) found that human umbilical vein endothelial cells incubated with growth medium conditioned by activated human peripheral blood mononuclear leukocytes elaborate an extracellular matrix highly enriched in PG. Induction of this matrix deposition was accompanied by marked cell shape change (sprouting) and reorganization of cytoskeletal elements (Montesano et al., 1985). Oohira et al. (1983) also noted changes in the HSPG component in "sprouting" endothelial cell cultures as compared to normal

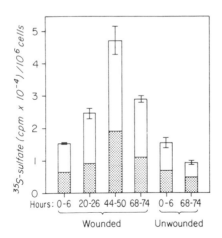

FIG. 8. Time course of $^{35}SO_4$ incorporation (6-hr pulses) in multiscratch-wounded monolayer cultures of bovine aortic endothelial cells at specified times following wounding. Peak incorporation is seen at 44–50 hr, the time period of maximal endothelial migration. All values represent the mean of triplicate cultures; bar, ±SD. Shaded areas represent proportion of total radioactivity present in the cell layer. Reproduced with permission from Kinsella and Wight, J. Cell Biol. 102, 679–689, 1986.

confluent endothelial cultures. These studies indicate that PG may be involved in maintaining a confluent endothelial morphology which could contribute to the nonthrombogenic nature of the endothelial cell surface. These studies have relevance to arterial wall biology, since endothelium participates in a wide range of normal and pathological processes including inflammation and lymphocyte trafficking (Jaffe, 1984). The presence of mononuclear leukocytes in lesions of atherosclerosis suggest that these cells may, in part, regulate the PG composition of the arterial wall. The physical and chemical nature of the PG changes induced by interleukins in endothelial cells has yet to be defined.

The HSPG synthesized by cultured endothelial cells probably have several functions which depend on their specific location. Ultrastructural histochemistry has suggested that heparan sulfate resides on the luminal surface of capillary endothelial cells (Simonescu *et al.*, 1981) and is also associated with the basement membrane beneath endothelial cells (Clowes *et al.*, 1984; Huang *et al.*, 1984; Charonis *et al.*, 1983; Kanwar and Farquhar, 1979). Studies indicate that luminal surface HSPG are involved in maintaining blood fluidity at the tissue interface by binding both coagulant and anticoagulant factors (see Marcum *et al.*, this volume). In addition, HSPG located at the luminal surface of endothelial cells affects lipoprotein metabolism. For example, it has been postulated for some time that lipoprotein lipase is bound to the surface of endothelium via interaction with a heparinlike molecule, since administration of heparin releases this enzyme into the circulation (Robinson and Jennings, 1965; Ho *et al.*, 1967). Two groups have used cultured bovine endothelial cells to demonstrate that exogenous lipoprotein lipase can no longer bind to endothelial cells that have been treated with enzymes specific for the degradation of heparin–heparan sulfate (Cheng *et al.*, 1981; Shimada *et al.*, 1981). In addition, Williams *et al.* (1983) have shown that lipoprotein lipase binds to intact endothelium *in vivo* in a specific, saturable manner which is reversible by addition of exogenous heparin. Klinger *et al.* (1985) subsequently used a lipoprotein lipase–Sepharose affinity column to isolate a HSPG from rat brain and found it to be a membrane-associated PG of $M_r = 220,000$ with GAG chains of $M_r = 14,000-15,000$. It has been previously reported (Bengtsson *et al.*, 1980) that chemically modified heparin preparations differing in the proportions of N-sulfate and N-acetyl groups display differing affinities for lipoprotein lipase, which suggests that relatively subtle variations in heparan sulfate structure may affect enzyme binding. It remains to be shown whether a specific class of membrane-bound HSPG present on endothelial cells is responsible for binding of lipoprotein lipase. The role of endothelial cell-bound lipoprotein lipase in the atherogenic process has been reviewed by Zilversmit (1973). It was suggested that triglyceride-rich and cholesterol-containing pre-β-lipoproteins and chylomicrons of the plasma could be adsorbed at arterial foci in proportion to local concentrations of sulfated polysaccharides (i.e., HSPG) at the endothelial cell surface. While adsorbed, the large complexes would be subjected to lipolysis, giving rise to cholesterol-rich β-lipoprotein or chylomicron remnants. Because the cross-linking effect of heparin and calcium at physiological salt concentration is weaker for lipoprotein than it is for chylomicrons or pre-β, much of the β-lipoproteins may be released into the bloodstream; however, the cholesterol-rich degradation products may stay bound to the intimal surface long enough to be incorporated into the blood vessel wall. Thus, high concentrations of plasma β-lipoproteins in some patients might be the consequence of an atherogenic lipolytic process rather than the cause of atherosclerosis. Although this scheme remains hypothetical, it illustrates, once more, the potential importance of sulfated PG in affecting lipid concentration within the vessel wall. The ability of endothelial cells to transport lipid-rich particles has been documented by Kenagy et al. (1984). It should be mentioned in this regard that LDL can be modified by endothelial cells so that it becomes recognized by the modified LDL receptor of the macrophage (Henriksen et al., 1983). It would be of interest to know what role PG play in this modification.

Finally, HSPG present on the abluminal surface of endothelial cells may be involved in the attachment of this cell type to its substratum (i.e., basement membrane). A number of studies have shown HSPG to be involved in the attachment of a variety of cells to specific substrata (Laterra et al., 1983; see Rapraeger and Bernfield, this volume). The fact that HSPG isolated from various cell types is capable of binding to hydrophobic gels and inserting into liposomes (Kjellén et al., 1981; Rapraeger and Bernfield, 1983) is consistent with the idea that HSPG is an intercalated membrane component. Colocalization of HSPG with cytoskeletal elements such as actin (Woods et al., 1985) suggests that this class of PG may regulate cell adhesion through association with other extracellular matrix components as well as with the cytoskeleton. Recent studies by Gill et al. (1986) have shown that treatment of attached cultured bovine aortic endothelial cells with *Flavobacterium heparinum* heparatinase, which remove all cell surface HSPG, did not affect cell attachment, but treatment of endothelial cells in suspension with the enzyme prevented reattachment. Thus, these data suggest that HSPG is involved in the process of cell attachment but may not be necessary for maintenance of attachment once it has taken place. Other experiments by Gordon *et al.* (1984) show that late-passage human umbilical vein endothelial cells (passage 27) detach more readily when subjected to brief protease treatment than early-passage cells (passage 4). Analysis of the attachment sites revealed considerably less PG present in the late-passage attachment sites as compared to early-passage attachment sites. Such differences may influence the relative degree of adhesion of endothelial cells to their substrata.

VI. CONCLUDING REMARKS

This chapter has described our current knowledge of the structure, location, and function of arterial proteoglycans (PG). This field has grown enormously since 1980, and it is becoming clear that PG within the blood vessel wall contribute to many of the key events or processes that form the basis for development and disease in this tissue. With the current techniques at hand, it will be important to continue to explore the intricate details of the structure of arterial PG, as a means for providing further clues to their function. For example, we will certainly discover subspecies or subpopulations of the major arterial PG that have specific functions. We will need to know how these subspecies arise. Are there genetic or epigenetic influences involved? We will need to probe the mechanisms by which PG interact with other macromolecules within the vessel wall and how this interaction influences the biological activity of the molecules involved. Can PG be classified as "regulatory molecules," acting as receptors, inhibitors, or stimulators of events such as cell growth and migration (chemotaxis?) and/or modifiers of enzymatic activity? From a number of studies, it appears that PG accumulation within the arterial intima predisposes the artery to the major complications of atherosclerosis such as lipid accumulation, calcification, and thrombosis. We need to know more about the role of PG in these events and why PG accumulate. Studies are needed to define those steps involved in the biosynthesis, secretion, and turnover of PG by arterial cells. The future of arterial PG research is indeed challenging, and it represents a critical component in the effort to understand those factors that contribute to cardiovascular disease—the chief cause of death in the United States and Europe.

ACKNOWLEDGMENTS

The authors wish to thank Mrs. Carol Hansen for the typing of this manuscript, Ms. Anita Coen for editing, and Ms. Melinda Riser for photography. The authors also express sincere appreciation to Lakshmi Subbaiah, Stephanie Lara, Sue Perigo, Kathi

Braun, Henderson Mar, and Stephen MacFarlane for their technical expertise in the original research that was reported in this chapter. This research was supported by the National Institutes of Health (Grant HL18645), an American Heart Association Grant-in-Aid, and a grant from R. J. Reynolds Inc. TNW is an Established Investigator of the American Heart Association and is supported in part by the Washington State Heart Association.

REFERENCES

Alavi, M., and Moore, S. (1985). *Exp. Mol. Pathol.* **42**, 389–400.
Arumugham, R., and Anderson, J. C. (1984). *Biochim. Biophys. Acta* **797**, 128–135.
Ausprunk, D., Boudreau, C., and Nelson, D. (1981). *Am. J. Pathol.* **105**, 353–366.
Bartholomew, J. S., and Anderson, J. C. (1983a). *Histochem. J.* **15**, 941–951.
Bartholomew, J. S., and Anderson, J. C. (1983b). *Histochem. J.* **15**, 1177–1190.
Basu, S. K., Brown, M. S., Ho, Y. K., and Goldstein, J. L. (1979). *J. Biol. Chem.* **254**, 7141–7146.
Bengtsson, G., Olivecrona, T., Höök, M., Riesenfeld, J., and Lindahl, U. (1980). *Biochem. J.* **189**, 625–633.
Berenson, G. S., Radhakrishnamurthy, B., Srinivasan, S. R., Vijayagopal, P., Dalferes, E. R., and Sharma, C. (1984). *Exp. Mol. Pathol.* **41**, 267–287.
Boucek, R. J., Nobel, N. L., and Wells, D. E. (1976). *Circ. Res.* **39**, 828–832.
Breton, M., Picard, J., and Berrou, E. (1981). *Biochemie* **63**, 515–525.
Brown, M. S., Goldstein, J. L., Kreiger, M., Ho, Y. K., and Anderson, R. G. W. (1979). *J. Cell Biol.* **82**, 598–613.
Brown, D. M., Klein, D. J., Michael, A. F., and Oegema, T. R. (1982). *Diabetes* **31**, 418–425.
Bumol, T. F., and Reisfeld, R. A. (1982). *Proc. Natl. Acad. Sci. U.S.A.* **79**, 1245–1249.
Bumol, T. F., Walker, L. E., and Reisfeld, R. A. (1984). *J. Biol. Chem.* **259**, 12733–12741.
Buonassissi, V. (1973). *Exp. Cell Res.* **76**, 363–368.
Buonassissi, V., and Coburn, P. (1983). *Biochim. Biophys. Acta* **760**, 1–12.
Camejo, G. (1982). *Adv. Lipid Res.* **19**, 1–53.
Camejo, G., Ponce, E., Lopez, F., Starosta, R., Hurt, E., and Romano, M. (1983). *Atherosclerosis* **49**, 241–254.
Camejo, G., Lopez, A., Lopez, F., and Quinones, J. (1985). *Atherosclerosis* **55**, 93–105.
Castellot, J. J., Jr, Faureau, L. V., Karnovsky, M. J., and Rosenberg, R. D. (1982). *J. Biol. Chem.* **257**, 11256–11260.
Chang, Y., Yanagishita, M., Hascall, V. C., and Wight, T. N. (1983). *J. Biol. Chem.* **258**, 5679–5688.
Charonis, A. S., Tsilibary, P. C., Kramer, R. H., and Wessig, S. L. (1983). *Microvasc. Res.* **26**, 108–115.
Chemnitz, J., and Collatz-Christensen, B. C. (1983). *Acta Pathol. Microbiol. Immunol. Scand. Sect. A* **91**, 477–482.
Chen, K., and Wight, T. N. (1984). *J. Histochem. Cytochem.* **32**, 347–357.
Cheng, C. F., Oosta, G. M., Benoadoun, A., and Rosenberg, R. D. (1981). *J. Biol. Chem.* **256**, 12893–12898.
Christner, J., and Baker, J. F. (1986). Submitted for publication.
Christner, J. E., Caterson, B., and Baker, J. R. (1980). *J. Biol. Chem.* **255**, 7102–7105.
Cleary, E. G., and Baker, J. R. (1985). *Arteriosclerosis*, in Press.
Clowes, A. W., Clowes, M. M., Gown, A. M., and Wight, T. N. (1984). *Histochemistry* **80**, 379–384.

Cohen, M. P., and Surma, M. L. (1981). *J. Lab. Clin. Med.* **98**, 715–722.

Coltoff-Schiller, B., and Goldfisher, S. (1981). *Am. J. Pathol.* **105**, 232–240.

Connolly, E. C., Elevack, L. R., and Auxman, A. (1983). *Mayo Clin. Proc.* **58**, 249–254.

Cöster, L., and Fransson, L. A. (1981). *Biochem. J.* **193**, 143–153.

Crane, W. A. (1962a). *J. Pathol. Bacteriol.* **83**, 183–193.

Crane, W. A. (1962b). *J. Pathol. Bacteriol.* **84**, 113–122.

Damle, S. P., Cöster, L., and Gregory, J. D. (1982). *J. Biol. Chem.* **257**, 5523–552.

Diehl, T., Scott-Burden, T., and Gevers, W. (1983). *Biochem. Int.* **6**, 29–41.

Eisenstein, R., Larsson, S. E., Kuettner, K. E., Sorgente, N., and Hascall, V. C. (1975). *Atherosclerosis* **22**, 1–17.

Falcone, D., Hajjar, D. P., and Minick, C. R. (1980). *Am. J. Pathol.* **99**, 81–104.

Falcone, D. J., Mateo, N., Shio, H., Minick, C. R., and Fowler, S. D. (1984). *J. Cell Biol.* **99**, 1266–1274.

Fogelman, A. M., Schechter, I., Seager, J., Hokom, M., Child, J. S., and Edwards, P. A. (1980). *Proc. Natl. Acad. Sci. U.S.A.* **77**, 2214–2218.

Fransson, L. A., Nieduszynski, I. A., Phelps, C. F., and Sheehan, J. K. (1979). *Biochim. Biophys. Acta* **586**, 179–188.

Fritze, L. M., Reilly, C. F., and Rosenberg, R. D. (1985). *J. Cell Biol.* **100**, 1041–1049.

Fujiwara, S., Wiedemann, H., Timpl, R., Lustig, A., and Engel, J. (1984). *Eur. J. Biochem.* **143**, 145–157.

Gallagher, J. T., Gasiunas, N., and Schor, S. L. (1983). *Biochem. J.* **215**, 107–116.

Gamse, G., Fromme, H. G., and Kresse, H. (1978). *Biochem. Biophys. Acta* **544**, 514–528.

Garrigues, H. J., Lark, M. W., Lara, S. L., Hellström, I., Hellström, K. E., and Wight, T. N. (1986). *J. Cell Biol.* **103**, 1699–1710.

Gill, P. J., Silbert, C. K., and Silbert, J. E. (1986). *Biochemistry* **25**, 405–410.

Glossl, J., Beck, M., and Kresse, H. (1984). *J. Biol. Chem.* **259**, 14144–14150.

Goldstein, J. L., Ho, Y., Basu, S. K., and Brown, M. S. (1979). *Proc. Natl. Acad. Sci. U.S.A.* **76**, 333–337.

Gordon, P. B., Levitt, M. A., Jenkins, C. S. P., and Hatcher, V. B. (1984). *J. Cell. Physiol.* **121**, 467–475.

Harris, S. A., Gajdusek, C., Schwartz, S. M., and Wight, T. N. (1981). *Fed. Proc. Fed. Am. Soc. Exp. Biol.* **40**, 623(A).

Hascall, G. (1980). *J. Ultrastruct. Res.* **70**, 369–380.

Hassell, J. R., Robey, P. G., Barrach, H. J., Wilczek, J., Rennard, S. I., and Martin, G. R. (1980). *Proc. Natl. Acad. Sci. U.S.A.* **77**, 4494–4498.

Hassell, J. R., Hascall, V. C., Ledbetter, S., Caterson, B., Thonar, E. J. M., Nakazawa, K., and Krachmer, J. (1984). *In Proc. 8th Symp. Ocular Visual Dev.* (S. Hilfer and J. Sheffield, ed.) pp. 101–114, Springer-Verlag, Berlin and New York.

Hassell, J. R., Leyshon, W. C., Ledbetter, S. R., Tyree, B. G., Kimata, K., and Kleinman, H. K. (1985). *J. Biol. Chem.* **260**, 8098–8105.

Hedman, K., Christner, J., Julkunen, I., and Vaheri, A. (1983). *J. Cell Biol.* **97**, 1285–1293.

Heinegård, D. K., Björne-Persson, A., Cöster, L., Franzén, A., Gardell, S., Malmström, A., Paulsson, M., Sandfalk, R., and Vogel, K. (1985). *Biochem. J.* **230**, 181–194.

Hellström, I. H., Garrigues, H. J., Cabasco, L., Mosely, G. H., Brown, J. P., and Hellström, K. E. (1983). *J. Immuno.* **130**, 1467–1472.

Henriksen, T., Mahoney, E. M., and Steinberg, D. (1983). *Arteriosclerosis* **3**, 149–159.

Ho, S. J., Ho, R. J., and Meng, C. H. (1967). *Am. J. Physiol.* **212**, 284–290.

Hoff, J. F., and Wagner, W. D. (1986). *Arteriosclerosis*, in press.

Hollander, W., Kramsch, D., Farmelant, M., and Madoff, I. M. (1968). *J. Clin. Invest.* **47,** 1221–1229.

Hollmann, J., Thiel, J., Schmidt, A., and Buddecke, E. (1986). *Exp. Cell Res.* **167,** 484–494.

Höök, M., Lindahl, U., and Iverius, P. H. (1974). *Biochem. J.* **137,** 33–43.

Horn, M. C., Breton, M., Deudon, E., and Picard, J. (1983). *Biochim. Biophys. Acta* **755,** 95–105.

Huang, W., Richardson, M., Alavi, M. Z., Julian, J., and Moore, S. (1984). *Atherosclerosis* **51,** 59–74.

Iozzo, R. V., Bolender, R. P., and Wight, T. N. (1982). *Lab. Invest.* **47,** 124–138.

Iverius, P. H. (1972). *J. Biol. Chem.* **247,** 2607–2613.

Jaffe, E. A., ed. (1984). "Biology of Endothelial Cells." Nijhoff, The Hague.

Joyner, W. L., Mayhan, W. G., Johnson, R. L., and Phares, C. K. (1981). *Diabetes* **30,** 93–100.

Kanwar, Y. S., and Farquhar, M. G. (1979). *Proc. Natl. Acad. Sci. U.S.A.* **76,** 1363–1307.

Kanwar, Y. S., Linker, A., and Farquhar, M. G. (1980). *J. Cell Biol.* **86,** 688–693.

Kanwar, Y. S., Hascall, V. C., and Farquhar, M. C. (1981). *J. Cell Biol.* **90,** 527–532.

Kapoor, R., Phelps, C. F., Cöster, L., and Fransson, L. Å. (1981). *Biochem. J.* **197,** 259–268.

Kapoor, R., Phelps, C. F., and Wight, T. N. (1986). *Biochem. J.* **240,** 575–583.

Kenagy, R., Bierman, E. L., Schwartz, S., and Albers, J. J. (1984). *Arteriosclerosis* **4,** 365–371.

Kielian, M. C., and Cohn, Z. A. (1982). *J. Cell Biol.* **93,** 875–882.

Kielian, M. C., Steinman, R. M., and Cohn, Z. A. (1982). *J. Cell Biol.* **93,** 866–874.

Kinsella, M. G., and Wight, T. N. (1985). *J. Cell Biol.* **101,** 338a.

Kinsella, M. G., and Wight, T. N. (1986a). *J. Cell Biol.* **102,** 679–689.

Kinsella, M. G., and Wight, T. N. (1986b). *Biochemistry.* Submitted for publication.

Kjellén, L., Pettersson, I., and Höök, M. (1981). *Proc. Natl. Acad. Sci. U.S.A.* **78,** 5371–5375.

Kjellén, L., Bielefeld, D., and Höök, M. (1982). *Diabetes* **32,** 337–342.

Klinger, M. M., Margolis, R. U., and Margolis, R. K. (1985). *J. Biol. Chem.* **260,** 4082–4090.

Kobayashi, S., Opuri, K., Kobayashi, K., and Okayama, M. (1983). *J. Biol. Chem.* **258,** 12051–12057.

Kramer, R. H., Vogel, K. G., and Nicolson, G. L. (1982). *J. Biol. Chem.* **257,** 2678–2686.

Lark, M., Hellström, I., Hellström, K. E., and Wight, T. N. (1985). *J. Cell Biol.* **101,** 337a.

Lark, M. W., and Wight, T. N. (1986). *Arteriosclerosis* **6,** 638–650.

Laterra, J., Silbert, J. E., and Culp, L. (1983). *J. Cell Biol.* **96,** 112–123.

Ledbetter, S. R., Tyree, B., Hassell, J. R., and Horigan, F. A. (1985). *J. Biol. Chem.* **260,** 8106–8113.

Leung, D. Y. M., Glagov, S., and Mathews, M. B. (1976). *Science* **191,** 475–477.

Mahley, R. W., Innerarity, T. L., Weisgraber, K. H., and Oh, S. Y. (1979). *J. Clin. Invest.* **64,** 743–750.

Mangkornkanok-Mark, M., Eisenstein, R., and Bahu, R. M. (1981). *J. Histochem. Cytochem.* **29,** 547–552.

Martin, G. R. (1980). *Proc. Natl. Acad. Sci. U.S.A.* **77,** 4494–4498.

Merrilees, M., and Scott, L. (1982). *In Vitro* **18,** 900–910.

Merrilees, M., and Scott, L. (1985). *Lab. Invest.* **52,** 409–419.

Merrilees, M. J., Merrilees, M. A., Birnbaum, P. S., Scott, L. J., and Flint, M. H. (1977). *Atherosclerosis* **27,** 259–264.

Montesano, R., Mossaz, A., Ryser, J. E., Orci, L., and Vassalli, P. (1984). *J. Cell Biol.* **99,** 1706–1715.

Montesano, R., Orci, L., and Vassalli, P. (1985). *J. Cell. Physiol.* **122,** 424–434.

Mourao, P. A. S., and Bracamonte, C. A. (1984). *Atherosclerosis* **50,** 133–146.

Mourao, P. A. S., Pillai, S., and DiFerrante, N. (1981). *Biochim. Biophys. Acta* **674,** 178–187.

Murray, E. (1983). Dissertation thesis, Univ. of Cape Town, South Africa.

Nilsson, J., Ksiazek, T., Thyberg, J., and Wasteson, Å. (1983). *J. Cell Sci.* **64,** 107–121.

Oegema, T. R., Jr., Hascall, V. C., and Dziewiatkowski, D. D. (1975). *J. Biol. Chem.* **250,** 6151–6159.

Oegema, T. R., Jr., Hascall, V. C., and Eisenstein, R. (1979). *J. Biol. Chem.* **254,** 1312–1318.

Oohira, A., Wight, T. N., McPherson, J., and Bornstein, P. (1982). *J. Cell Biol.* **92,** 357–367.

Oohira, A., Wight, T. N., and Bornstein, P. (1983). *J. Biol. Chem.* **258,** 2014–2021.

Oosta, G. M., Faureau, L., Beeler, D. L., and Rosenberg, R. D. (1982). *J. Biol. Chem.* **257,** 11249–11255.

Pearson, C. H., and Gibson, G. J. (1982). *Biochem. J.* **201,** 27–37.

Pietila, K. (1982). *Atherosclerosis* **42,** 67–75.

Pietila, K., and Jaakkola, O. (1984). *Atherosclerosis* **50,** 183–190.

Pietila, K., Moilanen, T., and Nikkari, T. (1980). *Artery* **7,** 509–518.

Pillai, S., Donnelly, P. V., Mourao, P. A. S., DiFerrante, N., and Eskiv, S. G. (1980). *Biochem. Int.* **1,** 55–63.

Radhakrishnamurthy, B., Jeansonne, N., and Berenson, G. S. (1986). *Biochem. Biophys. Acta,* in press.

Rapraeger, A. C., and Bernfield, M. (1983). *J. Biol. Chem.* **258,** 3632–3636.

Reynertson, R., Parmley, T., Rodén, L., and Oparil, S. (1986a). *Coll. Rel. Res.* **6,** 77–103.

Reynertson, R., and Rodén, L. (1986b). *Coll. Rel. Res.* **6,** 103–120.

Richardson, M., Ihnatowycz, I., and Moore, S. (1980). *Lab. Invest.* **43,** 509–516.

Richardson, M., Gerrity, R. G., Alavi, M. Z., and Moore, S. (1981). *Arteriosclerosis* **7,** 369–379.

Robinson, D. S., and Jennings, M. A. (1965). *J. Lipid Res.* **6,** 222–227.

Rohrbach, D. H., Hassell, J. R., Leinman, H. K., and Martin, G. R. (1982). *Diabetes* **31,** 185–188.

Rohrbach, D. H., Wagner, C. W., Star, V. L., Martin, G. S., Brown, K. S., and Won-Yoon, J. (1983). *J. Biol. Chem.* **258,** 11672–11677.

Rosenberg, L. C., Choi, H. U., Tang, L. H., Johnson, T. L., Pal, S., Webber, C., Reiner, A., and Poole, A. R. (1985). *J. Biol. Chem.* **260,** 6304–6313.

Ross, A. H., Cossu, G., Herlyn, M., Bell, J. R., Steplewski, Z., and Koprowski, H. (1983). *Arch. Biochem. Biophys.* **225,** 370–383.

Ross, R. (1986). *New Engl. J. Med.* **314,** 468–500.

Rowe, H. A., and Wagner, W. D. (1985). *Arteriosclerosis* **5,** 101–109.

Ruderman, N. B., and Haudenschild, C. (1984). *Prog. Cardiovasc. Dis.* **26,** 373–412.

Sajdera, S. W., and Hascall, V. C. (1969). *J. Biol. Chem.* **244,** 77–87.

Salisbury, B. G. J., and Wagner, W. D. (1981). *J. Biol. Chem.* **256,** 8050–8057.

Salisbury, B. G. J., Hajjar, D. P., and Minick, C. R. (1985a). *Exp. Mol. Pathol.* **42,** 306–319.

Salisbury, B. G. J., Falcone, D. J., and Minick, C. R. (1985b). *Am. J. Pathol.* **120,** 6–11.

Sandell, L. J. (1987). In "Biology of the Extracellular Matrix: Biology of Proteoglycans" (T. Wight and R. Mecham, eds.), pp. 27–57. Academic Press, Orlando, Florida.

Schechter, I., Fogelman, A. M., Haberland, M. E., Seager, J., Hokom, M., and Edwards, P. A. (1981). *J. Lipid Res.* **22**, 63–71.

Schmidt, A., Prager, M., Selmke, P., and Buddecke, E. (1982a). *Eur. J. Biochem.* **125**, 95–101.

Schmidt, A., Grünwald, J., and Buddecke, E. (1982b). *Atherosclerosis* **45**, 299–310.

Schmidt, A., VonTeutul, A., and Buddecke, E. (1984). *Hoppe-Seylers Z. Physiol. Chem.* **365**, 445–456.

Scott, J. E., and Orford, C. R. (1981). *Biochem. J.* **197**, 213–216.

Scott-Burden, T., Murray, E., Diehl, T., and Geven, W. (1983). *Hoppe-Seylers Z. Physiol. Chem.* **364**, 61–70.

Seethanathan, P., Taylor, P., and Ehrlich, K. (1980). *Experientia* **36**, 279–280.

Shimada, K., Gill, P. J., Silbert, J. E., and Douglas, W. H. J. (1981). *J. Clin. Invest.* **68**, 995–1002.

Simionescu, M., Simionescu, N., Silbert, J., and Palade, G. (1981). *J. Cell Biol.* **90**, 614–621.

Sirek, O. V., Sirek, A., and Cukerman, E. (1980). *Blood Vessels* **17**, 271–275.

Sirek, O. V., Sirek, A., and Cukerman, E. (1981). *Diabetologia* **21**, 154–159.

Srinivasan, S. R., Yost, C., Bhandaru, R. R., Radhakrishnamurthy, B., and Berenson, G. S. (1982). *Atherosclerosis* **43**, 289–301.

Srinivasan, S. R., Vijayagopal, P., Dalferes, E. R., Jr., Abbate, B., Radhakrishnamurthy, B., and Berenson, G. S. (1984). *Biochim. Biophys. Acta* **793**, 157–168.

Traber, M. G., and Kayden, H. J. (1979). *Proc. Natl. Acad. Sci. U.S.A.* **77**, 333–337.

Vijayagopal, P., Radhakrishnamurthy, B., Srinivasan, S. R., McMurtrey, J., and Berenson, G. S. (1980). *Artery* **6**, 458–470.

Vijayagopal, P., Srinivasan, S. R., Radhakrishnamurthy, B., and Berenson, G. S. (1981). *J. Biol. Chem.* **256**, 8234–8241.

Vijayagopal, P., Srinivasan, S. R., Radhakrishnamurthy, B., and Berenson, G. S. (1983). *Biochem. Biophys. Acta* **758**, 70–83.

Vijayagopal, P., Radhakrishnamurthy, B., Srinivasan, S. R., and Berenson, G. S. (1985a). *Biochem. Biophys. Acta* **839**, 110–118.

Vijayagopal, P., Srinivasan, S. R., Jones, K. M., Radhakrishnamurthy, B., and Berenson, G. S. (1985b). *Biochim. Biophys. Acta* **837**, 251–261.

Vogel, K. G., and Heinegård, D. (1985). *J. Biol. Chem.* **260**, 9298–9306.

Vogel, K. G., Sapien, R. E., and Pitcher, D. E. (1981). *J. Cell Biol.* **91**, 147a.

Vogel, K. G., Paulsson, M., and Heinegård, D. (1984). *Biochem. J.* **223**, 587–597.

Wagner, W. D., Connor, J. R., and Muldoon, E. (1982). *Biochim. Biophys. Acta* **717**, 132–143.

Wagner, W. D., Rowe, H. A., and Conner, J. R. (1983a). *J. Biol. Chem.* **258**, 11136–11142.

Wagner, W. D., Hardingham, T., and Edwards, I. (1983b). *Arteriosclerosis* **3**, 417a.

Wight, T. N. (1980). *Prog. Hemostasis Thomb.* **5**, 1–39.

Wight, T. N. (1985). *Fed. Proc. Fed. Am. Soc. Exp. Biol.* **44**, 381–385.

Wight, T. N., and Hascall, V. C. (1983). *J. Cell Biol.* **96**, 167–176.

Wight, T. N., and Ross, R. (1975). *J. Cell Biol.* **67**, 660–674.

Wight, T. N., Curwen, K. D., Litrenta, M. M., Alonso, D. R., and Minick, C. R. (1983). *J. Pathol.* **113**, 156–164.

Wight, T. N., Kinsella, M. G., and Potter-Perigo, S. (1985). In "Extracellular Matrix: Structure and Function" (H. Reddi, ed.), pp. 321–332. Liss, New York.

Wight, T. N., Kinsella, M. G., Lark, M. W., and Potter-Perigo, S. (1986). "Function of the Proteoglycans" (V. C. Haxall, ed.). Wiley, New York.

Williams, M. P., Streeter, H. B., Wusteman, F. S., and Cryer, A. (1983). *Biochim. Biophys. Acta* **756,** 83–91.

Wilson, B. S., Imai, K., Natale, P. G., and Ferrone, S. (1981). *Int. J. Cancer* **28,** 293–300.

Wilson, B. S., Ruberto, G., and Ferrone, S. (1983). *Cancer Immunol. Immunother.* **14,** 196–201.

Winterbourne, D. J., Schor, A. M., and Gallagher, J. T. (1983). *Eur. J. Biochem.* **135,** 271–277.

Woods, A., Couchman, J. R., and Höök, M. (1985). *J. Biol. Chem.* **260,** 10872–10879.

Wosu, L., Parisella, R., and Kalant, N. (1983). *Atherosclerosis* **48,** 205–220.

Wosu, L., McCormick, S., and Kalant, N. (1984). *Can. J. Biochem. Cell Biol.* **62,** 984–990.

Zilversmit, D. B. (1973). *Circ. Res.* **33,** 633–638.

Heparan Sulfate Species and Blood Vessel Wall Function

James A. Marcum
Christopher F. Reilly

Massachusetts Institute of Technology, Cambridge, Massachusetts, and the Harvard Medical School and Beth Israel Hospital, Boston, Massachusetts 02139

and

Robert D. Rosenberg

Massachusetts Institute of Technology, Cambridge, Massachusetts, and the Harvard Medical School and Beth Israel Hospital, and Dana Farber Cancer Institute, Boston, Massachusetts 02139

I. Introduction

Heparan sulfates have been postulated to play an important role in a variety of biological systems. Since the mid-1970s, there has been considerable progress in uncovering the manner by which these mucopolysaccharides and proteoglycans help to maintain normal blood vessel wall function. In this chapter we review recent biochemical and cell biologic studies that have significantly advanced our understanding of how these glycosaminoglycans act as critical control elements in suppressing thrombosis and atherosclerosis within the cardiovascular tree.

II. The Role of Heparan Sulfate Species as a Regulator of Coagulation Mechanism Activity

At the turn of the century, thrombin was observed to lose activity gradually upon addition to defibrinated plasma. On the basis of these data, it was thought that a specific inactivator of this enzyme, antithrombin, was present within the blood (1, 2). In 1916, McLean (3) isolated a substance from the liver and demonstrated its potent anticoagulant properties. Howell and Holt (4) later called this material

301

heparin. The anticoagulant effect of this substance on purified pro-coagulants was clarified by Brinkhous *et al.* (5), who showed that heparin was effective as an anticoagulant only in the presence of a plasma component that they termed heparin cofactor. Experiments conducted in the laboratories of Waugh (6) and Seegers (7) suggested that the plasma antithrombin activity and plasma heparin cofactor activity are intimately related. Abildgaard (8) isolated an α_2-globulin from human plasma that functioned in both capacities. However, Rosenberg and Damus (9) provided the first direct evidence that plasma antithrombin activity and the plasma heparin cofactor activity reside in the same molecule. The sections below describe recent advances in our understanding of the heparin (heparan sulfate)–antithrombin system and outline the probable physiologic role of this mechanism within the cardiovascular tree.

A.   Structure and Mode of Action of Antithrombin

Human antithrombin has a molecular wieght of 58,000 as well as an isoelectric point of 5.11 (9). The shape of the protease inhibitor approximates that of a prolate ellipsoid with an axial ratio of ~5 (10). The concentration of antithrombin within human plasma is ~150 μg/ml (11). The complete primary structure of this protease inhibitor has been reported by Petersen *et al.* (12). Recently, the human gene that codes for antithrombin has been identified, and the amino acid sequence derived from the cDNA for the protease inhibitor agrees with

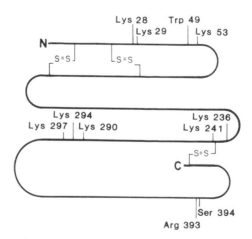

FIG. 1. Critical sites within antithrombin.

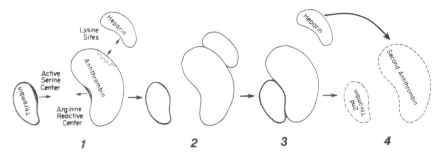

FIG. 2. The overall mechanism of action of heparin and antithrombin.

the primary structure of the molecule (13). The location of functionally important domains such as the enzyme-binding region, the potential heparin-binding sites, the conformation-sensitive tryptophan residue, and the S—S cross-links are shown in Fig. 1.

The mechanism by which thrombin is inhibited by antithrombin was uncovered by Rosenberg and Damus (9). These investigators showed that the protease inhibitor neutralizes the enzyme by forming a 1:1 stoichiometric complex between the two components via a reactive site (arginine)–active center (serine) interaction. Complex formation occurs at a relatively slow rate in the absence of heparin. However, in the presence of mucopolysaccharide, which binds to lysyl residues on antithrombin, enzyme–inhibitor interactions are accelerated dramatically (9). The heparin-induced acceleratory phenomenon appears to be due to an allosteric alteration in the position of a critical arginine residue of the protease inhibitor, so that this amino acid moiety is more readily available for interaction with thrombin (Fig. 2).

The above model of antithrombin action has been confirmed and extended by a variety of recently published structural data. Bjork *et al.* (14) have provided evidence that the arginine reactive site of antithrombin is located at Arg^{393}-Ser^{394}, near the carboxy terminal of this protein. Ferguson and Finley (15) have demonstrated that the reduction of an S—S bridge at Cys^{239}-Cys^{430} results in a dramatic decrease in the ability of antithrombin to bind heparin, but does not alter the capacity of the protease inhibitor to inactivate factor Xa or thrombin in the absence of mucopolysaccharide. In addition, the arginine reactive site lies within the S—S loop, and there appears to be a cluster of lysyl residues in this region which have been implicated as a binding site for heparin (see below).

Several groups of investigators (16, 17) have also claimed that a critical tryptophan residue is located within the heparin-binding site of antithrombin. Their conclusion is based on the fact that modification of Trp[49] prevents antithrombin from binding to heparin, produces a dramatic reduction in the ability of the mucopolysaccharide to accelerate thrombin–protease inhibitor interactions, but does not alter the capacity of the protein to inactivate the enzyme in the absence of the complex carbohydrate. Karp *et al.* (18), utilizing Trp[49]-modified antithrombin, have shown with quantitative binding techniques that this chemical alteration resulted in only a modest reduction (~10-fold) in the avidity of protease inhibitor for heparin. However, when the chemically altered protease inhibitor is saturated with the mucopolysaccharide, quantiative kinetic methods revealed a much larger decrease (~500-fold) in the heparin-dependent acceleration of thrombin–antithrombin and factor Xa–antithrombin interactions. Based on this experimental information, it appears likely that Trp[49] lies close to, but not within, a major binding region of antithrombin.

The coagulation cascade is composed of a series of linked proteolytic reactions (19). At each step, a zymogen is converted to its corresponding serine protease which is responsible for a subsequent zymogen–protease transition. The purified hemostatic enzymes of the intrinsic coagulation cascade (factors IXa, Xa, XIa, and XIIa) are neutralized by antithrombin in a manner similar to that outlined above for thrombin. Heparin is also able to accelerate dramatically each of these protease/protease inhibitor interactions (20–22). The heparin–antithrombin system described above constitutes the major pathway for neutralization of activated factors of the intrinsic coagulation cascade within plasma except possibly for factor XIIa, where C′ inhibitor may function as its natural inhibitor. In addition, factor VIIa and protein Ca are only slowly inactivated by the heparin–antithrombin system (23, 24).

Jordan *et al.* (25, 26), utilizing fluorescence spectroscopy and polarization fluorescence spectroscopy, demonstrated that heparin ($M_r \cong$ 6500) binds antithrombin with a stoichiometry of 1:1 and a dissociation constant of 5.74×10^{-8} M. These investigators also determined the binding of mucopolysaccharide to thrombin as well as factors IXa and Xa. Their results indicated that the stoichiometry of the heparin–thrombin interaction is 2:1 with two equivalent dissociation constants of 8×10^{-7} M, the stoichiometry of heparin–factor IXa interactions is 1:1 with a dissociation constant of 2.58×10^{-7} M, and the stoichiometry of the heparin–factor Xa interaction is 1:1 with a dissociation constant of 8.73×10^{-6} M.

The above investigators also determined experimentally the initial velocities of factors IXa and Xa as well as thrombin neutralization by antithrombin as a function of heparin ($M_r \simeq 6500$) concentration (25, 26). The results indicate that heparin-dependent enhancement in the rates of neutralization of these proteases by antithrombin requires binding of the mucopolysaccharide to the protease inhibitor but not necessarily to the enzyme. Comparisons of the various kinetic constants suggest that direct binding of heparin to antithrombin is responsible for a ~ 1000-fold acceleration of enzyme–inhibitor complex formation. The interaction between thrombin or factor IXa and mucopolysaccharide bound to antithrombin provides an additional 4- to 15-fold enhancement in the rate of enzyme neturalization. The binding of factor Xa to mucopolysaccharide bound to antithrombin does not augment the velocity of factor Xa inactivation. Therefore, the acceleration of antithrombin action is due mainly to the binding of heparin to the protease inhibitor, whereas the binding of heparin to the hemostatic enzymes accounts for either none (factor Xa) or 1–2% of the total rate enhancement (factor IXa and thrombin).

Recently, several groups (27, 28) have shown that heparin molecules of larger size accelerate certain of the hemostatic enzyme–protease inhibitor interactions to a greater degree than that outlined above. These results may reflect multiple binding sites of heparin for antithrombin as well as more potent approximation phenomena. In addition, it was noted throughout all of the above studies that the neutralization of hemostatic enzymes by the heparin–antithrombin complex results in the release of the mucopolysaccharide from the protease inhibitor on a 1:1 molar basis and that the binding of heparin to the hemostatic enzyme–antithrombin complex is 100–1000 times weaker than the interaction of mucopolysaccharide with free protease inhibitor (27). These data indicate that heparin catalyzes the above interactions by initiating multiple rounds of protease–protease inhibitor complex formation (Fig. 2).

B. Molecular Basis of Heparin–Antithrombin Interactions

Commercial heparin is one of the two mucopolysaccharides which are capable of binding antithrombin, activating the protease inhibitor, and accelerating the neutralization of hemostatic enzymes. Heparan sulfate is the other of these mucopolysaccharides (see below). Until recently, the nature of the critical groups on heparin which are responsible for this unique property were enigmatic. In 1976, Lam et al. (29) showed that only a small fraction of the mucopolysaccharide binds the protease inhibitor and is responsible for virtually all of the anticoagulant activity of the complex carbohydrate. These findings were rapidly confirmed by two other groups (30, 31) and allowed subsequent definition of the regions on heparin which are involved in its unique anticoagulant function.

Oosta et al. (32) randomly cleaved highly active heparin with chemical techniques and isolated mucopolysaccharide fragments of varying size which bind tightly to the protease inhibitor. Oligosaccharides of 8–16 sugar units accelerate factor Xa–antithrombin interactions but do not catalyze the rates of neutralization of other hemostatic enzymes by the protease inhibitor (domain 1). Heparin fragments of 16 residues or more accelerate thrombin–antithrombin as well as factor Xa–antithrombin interactions by directly activating antithrombin (domain 2). Mucopolysaccharide chains of 22 residues or more contain additional structural element(s) required for approximating free thrombin with protease inhibitor (domain 3). These latter segments of the mucopolysaccharide are also needed to accelerate factor IXa–antithrombin and factor XIa–antithrombin interactions (32). Thus, heparin contains several discrete structural regions which are responsible for its ability to complex with antithrombin and to modulate the biologic activities of the protease inhibitor.

Rosenberg et al. (33, 34), Lindahl et al. (35), and Choay et al. (36) have utilized highly active heparin as well as fragments of the mucopolysaccharide to further define domain 1. The region contains a critical tetrasaccharide sequence composed of IdA(GlcA)→GlcNAc-6-O-SO_3→GlcA→GlcN-SO_3-3-O-SO_3. Anticoagulantly inactive heparin does not contain this unique structure and consequently does not bind antithrombin. These various investigators (37–39) have also demonstrated that an octasaccharide which contains the unique tetrasaccharide region constitutes domain 1 of the heparin molecule. Direct studies of the interaction of this octasaccharide with the protease inhibitor revealed that it is responsible for about 8.7 to 10.2 kcal/mol of binding energy (40). The contributions of individual residues of the domain 1 have been evaluated by comparing the avidity of synthetic oligosaccharides as well as deaminative- and enzyme-cleavage fragments of the natural octasaccharide for antithrombin.

Based on the above experiments (40), the relative importance of individual monosaccharides within domain 1 have been estimated and are provided at the bottom of Fig. 3. The contribution of the nonreducing-end IdA or GlcA (residue 1) and GlcA (residue 3) to the binding energy of the octasaccharide is minimal, while the contributions of the 6-O-SO_3 of GlcNAc (residue 2) and the 3-O-SO_3 of GlcNSO₃-6-O-SO₃ (residue 4) to the binding energy of the octasaccharide are each about 4 to 5 kcal/mol. Interestingly, the 3-O-SO_3 group of residue 4 is functionally linked to the 6-O-SO$_3$ group of residue 2 such that both of these moieties are required for the binding of octasaccharide to the protease inhibitor. Indeed, the absence of either of the above sulfate moieties leads to the same 4–5 kcal/mol loss in binding energy. The apparent lack of contribution by residues 1 and 3 to the interaction of the octasaccharide with antithrombin does not indicate that these moieties are without importance. It is likely that these two nonsulfated uronic acid units function as critical spacers for the 6-O-SO$_3$ of residue 2 and the 3-O-SO$_3$ of residue 4. The contributions of the 2-O-SO$_3$ of IdA (residue 5) and the N-SO$_3$ of Glc-6-O-SO$_3$ (residue 6) to the binding of octasaccharide to antithrombin are 1.7 kcal/mol and 2.8 kcal/mol, respectively, whereas those of residues 7 and 8 are ~0.6 kcal/mol.

The mechanism by which domain 1 induces conformational changes within antithrombin as well as acceleration of factor Xa–protease inhibitor interactions has been examined. Rosenberg and Damus (9) initially demonstrated that heparin binds to antithrombin via a limited number of lysyl residues. Based on available NMR and X-ray data of the repeating units of heparin, Villanueva (41) has suggested that the three lysine residues (Lys[290], Lys[294], Lys[297]) within a α-helix of the protease inhibitor (286–297) could be matched for maximal interaction with the 6-O-SO$_3$ of residue 2, the N-SO$_3$ of residue 4, and the N-SO$_3$ of residue 6 of the octasaccharide. Atha $et\ al.$ (40, 42) have recently provided evidence that the sulfate groups of residues 2 and 6 are important for binding this region of heparin to antithrombin, but also revealed that the 3-O-SO$_3$ group of residue 4 is critical for complex formation (40). Although this latter moiety appears to be out of range of Lys[294], it is possible that the 3-O-SO$_3$ group of residue 4 might interact with a different helical segment in the N-terminal region of antithrombin and thereby induce a major alteration in the conformation of the protease inhibitor as evidenced by the spectral alterations outlined below.

The above hypothesis is consistent with results obtained by other investigators in which modification of Trp[49] near the N-terminal end of antithrombin blocks the heparin-enhanced inhibition of factor Xa (16, 17) and would also explain fluorescence transfer studies which indicate that one or two lysine residues such as Lys[28], Lys[29], and/or Lys[53] located near Trp[49] are essential for polysaccharide binding (43). If such is the case, the interactions of the 6-O-SO$_3$ of residue 2 and the 3-O-SO$_3$ of residue 4 with separate domains on antithrombin could induce a conformational change in the protease inhibitor involving the region about Trp[49] that would functionally link the binding of these

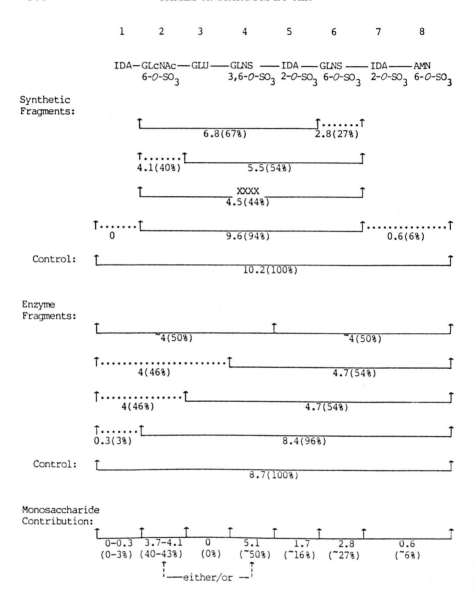

FIG. 3. Contribution of individual residues (numbers 1–8 across top) within the anti-thrombin-binding octasaccharide. IdA, Iduronic acid; GLU, glucuronic acid; GlcNAc, *N*-acetylglucosamine; GLNS, N-sulfated glucosamine; AMN, anhydromannitol. The numbers below solid brackets are the directly determined binding energies for the fragments in kilocalories per mole. The numbers below the dotted brackets have been calculated from the differences in the binding energies of the various fragments and are used to compute the relative contributions on the monosaccharides as given below. The numbers

two residues. This transition could also lead to the repositioning of the Arg[393]-Ser[394] locale or the formation of a secondary enzyme complexing site which is ultimately responsible for the acceleration of factor Xa neutralization.

Studies employing UV circular dichroism spectorscopy suggest that domain 2 of heparin interacts with a separate area of antithrombin and triggers a conformational transition of the protease inhibitor distinct from that induced by domain 1. To this end, Stone et $al.$ (44) examined the interaction of antithrombin with oligosaccharides ($\sim$8 to 18 monosaccharide units) as well as with heparin ($M_r \sim$ 6500 and 22,000) and observed two major chiral absorption spectra. The first, seen when oligosaccharides of $\sim$8 to 14 residues interacted with the protease inhibitor, showed spectral transitions typical of the domain 1–antithrombin interactions. The second, seen when oligosaccharides of $\sim$18 or more monosaccharide units bound to the protease inhibitor, are similar to the circular dichroism spectra produced by domain 1– protease inhibitor complexes, except for alterations within 292–282 nm and 275–255 nm ranges. The subtraction of the first spectra from the second revealed a shallow negative band between 300 and 275 nm, with potential negative minima at 290 and 283 nm as well as a deep negative band between 275 and 255 nm with possible negative minima at 268 and 262 nm. Conformational changes about a disulfide bridge(s) could account for this chiral absorption profile. These additional transitions are strongly correlated with the binding of domain 2 of the mucopolysaccharide to antithrombin. Indeed, the extension of the 14-monomer oligosaccharide by four residues permits this species to accelerate thrombin–antithrombin interactions as well as to induce the second type of spectral transition. These observations provide evidence that domain 2 of the heparin exerts its effect by binding to an area of antithrombin distinct from that of domain 1 and inducing an additional set of conformational changes within the protease inhibitor that

---

in parentheses are the percentages of the total binding energy of octasaccharide. Residue 2 is from tetra- and pentasaccharides. Residue 4 is N-sulfated glucosamine 6-O-sulfate in the non-3-O-sulfated synthetic pentasaccharide as indicated by XXX. Residue 5 is unsaturated in the reducing-end tetrasaccharide produced by using heparinase from *Flavobacterium heparinum*. The binding energies of the synthetic tetra- and pentasaccharide fragments were taken from Table II in this study (42). The binding energies of oligosaccharides produced from enyzmatic degradation (i.e., enzyme fragments) of the octasaccharide were taken from Table I of Atha et $al.$ (42). The relative contributions of residues 2 and 4 (either/or at bottom of figure) can be linked to the conformational change in antithrombin such that the loss of either results in essentially a complete loss of binding energy expected for the nonreducing-end tetrasaccharide.

are critical for the acceleration of thrombin–antithrombin interactions. It has been proposed that domain 2 of the heparin molecule binds to additional lysyl residues such as Lys^{236}, Lys^{241}, or other amino acid groups within the Cys^{239}–Cys^{430} loop. These interactions could induce a torsion of the S—S bridge which might either reposition the Arg^{393}–Ser^{394} locale or trigger the formation of a secondary binding site such that thrombin, factor IXa, and factor XIa can complex more quickly with the protease inhibitor.

C. Isolation of Anticoagulantly Active Heparan Sulfate from Vascular Tissue

The previous sections document the detailed structure–function relationship of the heparin–antithrombin system and suggest that these natural components could be important in the physiologic regulation of the hemostatic mechanism. Until recently, mast cells were thought to be the sole site of synthesis of anticoagulantly active heparin (45). Given that these cellular elements are located in the subendothelium, one was forced to hypothesize that biologically active mucopolysaccharides are discharged into the blood by mast cells which have been activated within the vessel wall. Unfortunately, the evidence for significant blood levels of endogenous anticoagulantly active heparin remains unconvincing.

In 1973, Damus *et al.* (21) postulated that the resistance of the endothelium to thrombogenesis could be due to heparin or heparinlike molecules present on the surface of the vascular lumen. The early studies of Teien *et al.* (46) showed that heparan sulfate (a heparinlike species with increased amounts of GlcA and GlcNAc-6-O-SO_3) is found within the aorta and exhibits trace amounts of anticoagulant activity. However, the manner by which the above component expresses this biologic activity and its precise location within vascular tissue were not determined. Marcum and Rosenberg (47) later isolated heparan sulfate from enriched endothelial cell preparations derived from calf cerebral microvessels as well as aortas, and affinity-fractionated these mucopolysaccharides into two populations by employing immobilized antithrombin. The aortic and cerebral microvascular heparan sulfates which complexed with the protease inhibitor exhibited a dramatic enhancement in the specific thrombin-inhibitory and factor Xa-inhibitory activities. These two fractions from aortas and cerebral vessels constituted 0.3 and 4.2%, respectively, of the chemical mass of the complex carbohydrate, and accounted for 60 and 75%, respectively, of the anticoagulant activity of the starting material.

Careful examination of the above vascular products revealed the existence of a small population of metachromically staining granulated cells (<1% of the total nucleated cells) which resembled mast cells. These cellular elements could account for the presence of the biologically active mucopolysaccharides within these tissues. In an attempt to define the cellular source of anticoagulantly active heparan sulfate found within vascular tissue, Marcum et al. (48) isolated retinal microvessels by modifications of previously reported techniques (49) and demonstrated that these vascular products were free of mast cells as suggested by earlier studies (50). The anticoagulant activity of these preparations were quite significant. The biologic potency of the retinal microvascular product appears to be due to a set of heparan sulfate proteoglycans. On the one hand, ~60% of the anticoagulant activity of the above material binds specifically to immobilized antithrombin. Furthermore, the anticoagulant activity was reduced by 85% when a chemically altered antithrombin was utilized that had been modified at Trp[49], and the biologic potency of these components was completely destroyed with purified *Flavobacterium* heparinase, which specifically cleaves heparin and heparan sulfate. On the other hand, anticoagulant activity could not be measured in retinal microvascular preparations when the products were preincubated with 10% trichloroacetic acid (TCA) prior to proteolytic treatment. Proteoglycans are insoluble under these conditions. These data suggest that heparan sulfate proteoglycans with anticoagulant activity are probably synthesized by vascular tissue.

D. Endothelial Cells Synthesize Anticoagulantly Active Proteoheparan Sulfate

To demonstrate unequivocally that endothelial cells synthesize anticoagulantly active heparan sulfate, Marcum et al. (51–53) isolated microvascular endothelial cells from the rat and mouse epididymal fat pads as well as macrovascular endothelial cells from bovine aortic intima. Rat and bovine cellular elements were cloned from single endothelial cells, which excluded the possibility that the results obtained in these studies were due to a minor contamination of the cultures with an additional cell type. The mucopolysaccharides were extracted enzymatically from the above cells, and anticoagulant activity was determined by quantitating the acceleration of thrombin–antithrombin complex formation. The data revealed that endothelial cells derived from microvascular tissue produced $0.82–1.79 \times 10^{-3}$ USP units per 10^6 cells, whereas endothelial cells obtained from macrovascular tissue generated $2.25-5.93 \times 10^{-3}$ USP units per 10^6 cells. The biologic activity was completely eliminated by incubating the samples with purified *Flavobacterium* heparinase.

The anticoagulant activity of the endothelial cells obtained from rat and mouse epididymal fat pads was expressed to only a minor extent when Trp^{49}-modified antithrombin was substituted for the native protease inhibitor. These results indicate that the endothelial cell-derived heparan sulfate activates antithrombin in a manner similar to commercial heparin (see above). Furthermore, virtually all of these macromolecules could be harvested by a brief exposure of cells to dilute trypsin, which suggests that these components are located on the surface of the cellular elements.

Cloned bovine aortic and rat epididymal fat pad endothelial cells were cultured with $[^{35}S]Na_2SO_4$ and proteolyzed extensively with papain. Radiolabeled sulfated glycosaminoglycans were isolated by DEAE–Sephacel chromatography. The anticoagulantly active mucopolysaccharide emerged from the ion-exchange matrix prior to chondroitin sulfate, and could be completely degraded by *Flavobacterium* heparinase. These two molecular properties are characteristic of heparan sulfate (51). This mucopolysaccharide was then affinity-fractionated into two separate populations utilizing immobilized antithrombin. The heparan sulfate derived from bovine aortic and rat fat pad endothelial cells, which bound tightly to the protease inhibitor, represented ~1 and ~10%, respectively, of the starting mass and exhibited a specific activity of 1.16 USP units/10^6 ^{35}S-cpm and 0.72 USP units/10^6 ^{35}S-cpm, respectively. However, the heparan sulfate obtained from bovine aortic and rat fat pad cellular elements, which interacted minimally with the protease inhibitor, constituted ~99 and ~90%, respectively, of the mucopolysaccharide mass and possessed a specific anticoagulant potency of <0.0002 USP units/10^6 ^{35}S-cpm and <0.002 USP units/10^6 ^{35}S-cpm, respectively. The high-affinity heparan sulfate from either species accounted for virtually all (>99%) of the starting anticoagulant activity.

The structures of affinity-fractionated heparan sulfates derived from cloned bovine aortic and rat epididymal fat pad endothelial cells were examined at the disaccharide level and compared to those of the mucopolysaccharides from the same sources with minimal affinity for the protease inhibitor. The data revealed that the affinity-fractionated heparan sulfates exhibited a significant increase in the amount of $GlcA{\rightarrow}AMN\text{-}3,6\text{-}O\text{-}SO_3$ and/or $GlcA{\rightarrow}GlcN\text{-}3\text{-}O\text{-}SO_3$ when compared to the corresponding mucopolysaccharides which possessed minimal affinity for the protease inhibitor (Table I). These disaccharides repre-

TABLE I

DISACCHARIDE COMPOSITION OF ENDOTHELIAL CELL HEPARAN SULFATE

	Percentage[b]	
Disaccharide[a]	Affinity fractionated, bovine	Depleted, bovine
GlcA—AMN 2-O-SO$_3$	5.5 ± 0.6	7.0 ± 1.5
GlcA—AMN 6-O-SO$_3$	11.8 ± 0.8	10.4 ± 1.0
IdA —AMN 6-O-SO$_3$	10.9 ± 1.5	8.7 ± 1.5
IdA —AMN 2-O-SO$_3$	56.0 ± 3.6	60.8 ± 3.1
GlcA—AMN 3-O-SO$_3$	5.5 ± 0.8	1.5 ± 0.7
IdA —AMN 2-O-SO$_3$ 6-O-SO$_3$	10.3 ± 1.3	11.6 ± 1.8
GlcA—AMN 3,6-O-(SO$_3$)$_2$	N.D.[c]	N.D.[c]

[a] Abbreviations: GlcA, glucuronic acid; IdA, iduronic acid; AMN, anhydromannitol.
[b] Values are mean ± SE and represent the averaged data of two separate experiments.
[c] N.D., None detected; <1%.

sent unique markers for the presence of domain 1 of heparinlike molecules which are able to bind and to activate antithrombin (see above). Furthermore, the above biochemical studies of affinity-fractionated mucopolysaccharides obtained from the cloned macrovascular and microvascular endothelial cells confirmed the heparan sulfate-like nature of these macromolecules. On the one hand, the low-pH nitrous acid treatment of these components revealed the presence of large amounts of N-acetyl groups (~40%). Heparan sulfate is known to contain significant numbers of these substituents, whereas heparin possesses few of these moieties. On the other hand, the relative distributions of the monosulfated and disulfated disaccharides of these glycosaminoglycans are characterized by an average of ~1.5 sulfate groups per disaccharide. Heparan sulfate contains ~1.0 sulfate groups per disaccharide, whereas heparin contains ~2.5 sulfate groups per disaccharide.

Cloned bovine aortic endothelial cells were also incubated with Na$_2$ ^{35}SO$_4$ and tritiated amino acids and completely solubilized with a

METABOLICALLY LABELED CELL EXTRACT

SEPHADEX G-50

EXCLUDED SAMPLE

DEAE-SEPHACEL

DEAE-I 90% / DEAE-II 10%
DEAE-II 20% / 80%

HEPARAN SULFATE
CHONDROITIN SULFATE

SEPHAROSE CL4B

PEAK CL4B-I
45% / 55%

PEAK CL4B-II
75% / 25%

PROTEOHEPARAN
PROTEOCHONDROITIN

HEPARAN CHAINS
CHONDROITIN CHAINS

OCTYL-SEPHAROSE

PEAK OS-I
2.0 × 10⁷

PEAK OS-II
3.0 × 10⁶

PEAK OS-III
4.5 × 10⁵

AT-AFFIGEL

AT-AFFIGEL

AT-AFFIGEL

AT-AFFIGEL

DEPLETED
1.5 × 10⁷
N.D.

HIGH AFFINITY
8.8 × 10³
<0.001

DEPLETED
2.0 × 10⁶
0.002

HIGH AFFINITY
3.8 × 10³
0.02

DEPLETED
1.5 × 10⁵
0.003

HIGH AFFINITY
1.9 × 10³
0.04

DEPLETED
1.3 × 10³
N.D.

HIGH AFFINITY
1.0 × 10³
0.009

RADIOACTIVITY
(³⁵S-cpm)

ANTICOAGULANT
ACTIVITY
(USP units)

RADIOACTIVITY
(³⁵S-cpm)

ANTICOAGULANT
ACTIVITY
(USP units)

NaCl (M)

DEAE-I
DEAE-II

³H
³⁵S
CPM × 10⁻⁶

FRACTION NUMBER

CL4B-I
CL4B-II

³H
³⁵S
CPM × 10⁻⁵

FRACTION NUMBER

TRITON X-100 (%)

OS-I
OS-II
OS-III

³H
³⁵S
CPM × 10⁻⁴

FRACTION NUMBER

guanidine–detergent solution (Fig. 4). Double-labeled macromolecules were isolated by DEAE–Sephacel chromatography (Fig. 4). The radiolabeled material (peaks DEAE-I and DEAE-II) were gel-filtered on Sepharose CL4B, and an initial peak (CL4B-I) of ^{3}H- and ^{35}S-labeled material eluted with a K_{av} of ~0.2, while a second peak (CL4B-II) of radiolabeled macromolecules eluted with a K_{av} ranging from 0.45 to 0.7. Treatment of peak CL4B-I with papain or with sodium hydroxide resulted in a shift in the K_{av} of the radioactivity from ~0.2 to ~0.5, suggesting that this fraction contained proteoglycans, whereas analysis of peak CL4B-II on Sepharose 4B revealed that this fraction coeluted with mucopolysaccharide chains obtained by proteolytic digestion of cellular elements (K_{av} ≈ 0.5). Incubation of the radiolabeled material from either fraction with mucopolysaccharidases of known substrate specificities indicated that peak CL4B-I contained a mixture of proteoheparan and proteochondroitin, whereas peak CL4B-II consisted of heparan and chondroitin chains (Fig. 4). Radiolabeled proteoglycans from peak CL4B-I were applied to octyl-Sepharose, and three peaks of ^{3}H- and ^{35}S-labeled material were eluted with a Triton X-100 gradient (Fig. 4). Peaks OS-II and OS-III, which probably represent native and partially degraded proteoglycans, emerged from the column at added-detergent concentrations of ~0.06 and ~0.26%, respectively, and possessed ~13 and ~2%, respectively, of the starting radioactive sulfate (Fig. 4). Peak OS-I, which probably constitutes a badly degraded proteoglycan missing its hydrophobic domain, did not bind to the matrix and accounted for ~85% of the initial ^{35}S radioactivity (Fig. 4).

The hydrophobic proteoglycans (peaks OS-II and OS-III) were then affinity-fractionated into two separate populations utilizing immobilized antithrombin (Fig. 4). The hydrophobic heparan sulfate proteoglycans which bound tightly to the protease inhibitor represented <1% of the starting material and exhibited a specific anticoagulant activity from ~5 to 21 USP units/10^6 ^{35}S-cpm. The heparan sulfate proteoglycans which interacted weakly with the protease inhibitor

FIG. 4. Isolation of radiolabeled proteoglycans from clone endothelial cells. Radiolabeled material obtained from metabolically labeled cloned endothelial cells and desalted on Sephadex G-50, was applied to DEAE–Sephacel, and bound material was eluted with a linear salt gradient. The above radiolabeled macromolecules (peaks DEAE-I and DEAE-II) were gel-filtered on Sepharose CL4B. Radiolabeled proteoglycans obtained from gel filtration (peak CL4B-I) were applied to octyl-Sepharose, and bound material was eluted with a Triton X-100 gradient. Proteoglycans (peaks OS-I, OS-II, and OS-III), as well as heparan sulfate chains (peak CL4B-II), were affinity-fractionated employing immobilized antithrombin (AT-Affigel). N.D., None detected.

constituted >99% of the starting material and possessed a specific anticoagulant potency from ~0.001 to 0.02 USP units/10^6 ^{35}S-cpm. The high-affinity heparan sulfate proteoglycans are responsible for >85% of the anticoagulant activity of the cloned bovine aortic endothelial cells (Fig. 4). Similar results have been obtained from studies conducted with cloned rat epididymal fat pad endothelial cells.

The above data suggest that all of the anticoagulantly active heparan sulfate chains may be attached to a small population of possibly unique hydrophobic core proteins. To examine this hypothesis, high-affinity hydrophobic proteoglycans as well as those with minimal ability to interact with the protease inhibitor were isolated from cloned rat epididymal fat pads as outlined above. These two populations of proteoglycans were treated with papain, and the resultant free mucopolysaccharide chains were rechromatographed on the immobilized antithrombin. The results showed that virtually all of the heparan sulfate chains of the proteoglycan which could interact strongly with the protease inhibitor bound tightly to the affinity matrix, whereas few, if any, of the glycosaminoglycan chains of the proteoglycan which could interact only weakly with the protease inhibitor bound to the affinity matrix. Thus a small population of proteoglycans contain all of the anticoagulantly active mucopolysaccharide, whereas the remaining larger population of proteoglycans bear glycosaminoglycan with minimal biologic potency (Fig. 5).

The hydrophobic nature of the anticoagulantly active proteoheparans implies that these molecular entities may represent an integral component of the cell membrane (54). To test this proposition, the

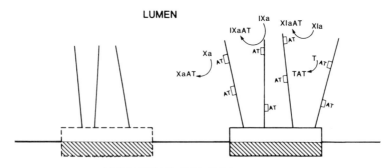

FIG. 5. Model of anticoagulantly active (solid-line figure) and inactive proteoheparan (dashed-line figure) proteoheparans positioned in cell membrane (hatched lines). AT, Antithrombin; TAT, Thrombin–antithrombin.

interaction of radiolabeled antithrombin with the surface of cloned bovine aortic endothelial cells was quantitated. At ~2 days postconflu-ence, these cells possess 5.8×10^4 protease inhibitor binding sites per cell with an apparent dissociation constant of 12.4 nM. These high-affinity antithrombin receptors are heparan sulfate membrane compo-nents, since complexing of the protease inhibitor is completely elimi-nated by pretreatment of endothelial cells with purified *Flavobac-terium* heparinase. The above results are in excellent agreement with previously conducted studies utilizing radiolabeled antithrombin and bovine aortic segments (55).

E. Physiological Role of Anticoagulantly Active Heparan Sulfate

The information summarized in the preceding section strongly sug-gests that anticoagulantly active heparan sulfate must be present on the luminal surface of the endothelium, which would thereby endow, in part, blood vessels with nonthrombogenic properties. Recently, sev-eral investigators have provided evidence both to support and to refute this hypothesis (56–58). Marcum *et al.* (52, 59) have attempted to re-solve this issue by directly demonstrating that intact blood vessels possess heparan sulfate, which accelerates the function of anti-thrombin.

To accomplish this goal, rat and mouse hindlimb preparations were perfused *in situ* with purified thrombin until a constant level of the enzyme was present in the effluent, at which time purified antithrom-bin was infused in the perfusion stream. The amount of thrombin–antithrombin complex formed within the vasculature was estimated by a specific radioimmunoassay for the product. The rate of enzyme–inhibitor complex formation in the rat and mouse hindlimb prepara-tions was enhanced by as much as 10-fold and 15-fold, respectively, as compared with the amount of interaction product generated in the absence of the mucopolysaccharide (Table II). The antithrombin-accel-erating activity detected in the hindlimb vasculature appears to be due to heparan sulfate. A physically homogeneous preparation of *Fla-vobacterium* heparinase which did not exhibit proteolytic or chondroitinase activities was recirculated through the hindlimb vas-culature prior to perfusion of the hemostatic components. The amount of enzyme–inhibitor complex generated within the animal was re-duced to uncatalyzed levels (Table II). The anatomic location of these mucopolysaccharides within the hemicorpus preparation has also been investigated. Buffer was recirculated through the hindlimb prepara-

TABLE II

In Situ Generation of Thrombin–Antithrombin Complex
in the Hindlimb Preparation

Proteins perfused through the hindlimb preparation[a]	Thrombin–antithrombin complex[b]		
		Mouse	
	Rat	+/+	W/W^v
Thrombin followed by native antithrombin	0.94 ± 0.17 (N = 13)	0.48 ± 0.04 (N = 13)	0.48 ± 0.05 (N = 14)
Thrombin followed by modified antithrombin	0.22 ± 0.04 (N = 4)	0.05 ± 0.03 (N = 6)	0.05 ± 0.03 (N = 6)
Thrombin after treatment with purified heparinase followed by antithrombin	0.08 ± 0.01 (N = 4)	0.05 ± 0.02 (N = 5)	0.05 ± 0.01 (N = 7)
Uncatalyzed amount	0.09	0.03	0.03

[a] Thrombin was perfused through hindlimb vasculature at an initial concentration of 5.4 nM, and antithrombin was perfused through rat and mouse vasculature at initial concentrations of 0.18 µM and 0.09 µM, respectively.

[b] Values are mean ± SE. Enzyme–inhibitor complex formation within the cannuli was subtracted from the original data to obtain the amount of complex generated within the vasculature of the rat (pmol/20 sec) and mouse (pmol/17 sec).

tion for extended periods of time, and the presence of mucopolysaccharide was ascertained by determining the acceleration of thrombin–antithrombin complex formation. Anticoagulant activity was not detected within the buffer, which implies that the anticoagulantly active molecules are tightly bound to the luminal surface of the endothelium (59).

To show that heparan sulfate present on the endothelium functions in a manner identical to commercial heparin, antithrombin modified at Trp^{49} was substituted for native protease inhibitor during studies employing the rat and mouse hindlimb perfusion system. This change in components resulted in a reduction of thrombin–antithrombin complex formation to uncatalyzed levels (Table II). These observations indicate that the heparan sulfate present on the endothelium potentiates thrombin–antithrombin complex formation via a mechanism dependent on the same critical tryptophan residue which is required for heparin-induced acceleration of the protease inhibitor (see above).

To assess the contribution of mast cells to the maintenance of blood fluidity, the hindlimb vasculature of mast cell-deficient mice (W/W^v) and littermates containing normal levels of mast cells (+/+), were perfused with purified human thrombin and antithrombin as outlined above (52). Thrombin–antithrombin complex generation within the vasculature of W/W^v and +/+ mice was enhanced to a comparable extent over the uncatalyzed rate (Table II). These observations suggest that the heparan sulfate produced by the endothelium are wholly responsible for the acceleration of antithrombin action and that mast cell heparin is not involved in maintaining the nonthrombogenic properties of blood vessels.

   In conclusion, we believe that a small fraction of plasma antithrombin is normally bound to a specific population of heparan sulfate proteoglycans synthesized by macrovascular and microvascular endothelial cells. This permits the protease inhibitor to be selectively activated at blood surface interfaces where enzymes of the intrinsic coagulation cascade are commonly generated. Thus, antithrombin is critically placed to neutralize these hemostatic enzymes and thereby protect natural surfaces against thrombus formation. Furthermore, the catalytic nature of this specific set of heparan sulfate proteoglycans ensures the continual regeneration of the nonthrombotic properties of the endothelial cell layer (Fig. 5). Alterations in the synthesis and/or placement of the anticoagulantly active heparan sulfate proteoglycans on the surface of microvascular and macrovascular endothelial cells could be responsible for arterial and venous thrombotic disease in humans.

III. Role of Heparan Sulfate as a Modulator of Smooth Muscle Proliferation

   The healthy arterial wall consists of the intimal endothelial cells which line the lumen and the underlying medial smooth muscle cells (SMC), which remain in a quiescent growth state. Several investigators have observed that within minutes of experimental desquamation of the endothelium, platelets adhere to the denuded surface and release their α-granule content. Platelet proteins such as platelet-derived growth factor (PDGF) appear on the surface of the exposed SMC within the denuded area and are internalized within the first hour (60–63). The underlying SMC apparently respond to this mitogen as well as to factors released from endothelial cells and macrophages present at the site of injury and migrate from the media into the intima. Thereafter, the SMC proliferate to form eventually a myointimal plaque (64–67). At the periphery of the damaged site, uninjured endothelial cells multiply and migrate until the damaged area is reendothelialized. Provided that the site of injury is sufficiently small, endothelial cells will completely cover this area and the SMC growth will cease. It is currently believed that repeated endothelial cell injury might lead to cycles of SMC hyperplasia with subsequent lipid deposition and the eventual development of the atherosclerotic lesion. Until a few years ago, little was known about endogenous components that might prevent the action of growth factors on SMC and hence suppress the pathologic proliferation of SMC. In the sections below, we describe recent progress in elucidating the role that heparinlike molecules play in modulating the growth of SMC and consider how these mucopolysaccharides are able to perform this function.

A. *Effects of Heparin on the in Vivo Proliferation of Smooth Muscle Cells*

Clowes and Karnovsky were the first to demonstrate that administration of commercially available porcine mucosal heparin to animals following denudation of intimal endothelial cells markedly reduced the subsequent SMC proliferation (68). These investigators ligated the right carotid arteries of Sprague–Dawley rats. Air was infused into the ligated section and immediately afterwards, blood flow was reestablished. This procedure resulted in complete desiccation and desquamification of the endothelium within the ligated area. Experimental animals were infused 24 hr later with 50 or 100 units/kg/hr of heparin while control animals were infused continuously with Ringer's solution. Animals were sacrificed at later times and the carotid arteries were sectioned at a point midway between the two ligatures. The isolated arterial segments were prepared for light, scanning, and transmission electron microscopy.

Control as well as heparin-treated animals showed a thick covering of platelets adhering to the exposed subendothelial surface as well as endothelial cells approaching from the intact wound edges. At 10 and 14 days, intimal thickening due to medial SMC proliferation was apparent in the injured arteries. However, heparin-treated animals had a pronounced diminution of the SMC thickening. The maximal ratio of intima to media in 15 heparin-treated animals was 0.5 ± 0.1 whereas the same ratio in five control animals was 1.9 ± 0.5 (P = 0.007). Thus, heparin exhibited a significant effect on SMC proliferation but did not alter endothelial regeneration or platelet binding to the media.

Given that heparin concentrations used in these studies prolonged the clotting times, it was possible that the antiproliferative effect of the mucopolysaccharide toward SMC could be related to its ability to accelerate thrombin–antithrombin interactions. Thrombin induces PDGF release from platelets and is also a potential mitogen for SMC.

Therefore, heparin could function by preventing thrombin-dependent stimulation of SMC growth.

We have previously separated anticoagulantly active heparin species from nonanticoagulantly active mucopolysaccharide by affinity chromatography utilizing immobilized antithrombin (see above). These two forms of heparin allowed us to establish whether the antiproliferative activity of the mucopolysaccharide is related to its anticoagulant potential (69). To this end, the carotid arteries of 22 rats were air-dried as described above. After 24 hrs, the various heparin fractions were infused (0.5–0.65 mg/kg/hr), and the carotid arteries were examined 14 days later (Fig. 6). Our data indicated that nonanticoagulantly active heparin was equally effective as anticoagulantly active mucopolysaccharide at inhibiting intimal SMC proliferation. The level of growth inhibition was 77% of that observed in the absence of mucopolysaccharide. Thus, the antiproliferative effect of heparin is unrelated to its anticoagulant activity.

B. Effects of Heparin on the in Vitro Proliferation of Smooth Muscle Cells

We have also studied the effects of heparin on the *in vitro* proliferation of SMC isolated from bovine and rat aortas (70). SMC were plated at a density of 6000 cells/ml, growth-arrested in a medium supplemented with low concentrations of serum (0.4%) or platelet-poor plasma (2.0%), and released from the growth-arrested state by feeding them with medium supplemented with 10% serum in the presence or absence of heparin. Cell number determinations performed 7 days later revealed that heparin, at doses between 1 and 10 μg/ml, inhibited cell proliferation by 60–90%. Nonanticoagulantly active heparin also suppressed the growth of the cultured SMC. We have also shown that the growth-inhibitory effect is not observed in the presence of glycosaminoglycans other than heparin or heparinlike molecules. For example, chondroitin sulfate, dermatan sulfate, and hyaluronic acid exhibited negligible antiproliferative potency.

C. Synthesis of Antiproliferative Species of Heparan Sulfate by Vascular Endothelial Cells and Smooth Muscle Cells

Given that SMC proliferation ceases after endothelial cell regrowth at sites of vascular injury, we thought it likely that endothelial cells could secrete a heparinlike substance (heparan sulfate) which might

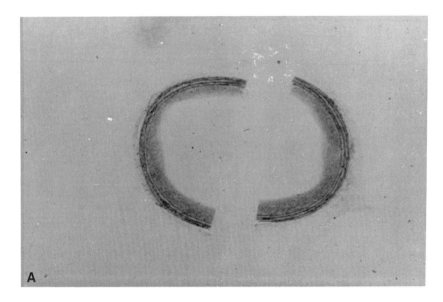

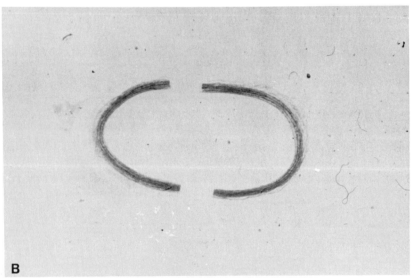

FIG. 6. Histological sections of injured right carotids in control (A) and in heparin-treated (B) animals at 14 days.

regulate SMC growth. To test this hypothesis, we examined bovine aortic endothelial cells (BAEC) in culture (71). Our data showed that conditioned medium from postconfluent but not exponentially growing BAEC inhibited the proliferation of growth-arrested SMC by approximately 70 to 80%. The inhibitory activity appeared to be due to a heparinlike substance. This conclusion was supported by the fact that a crude isolate of glycosaminoglycans (TCA-soluble, ethanol-precipitable material) from BAEC-conditioned medium inhibited SMC growth, whereas glycosaminoglycans isolated from unconditioned medium had no effect on SMC proliferation. In addition, neither exposure of the conditioned medium to high temperatures for prolonged periods of time, nor treatment with enzymes such as trypsin, chymotrypsin, hyaluronidase, or chondroitin sulfate ABC lyase destroyed the growth-inhibitory activity. However, the antiproliferative potency was abolished by exposure to purified *Flavobacterium* heparinase (Table III). This highly specific enzyme cleaves heparin and heparan sulfate into tetrasaccharide and disaccharide units, which we have shown do not possess growth-inhibitory activity.

While engaged in these studies, we noticed that the inhibitory activity was present in the medium of BAEC conditioned in serum but absent in medium conditioned in plasma. This indicated to us that a platelet product was involved in the release of the heparinlike inhib-

TABLE III

BIOCHEMICAL PROPERTIES OF THE BAEC-CONDITIONED MEDIUM INHIBITORY ACTIVITY[a]

	Maximum (%) Inhibition	
Treatment	Calf SMC	Rat SMC
None	84 ± 8	58 ± 5
90°C, 5 min	93 ± 4	62 ± 5
60°C, 30 min	76 ± 9	53 ± 10
Trypsin (30 μg/ml, 2 hr, 37°C)	82 ± 5	60 ± 10
Chymotrypsin (30 μg/ml, 2 hr, 37°C)	85	54
Hyaluronidase (15 U/ml, 90 min, 37°C)	94 ± 4	50
Chondroitin SO₄ ABC lyase (1 U/ml, 90 min, 37°C)	88 ± 7	58 ± 2
Heparinase (10 U/ml, 90 min, 37°C)	19	9 ± 6

[a] Conditioned medium from BAEC was subjected to the indicated treatments and then assayed for its ability to inhibit growth-arrested calf and rat SMC when mixed 1:1 with RPMI + FCS (final concentration 20% FCS). SD are given in those instances when the treatments were performed three or more times. In other cases, the treatments were performed twice.

itor from the surface of the endothelial cells. This observation prompted us to develop a procedure for isolating a lysosomal endoglycosidase
from extracts of outdated human platelets (72). Utilizing standard
chromatographic techniques, the platelet heparitinase was purified
about 240,000-fold with an overall yield of 5.6%. The final product is
physically homogeneous as judged by disk gel electrophoresis at acidic
pH as well as gel filtration chromatography and exhibits an apparent
molecular weight of ~134,000 (72).

The biologic potency of the endoglycosidase was examined as a function of pH. The data show that the platelet heparitinase is maximally
active from pH 5.5 to 7.5. The substrate specificity of the platelet
endoglycosidase was determined by identifying susceptible linkages
within the heparin molecule that can be cleaved by the above component. Our studies indicate that this enzyme is only able to hydrolyze
glucuronsyl–glucosamine linkages. Furthermore, investigation of the
structure of the disaccharide which lies on the nonreducing end of the
cleaved glucuronic acid residue suggests that N-sulfation of the glucosamine moiety or ester sulfation of the adjacent iduronic acid groups
are not essential for bond scission (72). We examined the ability of the
purified heparitinase to release heparan sulfate species from cultured
BAEC (73). Our results show that when endothelial cells were exposed
to serum-free medium containing 1 ng/ml of the purified platelet endoglycosidase, at least as much inhibitory activity was liberated as
was obtained with 0.4% serum (Fig. 7). Dose–response experiments
indicated that as little as 10 pg/ml of the enzyme were necessary to
liberate 50% of the inhibitory activity from endothelial cells (73).

We thought it likely that SMC might also synthesize heparinlike
molecules possessing inhibitory activity and, thereby, regulate their
own growth potential. This possibility was investigated by isolating
heparan sulfate from bovine SMC and determining the antiproliferative potencies of these mucopolysaccharides (74). Exponentially growing as well as postconfluent SMC were incubated with $Na_2\,^{35}SO_4$ and
the radiolabeled glycosaminoglycans were obtained from the cell surface, cell pellet, as well as culture medium. These molecular species
were freed of protein by extensive proteolytic digestion and the resultant glycosaminoglycan chains were chromatographed on DEAE–
Sephadex. As indicated in Fig. 8, this step resolved the ^{35}S-labeled
material into two peaks regardless of growth stage or cellular origin.
Peak I is heparan sulfate based on its susceptibility to digestion with
purified Flavobacterium heparinase. Peak II is chondroitin sulfate,
since it is completely degraded by chondroitinase ABC lyase. When
these components were bioassayed for their SMC growth-inhibitory

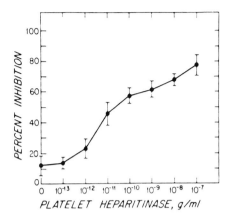

Fig. 7. Ability of platelet heparitinase to release inhibitory activity from bovine aortic endothelial cells. Confluent, primary cultures of endothelial cells were incubated in serum-free RPMI 1640 containing the indicated concentrations of platelet heparitinase for 4 hr at 37°C. The resulting medium was mixed with RPMI 1640 and fetal calf serum (final serum concentration 20%) and tested for its ability to inhibit SMC growth. Error bars, SD.

potential, we found that heparan sulfate from the surface, pellet, and medium of exponentially growing as well as postconfluent SMC exhibits antiproliferative activity, whereas the similarly designated fractions of chondroitin sulfate possess no such biologic potency. Therefore, the anionic charge of the heparan sulfate cannot be the sole reason for the growth-inhibitory potency of this component, since the DEAE–Sephadex chromatographic analyses clearly show that chondroitin sulfate had a higher average charge density than the heparan sulfate.

The heparan sulfate contained in the peak I fractions from DEAE–Sephadex were subjected to pronase digestion and then chromatographed on Sepharose 4B. The resultant purified heparan sulfate species from the surface, interior, and medium of postconfluent and exponentially growing SMC elute as one peak with a molecular weight of 35,000–40,000. The inhibitory potential of the various species was next established. Table IV indicates that heparan sulfate isolated from the surface of postconfluent SMC exhibits about eight times the antiproliferative potency as the corresponding material obtained from the surface of exponentially growing SMC. Heparan sulfate isolated from the cell pellet, or culture medium of SMC in either growth state possesses only minimal amounts of inhibitory activity (Table IV).

The chemical masses of the various fractions of heparan sulfate

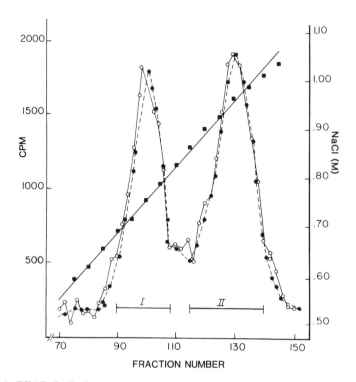

FIG. 8. DEAE–Sephadex chromatography of radiolabeled glycoshaminoglycans from the surface of postconfluent and exponentially growing SMC. The ^{35}S-labeled glycosaminoglycans obtained from the surface of postconfluent (dashed line) or exponentially growing (solid line) SMC by treatment with trypsin–EDTA were chromatographed on a column of DEAE–Sephadex A50. Concentrations of NaCl were determined based on conductance measurements. Elution profiles of medium and cell pellet glycosaminoglycans from postconfluent and exponentially growing SMC were similar to those shown.

were also determined by hexosamine analyses, and the specific antiproliferative activities of these glycosaminoglycans were calculated. The data showed that exponentially growing SMC synthesize about 1.5 to 3.0 times the amount of heparan sulfate found in the corresponding fractions of postconfluent SMC (Table IV). Thus, the large amounts of antiproliferative activity present on the surface of postconfluent SMC are not simply due to the augmented production of this glycosaminoglycan. Indeed, the heparan sulfate isolated from the surface of postconfluent SMC had a specific inhibitory activity which is 13 times that of the similarly designated mucopolysaccharide obtained from exponentially growing SMC. This highly active heparan sulfate is able to inhibit SMC proliferation dramatically when added at a level as low as 20 ng/ml, and hence its potency is >40 times greater than that of commercial heparin in suppressing the growth of these cells (Fig. 9). Thus, our data indicate that postconfluent SMC are uniquely able to synthesize a heparan sulfate with remarkably potent antiproliferative activity and place these components on their cell surface. This highly active heparan sulfate is likely to differ structurally to only a very minor extent when compared to mucopolysaccharides isolated from exponentially growing cells, since both types of glycosaminoglycans appear to have similar average charge densities and average molecular sizes as determined by identical elution profiles on DEAE–Sephadex and Sepharose 4B chromatography. Indeed, the above findings indicate that certain specific structural elements of heparan sul-

TABLE IV

CHEMICAL MASS AND ANTIPROLIFERATIVE ACTIVITIES OF HEPARAN SULFATE ISOLATED FROM EXPONENTIALLY GROWING AND POSTCONFLUENT SMCs[a]

Fraction	Growth stage	Antiproliferative activity (inhibitory units/10^6 cells)	Heparan sulfate ($\mu g/10^6$ cells)	Specific antiproliferative activity
Cell surface (exponentially growing)		2.97	0.818	3.63
Cell surface (postconfluent)		22.35	0.499	44.82
Cell pellet (exponentially growing)		1.27	0.640	1.98
Cell pellet (postconfluent)		1.21	0.464	2.61
Culture medium (exponentially growing)		1.80	0.680	2.63
Culture medium (postconfluent)		0.63	0.236	2.65

[a] The chemical mass of heparan sulfate obtained from the cell surface, cell pellet, and medium of exponentially growing as well as postconfluent SMCs was determined by glucosamine assay. The antiproliferative activity was calculated by comparing the amount of growth inhibition produced by each sample to that produced by known amounts of the heparin standard. Under these conditions, 1 μg of heparin per milliliter produced a 30–40% inhibition of growth relative to control cells which underwent four to five doublings during the experiment. Cell numbers in control wells typically reached 0.8–1.0 × 10^5 cells/cm^2. One inhibitory unit is equivalent to the amount of inhibition caused by 1 μg of heparin per milliliter. The specific antiproliferative activity of a given heparan sulfate is computed by dividing the growth-inhibitory activity of the sample by its chemical mass. The specific antiproliferative activity of heparin is equal to 1.

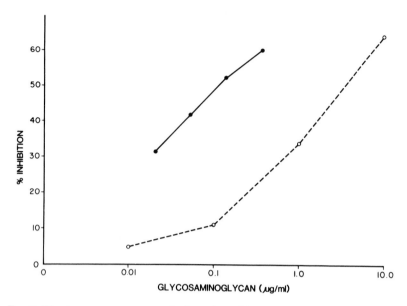

Fig. 9. The dose–response curves for heparin and heparan sulfate in the SMC growth inhibition assay. Each point in the heparin curve (dashed line) represents the average of at least 10 determinations in experiments using the same heparin standard. Each point in the heparan sulfate curve (solid line) represents the average of at least 4 determinations in experiments using three different preparations of heparan sulfate isolated from the surface of postconfluent SMC.

fate such as glycosidic bond configuration, sulfate position, or iduronic acid residues may be required for the observed antiproliferative effect (75).

Given that the surfaces of exponentially growing SMC possess heparan sulfate with minimal growth-inhibitory activity, we wondered whether the levels of this biologically potent mucopolysaccharide could represent residual highly active glycosaminoglycan generated by the primary postconfluent SMC utilized to seed our cultures. To test this hypothesis, we harvested the surface glycosaminoglycans present on SMC from the time of seeding to the period of postconfluence, isolated heparan sulfate by column chromatography, and ascertained the antiproliferative activity of the mucopolysaccharide per 10^6 cells. Our results indicate that exponentially growing SMC retain small amounts of residual highly active heparan sulfate from the surface of primary postconfluent SMC but can produce little, if any, of the biologically active component. Indeed, we would suggest that the

postconfluent SMC as compared to SMC at other stages of growth
probably differ by as much as several hundred fold in their ability to
synthesize heparan sulfate with antiproliferative potency.

It is also of interest to note that SMC possess large amounts of the
same lysosomal heparitinase which has been isolated from platelets
(L. Fritze and R. Rosenberg, unpublished observations). Hence this cell
type is able to synthesize all of components required for an autocrine
type of growth-inhibitory circuit.

D. Effects of Heparin on the in Vitro Kinetics of Smooth Muscle Cell Growth

Recent data from the laboratory of Morris Karnovsky have established that SMC possess specific high-affinity receptors for heparin
and that these cells are capable of internalizing the bound mucopolysaccharide (76). Using both ³H-labeled heparin and ¹²⁵I-labeled
heparin and following the extent of binding to rat as well as calf aortic
SMC, ~100,000 binding sites per cell with an apparent dissociation
constant of 1×10^{-9} M were found. The bound glycosaminoglycan was
internalized at 37°C but not at 4°C. Moreover, the rate of uptake of
heparin was biphasic, with an initial rapid entry of 50% of the mucopolysaccharide into the cell ($t_{0.5}$ = 20 min) followed by a slow, steady
internalization of the remainder over 1–2 days. Experiments performed with fluorescently labeled heparin established that the initial
internalization was a receptor-mediated endocytotic process. Moments
after binding, the FITC-labeled heparin appeared as a diffuse pattern
over the cell surface, but within minutes the mucopolysaccharide had
moved into intensely staining, punctate vesicles. By 15 min the
heparin was internalized, as neither trypsin nor Flavobacterium
heparinase treatment of the cells affected the fluorescent pattern.
After 1 hr, the large punctate vesicles had concentrated in the per-
inuclear region. Thus, heparin enters SMC by both a receptor-mediated mechanism and a slower endocytotic process. These results raise
the interesting possibility that heparin suppresses the growth of SMC,
not in the extracellular environment of the cell, but at specific intracellular sites.

To obtain more evidence for the above hypothesis, we have investigated the kinetics of heparin inhibition of SMC growth. The addition
of mucopolysaccharide to quiescent cells prior to their exposure to
serum results in a dramatic enhancement in the level of inhibition, as
measured 5 days later, compared to cultures in which the mucopolysaccharide is added coincident with exposure to serum. This augmenta-

tion in the final extent of growth suppression is time dependent, since addition of heparin to quiescent SMC for ≥48 hr had the greatest impact on the final level of proliferative activity while pretreatment for shorter periods of time resulted in a reduced effect. These data suggest that the time of exposure of SMC to heparin is critically important with regard to the inhibitory potential of the mucopolysaccharide. Further examination of the effects of heparin on the initial growth rate revealed that SMC must be preexposed to the mucopolysaccharide for ~48 hr in order to block the first cell cycle traverse and hence have the optimal effect on the final level of inhibition at 5 days. When SMC are pretreated with the glycosaminoglycan for shorter periods of time, the initial cell cycle traverse is not suppressed.

We also examined the effect of removing heparin from the cultures after a 48-hr pretreatment. In this situation, the SMC resumed proliferation at normal rates within 2 days and the final level of cell growth is minimally affected. Thus our data would indicate that SMC must be exposed to heparin for a minimum of 48 hr to observe any antiproliferative effect and that maximal inhibition requires a continuous exposure to the mucopolysaccharide. These results are consistent with a model in which heparin slowly enters SMC and accumulates at an internal site where it can block critical biochemical events required for proliferation. Furthermore, it would appear that the mucopolysaccharide undergoes constant degradation at this site and must be continuously replaced with newly internalized glycosaminoglycan.

In other studies, we added platelet factor 4 (PF4) to our culture system and determined its ability to block the action of heparin. This protein binds tightly to heparin with K_d of 5×10^{-9} M and prevents the specific interaction of heparin with the plasma proteinase inhibitor, antithrombin (27). PF4 at levels 80 times higher than normally found in our assay medium (5% CS-DME) (77), however, could not alter the inhibitory effects of heparin on SMC growth. Since this amount of PF4 should have been sufficient to bind heparin unless other binding species with higher affinities were present, it suggests that inhibition of SMC growth may result from specific high-affinity binding of heparin to as yet unidentified sites.

E. Biochemical Mechanisms by Which Heparin and Heparan Sulfate Inhibit the Proliferation of Smooth Muscle Cells

The *in vivo* and *in vitro* studies described above have established that the antiproliferative activity of heparin and heparan sulfate are unrelated to their anticoagulant potential (69, 70). However, the precise mechanism by which these mucopolysaccharides are able to inhibit SMC growth has remained somewhat enigmatic. To address this issue, several laboratories have examined the effects of heparin on protein synthesis and cell metabolism of cultured SMC. Majack and Bornstein (78) have demonstrated that heparin induces the synthesis of a 60,000 MW collagenlike protein which is associated with the cell layer of early-passage rat SMC. This molecule is produced 18–24 hr after heparin addition to the cultures and persisted for up to 72 hr after removal of the glycosaminoglycan. Other studies have noted the rapid appearance of 35,000, 37,000, and 39,000 MW proteins in the culture medium of heparin-treated SMC (79, 80). The levels of these components are maximal within 4–8 hr of heparin exposure and are unaffected by prior actinomycin D treatment. Thus, new synthesis of mRNA may not be necessary for their generation. Heparin also regulates the production of thrombospondin by SMC (81). The levels of this protein in the extracellular matrix are decreased substantially in cells previously exposed to heparin. Although none of these investigations have established a direct correlation between heparin regulation of these proteins and growth inhibition, it is noteworthy that all of these phenomena occur at doses of mucopolysaccharide which are maximally inhibitory (10 µg/ml). Moreover, it appears that glycosaminoglycans with little antiproliferative activity also possess negligible protein-modulating potential. The situation with respect to thrombospondin is especially intriguing, since antibodies to this cell matrix constituent inhibit DNA synthesis by SMC. Thus, heparin may prevent the normal deposition of this protein in the extracellular matrix and thereby could alter the proliferative response of SMC. The ability of mucopolysaccharides to regulate the production of collagenlike proteins and thrombospondin may also be of importance with regard to the ability of SMC to undergo migration, which is required to form the atherosclerotic lesion (82). Heparin may also influence other key metabolic events in SMC. Castellot et al. (83) have noted that the mucopolysaccharide induces a significant decrease in the transport of thymidine and uridine but has no effect on the transport of amino acids or glucose into SMC. This phenomena is observed within 2 hr of treating SMC with heparin and suggests that the mucopolysaccharide may have rather specific effects on membrane transport processes.

Ross and his collaborators (84) have demonstrated that SMC enter the G_0/G_1 growth state and are unable to proliferate when cultured in plasma-containing medium from which platelet release products are absent. The addition of serum or purified mitogens such as PDGF or epidermal growth factor (EGF) allows the cells to traverse G_1 to S and begin DNA synthesis (85, 86). Given that serum contains many growth factors including PDGF, EGF-like peptides of platelet origin (87, 88), as well as insulinlike growth factor 1 (IGF-1), proliferation of SMC induced by serum may be caused by the concerted actions of several mitogens (89, 90). It is known that growth factors such as PDGF and EGF bind to specific receptors on the surface of SMC with high affinity (86, 91–93). After receptor occupancy, a variety of intracellular changes occur including increased tyrosine kinase activity, phospholipid degradation, altered ion transport, specific mRNA synthesis, and specific protein synthesis (94, 95). Some or all of these secondary events may be critical for the subsequent initiation of DNA synthesis.

Rat aortic SMC treated with heparin (100 µg/ml) are blocked in G_0/G_1 as early as 28 hr after addition of mucopolysaccharide to the cultures (83). Moreover, the heparinlike species present in the conditioned medium of BAEC also halt SMC in G_0/G_1 but only after a 2- to 3-day exposure of the cells to the mucopolysaccharide (71). Given these previous findings, we thought it likely that heparin and heparan sulfate might prevent the mitogenic effects of specific growth factors involved in the G_0/G_1 to S transition of SMC by either directly preventing a serum mitogen from interacting with the cell or by interfering with a subsequent effect of the mitogen. In either situation, increased concentrations of a particular growth factor above that normally found in serum should reverse the antiproliferative action of the mucopolysaccharide.

Weinstein et al. (96) have demonstrated that vascular SMC proliferate in serum-free medium supplemented with various components including purified thrombin, transferrin, ovalbumin, multiplication-stimulating activity, insulin, PDGF, IGF-1, and EGF. Heparin could potentially modulate the response of the cell to any of these components. Therefore, we added a large excess of each of these constituents to our biologic inhibition assay and determined whether they altered the antiproliferative action of heparin. None of these components, with the exception of EGF, had any substantial effect on the phenomena. However, the addition of EGF to the serum-containing medium at concentrations of 50 ng/ml or 250 ng/ml decreased the growth-inhibitory action of heparin by 60–77%. Thus, EGF appears to be the only mitogen capable of reversing the antiproliferative effects of the mucopolysaccharide.

Given that the experiment described above was performed in culture medium containing 5% serum, SMC proliferation was independent of the exogenous EGF. Therefore, the possibility existed that EGF displaced heparin from a non-EGF-like mitogen and thereby allowed this latter factor to stimulate cell growth. To ensure that heparin specifically interferes with EGF-like components, it was essential to reverse the inhibitory capability of the mucopolysaccharide with EGF under conditions where this mitogen alone was responsible for growth. Toward this end, SMC were growth-arrested in 5% plasma-containing medium. We then compared the growth induced by EGF with that of PDGF. The addition of either mitogen to these cultures 2 days after plating, stimulated the cells to leave the G_0/G_1 state and undergo sustained cellular proliferation over the next 5 days. In these studies, EGF and PDGF were nearly as effective as 5% calf serum at stimulating SMC growth.

The establishment of a well-defined system where single mitogens were able to induce SMC growth allowed us to confirm more rigorously our prior observation that EGF but not PDGF could reverse the inhibitory effects of heparin. In a control series of experiments, the plasma-containing medium was supplemented with varying concentrations of PDGF (0.6–48 ng/ml) after the cultures had been pretreated with heparin for 2 hr or 48 hr. A linear augmentation in the rate of growth was observed as the levels of PDGF were increased to 5 ng/ml in the absence of mucopolysaccharide. If the SMC were pretreated with heparin for 2–48 hr, the PDGF-induced proliferation was greatly decreased regardless of the amount of mitogen used. We then added EGF (0.4–50 ng/ml) to the growth arrest medium and examined the extent of heparin inhibition due to pretreatment of SMC for 2 or 48 hr with mucopolysaccharide. Control cultures exhibited a linear increase in their rate of growth up to a mitogen concentrationof 4 ng/ml. When SMC were pretreated with heparin for only 2 hr prior to EGF addition, there was no observable antiproliferative effect of the mucopolysaccharide. In contrast, when heparin was added to the cultures 48 hr before EGF addition, the subsequent growth was significantly inhibited at all levels of mitogen.

The simplest interpretation of our data is that heparin exerts its inhibitory effect either by binding EGF or by blocking the EGF receptor on SMC. To investigate these possibilities, EGF was radioiodinated and the binding of mitogen to SMC was established in the presence and absence of mucopolysaccharide. Heparin concentrations of 100 µg/ml had no effect on the interaction of ^{125}I-labeled EGF with SMC. Thus, levels of mucopolysaccharide which afford maximal suppression of growth do not interfere with the interaction of EGF and its receptor. Given that the protective effect of EGF was observed when heparin was added 2 hr before the mitogen, but that this phenomena was absent when the mucopolysaccharide was added 48 hr before the mitogen, we wondered whether the glycosaminoglycan functioned by reducing the number and/or avidity of the EGF receptors on SMC in a time-dependent manner. To test this hypothesis, we examined EGF binding in exponentially growing SMC. After serum addition, the log phase SMC bound progressively more EGF, eventually binding 3-fold more as the cultures approached confluence. If heparin was added in conjunction with serum, the increase in binding was prevented and the growth-inhibited cultures bound an equivalent amount of mitogen when compared to quiescent SMC. The rise in EGF binding observed in control cultures was also abrogated in 48-hr pretreated SMC where maximal growth inhibition was observed. These cultures bound 7-fold less EGF than exponentially growing SMC and 2-fold less EGF than quiescent SMC. Thus, heparin in both pretreated and nonpretreated cultures suppresses the binding of EGF to SMC by a mechanism which closely parallels its growth-inhibitory action. Moreover, the enhanced inhibitory effects of the mucopolysaccharide in 48-hr pretreated cultures may be directly due to large decreases in EGF binding relative to quiescent SMC.

Heparin effects on EGF binding might simply reflect a general aspect of growth inhibition, that is, the resemblance of mucopolysaccharide-inhibited cells to quiescent cells. To ensure that the observed reductions were independent of the SMC growth state, subsequent experiments were performed with quiescent SMC before serum addition. Therefore, we examined the effects of pretreatment with varying concentrations of heparin on EGF binding in quiescent SMC and correlated the changes with subsequent cell growth upon mitogen addition. Cultures pretreated for 48 hr bound progressively less EGF as the heparin concentration increased, eventually binding 50% less EGF at ≥ 10 µg/ml. Parallel cultures were serum-stimulated after the pretreatment, and the effects on cell proliferation were determined. Growth inhibition increased rapidly above heparin levels of 1 µg/ml and was maximal at 10 µg/ml. Thus, heparin concentrations which maximally affect EGF binding on quiescent SMC also have the greatest impact on cell proliferation. A strong temporal correlation also exists between the antiproliferative effects of heparin and its action on EGF binding to SMC. SMC, pretreated for varying lengths of time, bound progressively less EGF as the time of pretreatment was increased. A 2-hr pretreatment had no effect on EGF binding, indicating that residual glycosaminoglycan present in the binding assay medium does not affect the results. Decreases in EGF binding were apparent at

24 hr and maximal after 48–72 hr. The length of heparin pretreatment also had a dramatic effect on the initial growth rate of SMC. Proliferation was minimally affected by the 2-hr pretreatment. With increasing times of preincubation, concomitant increases in growth inhibition were observed, reaching a maximum by 48 hr.

We have also investigated the effects of an inhibitory heparan sulfate species produced by postconfluent SMC on EGF binding potential (74). When SMC were pretreated with 75 ng/ml of this latter moiety for 48 hr, EGF binding decreased 45% relative to the binding in quiescent SMC. EGF binding in SMC pretreated with 10 µg/ml of heparin for 48 hr declined by 35%. The growth of the heparin-pretreated SMC was inhibited by 77%. Heparan sulfate, at levels 130-fold lower than that of heparin, inhibited the subsequent proliferation by 41%. We also determined whether glycosaminoglycans with negligible growth-regulatory activity induce changes in EGF binding (71). SMC pretreated with 100 µg/ml of equal amounts of chondroitin 4-sulfate and chondroitin 6-sulfate (Sigma) bound an equivalent amount of EGF as quiescent cultures. Since chondroitin sulfates are more negatively charged than heparan sulfate, it would suggest that effects of inhibitory glycosaminoglycans on EGF binding are not simply due to charge density. Rather, the ability of postconfluent SMC heparan sulfate to inhibit growth and regulate EGF binding when used at low levels suggest that unique structural elements may be present which endow these species with biologic activity.

To establish whether heparin specifically regulated EGF binding or influenced mitogen binding in general, we examined PDGF and IGF-1 binding in pretreated SMC. Experiments were performed at several ligand concentrations in order to measure both high- and low-affinity binding. Pretreatment for 2 hr or 48 hr with 10 µg/ml of heparin had no affect on either IGF-1 or PDGF binding at any ligand concentration. We have also analyzed our compiled EGF binding data by Scatchard analysis. The results indicate that heparin induces a 53–60% reduction in the numbers of high- and low-affinity EGF receptors without alteration of the dissociation constants. These results provide evidence that heparinlike molecules decrease the total number of EGF receptors on SMC over a 48-hr period without affecting the avidity of these receptors for their mitogen. Other agents capable of reducing EGF binding such as PDGF, diacylglycerol, and phorbol esters function in a different manner. These latter components decrease the affinity of the EGF receptor within minutes, presumably by inducing specific phosphorylation events (97–99). Thus, our data are somewhat more consistent with a specific effect of the mucopolysaccharide on the synthesis of the EGF receptor rather than on a postsynthetic inactivation event.

It should be noted that in SMC as well as other cell types, phorbol ester alteration of the EGF receptor is associated with subsequent reduction of EGF-dependent events including DNA synthesis (100, 101). On the other hand, in 3T3 cells PDGF effects on EGF binding apparently enhance cell sensitivity to EGF (102). The reason for this discrepant behavior is not known. In this context, we should emphasize that the addition of EGF at concentrations of 60 pM to 1000 pM to plasma-containing culture medium allows SMC to undergo one to four population doublings over 5 days (see above). Given that the levels of mitogen required for SMC growth are similar to the apparent dissociation constants calculated above, it seems quite likely that the reduction in the number of EGF receptors induced by heparin contributes to a decrease in the proliferative potential of the SMC. This surmise is bulwarked by the apparent relationship between the extent of the heparin-induced alteration in EGF receptor number and the subsequent effects on the growth of SMC. However, it is quite likely that heparin affects other steps within the pathways of mitogenic response in SMC which may act synergistically with the alteration in EGF binding. This is to be expected given the pleiotropic actions of growth mitogens such as PDGF and EGF.

In summary, we would propose a simple model for the possible role of heparan sulfate and endoglycosidase in the regulation of SMC growth within the vessel wall. In the normal artery, endothelial cells, macrophages, and/or platelets serve as sources for mitogenic factors such as PDGF, EGF, and IGF-1 which are necessary for the growth of medial SMC. However, SMC also generate a specific type of heparan sulfate with antiproliferative activity which is positioned at the surface of the cell. The endoglycosidase, which can liberate the growth-inhibitory mucopolysaccharide, is also available within the vessel wall. Under normal circumstances, a small amount of the antiproliferative heparan sulfate is cleaved from the surface of SMC, taken up by these cellular elements, and is then able to regulate the levels of EGF receptors and other important components of the mitogenic pathway such as thrombospondin. The net effect of the mitogenic factors and the intracellular actions of the above heparan sulfate permits a small amount of SMC growth to compensate for the death of these cellular elements (<0.1% per day) (Fig. 10).

During damage to the endothelium, platelets and macrophages would appear at the site of injury and release high concentrations of

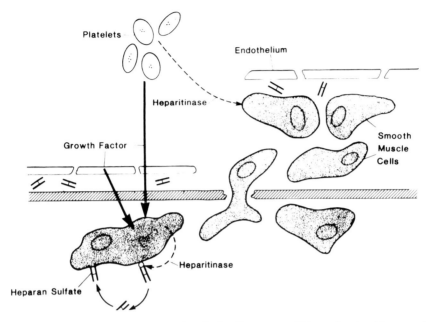

FIG. 10. Model depicting the postulated role of heparan sulfate, endogylcosidases, and mitogens in the regulation of SMC proliferation within the arterial wall.

growth factors. The SMC might be able to respond to these pathologic alterations by augmenting the synthesis, release, and internalization of the heparan sulfate with antiproliferative activity in a fashion identical to that noted when these cellular elements reach a postconfluent stage of cell growth. In this situation, the net balance between the elevated concentrations of mitogenic factors and the increased levels of free heparan sulfate with growth-inhibitory activity would ultimately determine whether SMC migrate to the luminal surface of the blood vessel wall and mount a proliferative response. The ability of heparan sulfate to modulate the synthesis and release of a specific collagen subtype as well as thrombospondin may be particularly critical with regard to the migration of SMC. Thus, this specific form of heparan sulfate would be positioned to act as a negative control element during the regulation of cell migration and proliferation in a similar manner to that postulated for certain proteins isolated from the surface of 3T3 cells and endothelial cells (103, 104) (Fig. 10). Alterations in the synthesis, release, or internalization of the growth-inhibitory forms of heparan sulfate by SMC may, in part, be responsible for the development of atherosclerotic lesions in humans.

References

1. Contejean, C. (1895). Recherches sur les injections intraveineuses de peptone et leur influence sur la coagulabilite' du sang chez le chien. *Arch. Physiol. Norm. Pathol.* **7**, 45–53.
2. Morowitz, P. (1968). "The Chemistry of Blood Coagulation." Thomas, Springfield, Illinois.
3. McLean, J. (1916). The thromboplastic action of cephalin. *Am. J. Physiol.* **41**, 250–257.
4. Howell, W. H., and Holt, E. (1918). Two new factors in blood coagulation: Heparin and pro-antithrombin. *Am. J. Physiol.* **47**, 328–341.
5. Brinkhous, K. M., Smith, H. P., Warner, E. D., and Seegers, W. H. (1939). The inhibition of blood clotting: An unidentified substance which acts in conjunction with heparin to prevent the conversion of prothrombin to thrombin. *Am. J. Physiol.* **125**, 683–687.
6. Waugh, D. F., and Fitzgerald, M. A. (1956). Quantitative aspects of antithrombin and heparin in plasma. *Am. J. Physiol.* **184**, 627–639.
7. Monkhouse, F. C., France, E. S., and Seegers, W. H. (1955). Studies on the antithrombin and heparin cofactor activities of a fraction absorbed from plasma by aluminum hydroxide. *Circ. Res.* **3**, 397–402.
8. Abildgaard, U. (1968). Highly purified antithrombin III with heparin cofactor activity prepared by disc gel electrophoresis. *Scand. J. Clin. Lab. Invest.* **21**, 89–91, 1968.
9. Rosenberg, R. D., and Damus, P. S. (1973). The purification and mechanism of action of human antithrombin–heparin cofactor. *J. Biol. Chem.* **248**, 6490–6505.
10. Nordenman, B., Nystrom, C., and Bjork, I. (1977). The size and shape of human and bovine antithrombin III. *Eur. J. Biochem.* **78**, 195–203.
11. Murano, G., Williams, L., Miller-Andersson, M., Aronson, D., and King, C. (1980). Some properties of antithrombin III and its concentration in human plasma. *Thromb. Res.* **18**, 259–262.
12. Petersen, E. E., Dudek-Wojciechowska, G., Sottrup-Jensen, L., and Magnusson, S. (1979). The primary structure of antithrombin III (heparin-cofactor). Partial homology between alpha$_1$-antitrypsin and antithrombin III. *In* "The Physiological Inhibitors of Coagulation and Fibrinolysis" (D. Collen, B. Wiman, and M. Verstraete, eds.), 43–54. Elsevier/North Holland, Amsterdam.
13. Prochownik, E. V., and Orkin, S. H. (1984). In vivo transcription of a human antithrombin III "Minigene." *J. Biol. Chem.* **259**, 15386–13592.
14. Bjork, I., Jackson, C. M., Journvall, H., Lavine, K. K., Nordling, K., and Salsgiver, W. J. (1982). The active site of antithrombin. *J. Biol. Chem.* **257**, 2406–2411.
15. Ferguson, W. S., and Finaly, T. H. (1983). Localization of disulfide bond in human antithrombin III required for heparin-accelerated thrombin inactivation. *Arch. Biochem. Biophys.* **221**, 304–307.
16. Blackburn, M. N., and Sibley, C. C. (1980). The heparin binding site of antithrombin III. *J. Biol. Chem.* **255**, 824–826.
17. Villanueva, G. B., Perret, V., and Danishefsky, I. (1980). Tryptophan residue at the heparin binding site in antithrombin III. *Arch. Biochem. Biophys.* **203**, 453–457.
18. Karp, G. I., Marcum, J. A., and Rosenberg, R. D. (1984). The role of tryptophan residues in heparin–antithrombin interactions. *Arch. Biochem. Biphys.* **233**, 712–720.

19. Davie, E. W., and Ratnoff, O. D. (1964). Waterfall sequence of intrinsic blood clotting. *Science* **145**, 1310–1312.

20. Rosenberg, J. S., McKenna, P., and Rosenberg, R. D. (1975). Inhibition of human factor IXa by human antithrombin–heparin cofactor. *J. Biol. Chem.* **250**, 8883–8888.

21. Damus, P. S., Hicks, M., and Rosenberg, R. D. (1973). Anticoagulant action of heparin. *Nature (London)* **246**, 355–357.

22. Stead, N., Kaplan, A. P., and Rosenberg, R. D. (1976). Inhibition of activated factor XII by antithrombin–heparin cofactor. *J. Biol. Chem.* **251**, 6481–6488.

23. Godal, H. C., Rygh, M., and Laake, K. (1974). Progressive inactivation of purified factor VII by heparin and antithrombin III. *Thromb. Res.* **5**, 773–775.

24. Broze, G. J., Jr., and Majerus, P. W. (1980). Purification and properties of human coagulation factor VII. *J. Biol. Chem.* **255**, 1242–1250.

25. Jordan, R. E., Oosta, G. M., Gardner, W. T., and Rosenberg, R. D. (1980). The binding of low-molecular-weight heparin to hemostatic enzymes. *J. Biol. Chem.* **255**, 10073–10080.

26. Jordan, R. E., Oosta, G. M., Gardner, W. T., and Rosenberg, R. D. (1980). The kinetics of hemostatic enzyme–antithrombin interactions in the presence of low-molecular-weight heparin. *J. Biol. Chem.* **255**, 10081–10090.

27. Jordan, R. E., Favreau, L. V., Braswell, E. H., and Rosenberg, R. D. (1982). Heparin with two binding sites for antithrombin or platelet factor 4. *J. Biol. Chem.* **277**, 400–406.

28. Hoylaerts, M., Owen, W. G., and Collen, D. (1984). Involvement of heparin chain length in the heparin-catalyzed inhibition of thrombin by antithrombin III. *J. Biol. Chem.* **259**, 5670–5677.

29. Lam, L. H., Silbert, J. E., and Rosenberg, R. D. (1976). The separation of active and inactive forms of heparin. *Biochem. Biophys. Re. Commun.* **69**, 570–577.

30. Hook, M., Bjork, I., Hopwood, J., and Lindahl, U. (1976). Anticoagulant activity of heparin: Separation of high-activity and low-activity heparin species by affinity chromatography on immobilized antithrombin. *Fed. Eur. Biochem. Soc. Lett.* **66**, 90–93.

31. Andersson, L. O., Barrowcliffe, T. W., Holmer, E., Johnson, E. A., and Sims, G. E. C. (1976). Anticoagulant properties of heparin fractionated by affinity chromatography on matrix-bound antithrombin III and by gel filtration. *Thromb. Res.* **9**, 575–581.

32. Oosta, G. M., Gardner, W. T., Beeler, D. L., and Rosenberg, R. D. (1981). Multiple functional domains of the heparin molecule. *Proc. Natl. Acad. Sci. U.S.A.* **78**, 829–833.

33. Rosenberg, R. D., Armand, G., and Lam, L. (1978). Structure–function relationship of heparin species. *Proc. Natl. Acad. Sci. U.S.A.* **75**, 3065–3069.

34. Rosenberg, R. D., and Lam, L. H. (1979). Correlation between structure and function of heparin. *Proc. Natl. Acad. Sci. U.S.A.* **76**, 1218–1222.

35. Lindahl, U., Backstrom, G., Höök, M., Thunberg, L., Fransson, L. A., and Linker, A. (1979). Structure of the antithrombin-binding site of heparin. *Proc. Natl. Acad. Sci. U.S.A.* **76**, 3198–3202.

36. Choay, J., Lormeau, J. C., Petitou, M., Sinay, P., Casu, B., Oreste, P., Torri, G., and Tatti, G. (1980). Anti-Xa active heparin oligosaccharides. *Thromb. Res.* **18**, 573–578.

37. Choay, J., Petitou, M., Lormeau J.-C., Sinay, P., Casu, B., and Gatti, G. (1983). Structural–activity relationship in heparin: A synthetic pentasaccharide with

high affinity for antithrombin III and eliciting high anti-factor Xa activity. *Biochem. Biophys. Res. Commun.* **116**, 492–499.

38. Atha, D. H., Stephens, A. W., Rimon, A., and Rosenberg, R. D. (1984). Sequence variation in heparin octasaccharides with high affinity for antithrombin III. *Biochemistry* **23**, 5801–5812.

39. Lindahl, U., Backstrom, G., and Thunberg, L. (1983). The antithrombin-binding sequence in heparin: Identification of an essential 6-*O* sulfate group. *J. Biol. Chem.* **258**, 9826–9830.

40. Atha, D. H., Lormeau, J. C., Petitou, M., Choay, J., and Rosenberg, R. D. (1985). Contribution of monosaccharide residues in heparin binding to antithrombin III. *Biochemistry* **24**, 6723–6729.

41. Villanueva, G. B. (1984). Predictions of the secondary structure of antithrombin III and the location of the heparin-binding site. *J. Biol. Chem.* **259**, 2531–2536.

42. Atha, D. H., Stephens, A. W., and Rosenberg, R. D. (1984). Evaluation of critical groups required for binding of heparin to antithrombin. *Proc. Natl. Acad. Sci. U.S.A.* **81**, 1030–1034.

43. Pecon, J. M., and Blackburn, M. N. (1984). Pyridoxylation of essential lysines in the heparin-binding site of antithrombin III. *J. Biol. Chem.* **259**, 935–938.

44. Stone, A. L., Beeler, D., Oosta, G., and Rosenberg, R. D. (1982). Circular dichroism spectroscopy of heparin–antithrombin interactions. *Proc. Natl. Acad. Sci. U.S.A.* **79**, 7190–7194.

45. Galli, S. J., Orenstein, N. S., Gill, P. J., Silbert, J. E., Dvorak, A. M., and Dvorak, H. F. (1984). *In* "The Mast Cell: Its Role in Health and Disease" (J. Pepys, and A. M. Edwards, eds.), p. 842. Pitman Medical, Kent, England.

46. Teien, A. N., Abildgaard, U., and Hook, M. (1976). The anticoagulant effect of heparan sulfate and dermatan sulfate. *Trhomb. Res.* **8**, 859–867.

47. Marcum, J. A., and Rosenberg, R. D. (1984). Anticoagulantly active heparin-like molecules from vascular tissue. *Biochemistry* **23**, 1730–1737.

48. Marcum, J. A., Fritze, L., Galli, S. J., Karp, G., and Rosenberg, R. D. (1983). Microvascular heparin-like species with anticoagulant activity. *Am. J. Physiol.* **245**, H725-H733.

49. Meezan, E., Brendel, K., and Carlson, E. C. (1974). Isolation of a purified preparation of metabolically active retinal blood vessels. *Nature* (London). **251**, 65–67.

50. Jarrott, B., Hjelle, J. T., and Spector, S. (1979). Association of histamine with cerebral microvessels in regions of bovine brain. *Brain Res.* **168**, 323–330.

51. Marcum J. A., and Rosenberg, R. D. (1985). Heparinlike molecules with anticoagulant activity are synthesized by cultured endothelial cell. *Biochem. Biophys. Res. Commun.* **126**, 365–372.

52. Marcum, J. A., McKenney, J. B., Galli, S. J., Jackman, R. W., and Rosenberg, R. D. (1986). Anticoagulantly active heparinlike molecules from mast cell-deficient mice. *Am. J. Physiol.* **250**, H879–H888.

53. Marcum, J. A., Atha, D. H., Fritze, L. M. S., Nawroth, P., Stern, D., and Rosenberg, R. D. (1986). Cloned bovine aortic endothelial cells synthesize anticoagulantly active heparan sulfate proteoglycan. *J. Biol. Chem.* **261**, 7507–7517.

54. Yanagishita, M., and Hascall, V. C. (1984). Proteoglycans synthesized by rat ovarian granulosa cells in culture: Isolation, fractionation, and characterization of proteoglycans associated with cell layer. *J. Biol. Chem.* **259**, 10260–10269.

55. Stern, D., Nawroth, P., Marcum, J. A., Handley, D., Rosenberg, R. D., Kisiel, W., and Stern, K. (1985). Interaction of antithrombin III with bovine aortic segments:

Role of heparin in binding and enhanced anticoagulant activity. *J. Clin. Invest.* **75,** 272–279.

56. Lollar, P., and Owen, W. G. (1980). Clearance of thrombin from the circulation in rabbits by high-affinity binding sites on the endothelium: Possible role in the inactivation of thrombin by antithrombin III. *J. Clin. Invest.* **66,** 1222–1330.

57. Busch, C., and Owen, W. G. (1982). Identification *in vitro* of an endothelial cell surface cofactor for antithrombin III: Parallel studies with isolated perfused rat hearts and microcarrier cultures of bovine endothelium. *J. Clin. Invest.* **69,** 726–729.

58. Lollar, P., MacIntosh, S. C., and Owen, W. G. (1984). Reaction of antithrombin III with thrombin bound to the vascular endothelium: Analysis in a recirculating perfused rabbit heart preparation. *J. Biol. Chem.* **259,** 4335–4338.

59. Marcum, J. A., McKenney, J. B., and Rosenberg, R. D. (1984). The acceleration of thrombin–antithrombin complex formation in rat hindquarters *via* naturally occurring heparin-like molecules bound to the endothelium. *J. Clin. Invest.* **74,** 341–350.

60. Ross, R., Glomset, B., Kariya, B., and Harker, L. (1974). Platelet and smooth muscle cell proliferation. *Proc. Natl. Acad. Sci. U.S.A.* **71,** 1207–1210.

61. Harker, L. A., Ross, R., Slichter, S., and Scott, C. (1976). The role of endothelial cell injury and platelet response in its genesis. *J. Clin. Invest.* **58,** 731–741.

62. Heldin, C. H., Wasteson, A., and Westermark, B. (1977). Partial purification and characterization of platelet factors stimulating the multiplication of normal human glial cells. *Exp. Cell Res.* **109,** 429–437.

63. Scher, C., Pledger, W., Martin, P., Antoniades, H., and Stiles, C. (1978). Transforming viruses directly reduce the cellular growth requirement for a platelet-derived growth factor. *J. Cell. Physiol.* **97,** 371–380.

64. Gajdusek, C., DiCorleto, P., Ross, R., and Schwartz, S. (1980). An endothelial cell-derived growth factor. *J. Cell Biol.* **85,** 467–472.

65. Friedman, R. J., Stemerman, M., Wenz, B., Moore, S., Galudie, J., Gent, M., Tiell, M., and Spaet, T. (1977). The effect of thrombocytopenia on experimental arteriosclerotic lesion formation in rabbits: Smooth muscle cell proliferation and re-endothelialization. *J. Clin. Invest.* **60,** 1191–1201.

66. Groves, H., Kinlough-Rathbone, R., Richardson, M., Moore, S., and Mustard, J. (1979). Platelet interaction with damaged rabbit aorta. *Lab. Invest.* **40,** 194–200.

67. Moore, S., Friedman, R., Singal, D., Galudie, J., Blajchman, M., and Roberts, R. (1976). Inhibition of injury induced thromboatherosclerotic lesions by antiplatelet serum in rabbits. *Thromb. Haemostasis* **35,** 70–81.

68. Clowes, A. W., and Karnovsky, M. J. (1977). Suppression by heparin of smooth muscle cell proliferation in injured arteries. *Nature (London)* **265,** 625–626.

69. Guyton, J., Rosenberg, R., Clowes, A., and Karnovsky, M. (1980). Inhibition of rat arterial smooth muscle cell proliferation by heparin. I. In vivo studies with anticoagulant and non-anticoagulant heparin. *Circ. Res.* **46,** 625–634.

70. Hoover, R., Rosenberg, R., Haering, W., and Karnovsky, M. J. (1980). Inhibition of rat arterial smooth muscle cell proliferation by heparin, Part II. In vitro studies. *Circ. Res.* **47,** 578–583.

71. Castellot, J., Addonizio, M., Rosenberg, R., and Karnovsky, M. J. (1981). Cultured endothelial cells produce a heparinlike inhibitor of smooth muscle cell growth. *J. Cell Biol.* **90,** 372–379.

72. Oosta, G., Favreau, L., Beeler, D., and Rosenberg, R. (1982). Purification and properties of human platelet heparitinase. *J. Biol. Chem.* **257,** 11249–11255.

73. Castellot, J., Favreau, L., Rosenberg, R., and Karnovsky, M. (1982). Inhibition of vascular smooth muscle cell growth by endothelial cell-derived heparin: Possible role of a platelet endoglycosidase. *J. Biol. Chem.* **257**, 11256–11260.

74. Fritze, L., Reilly, C., and Rosenberg, R. (1985). An antiproliferative heparan sulfate species produced by postconfluent smooth muscle cells. *J. Cell Biol.* **100**, 1041–1049.

75. Castellot, J., Beeler, D., Rosenberg, R., and Karnovsky, M. J. (1984). Structural determinants of the capacity of heparin to inhibit the proliferation of vascular smooth muscle cells. *J. Cell. Physiol.* **120**, 315–320.

76. Castellot, J., Wong, K., Herman, B., Hoover, R., Albertini, D., Wright, T., Caleb, B., and Karnovsky, M. (1985). Binding and internalization of heparin by vascular smooth muscle cells. *J. Cell. Physiol.* **124**, 13–20.

77. Levine, S., and Krentz, L. (1977). Development of a radioimmunoassay for human platelet factor 4. *Thromb. Res.* **11**, 673–686.

78. Majack, R., and Bornstein, P. (1985). Heparin regulates the collagen phenotype of vascular smooth muscle cells: Induced synthesis of an M_r 60,000 collagen. *J. Cell Biol.* **100**, 613–619.

79. Cochran, D., Castellot, J., and Karnovsky, M. (1985). Effect of heparin on vascular smooth muscle cells. II. Specific protein synthesis. *J. Cell. Physiol.* **124**, 29–36.

80. Majack, R., and Bornstein, P. (1984). Heparin and related glycosaminoglycans modulate the secretory phenotype of vascular smooth muscle cells. *J. Cell Biol.* **99**, 1688–1695.

81. Majack, R., Cook, S., and Bornstein, P. (1985). Platelet-derived growth factor and heparin-like glycosaminoglycans regulate thrombospondin synthesis and deposition in the matrix by smooth muscle cells. *J. Cell Biol.* **101**, 1059–1070.

82. Majack, R., and Clowes, A. (1984). Inhibition of vascular smooth muscle cell migration by heparin-like glycosaminoglycans. *J. Cell. Physiol.* **118**, 253–256.

83. Castellot, J., Cochran, D., and Karnovsky, M. (1985). Effect of heparin on vascular smooth muscle cells. I. Cell metabolism. *J. Cell. Physiol.* **124**, 21–28.

84. Ross, R., Glomset, J., Kariya, B., and Harker, L. (1974). Platelets and smooth muscle cell proliferation. *Proc. Natl. Acad. Sci. U.S.A.* **71**, 1207–1210.

85. Pledger, W., Stiles, C., Antoniades, H., and Scher, C. (1977). Induction of DNA synthesis in Balb/C 3T3 cells by serum components: Reevaluation of the commitment process. *Proc. Natl. Acad. Sci. U.S.A.* **74**, 4481–4485.

86. Bhargava, G., Rifas, L., and Makman, M. (1979). Presence of epidermal growth factor receptors and influence of epidermal growth factor on proliferation and aging in cultured smooth muscle cells. *J. Cell. Physiol.* **100**, 365–374.

87. Assoian, R., Grotendorst, G., Miller, D., and Sporn, M. (1984). Cellular transformation by coordinated action of three peptide growth factors from human platelets. *Nature (London).* **309**, 804–806.

88. Oka, Y., and Orth, D. (1983). Human plasma epidermal growth factor/B-urogastrone is associated with blood platelets. *J. Clin. Invest.* **72**, 249–259.

89. Loef, E., Wharton, W., Van Wyk, J., and Pledger, W. (1982). Epidermal growth factor (EGF) and somatomedin C regulate G1 progression in competent Balb/c-3T3 cells. *Exp. Cell Res.* **141**, 107–115.

90. Vogel, A., Raines, E., Kariya, B., Rivest, M., and Ross, R. (1978). Coordinate control of 3T3 cell proliferation by platelet-derived growth factor and plasma components. *Proc. Natl. Acad. Sci. U.S.A.* **75**, 2810–2814.

91. Heldin, C., Westermark, B., and Wasteson, A. (1982). Specific receptors for

platelet-derived growth factor on cells derived from connective tissue and glia. *Proc. Natl. Acad. Sci. U.S.A.* **78**, 3664–3668.

92. Williams, L., Tremble, P., and Antoniades, H. (1982). Platelet-derived growth factor binds specifically to receptors on vascular smooth muscle cells and the binding becomes nondissociable. *Proc. Natl. Acad. Sci. U.S.A.* **79**, 5867–5870.

93. Bowen-Pope, D., and Ross, R. (1982). Platelet-derived growth factor II. Specific binding to cultured cells. *J. Biol. Chem.* **257**, 5161–5171.

94. Stiles C. (1983). The molecular biology of platelet-derived growth factor. *Cell* **33**, 653–655.

95. Guroff, G., ed. (1983). "Growth and Maturation Factors," Vol. 1. Wiley, New York.

96. Weinstein, R., Stemerman, M., and Maciag, T. (1981). Hormonal requirements for growth of arterial smooth muscle cells in vitro: An endocrine approach to atherosclerosis. *Science* **212**, 818–820.

97. Bowen-Pope, D., DiCorleto, P., and Ross, R. (1983). Interactions between the receptors for platelet-derived growth factor and epidermal growth factor. *J. Cell Biol.* **96**, 679–683.

98. Davis R., and Czech, M. (1985). Tumor-promoting phorbol diesters cause the phosphorylation of epidermal growth factor receptors in normal human fibroblasts at threonine-654. *Proc. Natl. Acad. Sci. U.S.A.* **82**, 1974–1978.

99. McCaffrey, P., Freidman, B., and Rosner, M. (1984). Diacylglycerol modulates binding and phosphorylation of the epidermal growth factor receptor. *J. Biol. Chem.* **259**, 12502–12507.

100. Owen, N. (1985). Effect of TPA on ion fluxes and DNA synthesis in vascular smooth muscle cells. *J. Cell Biol.* **101**, 454–459.

101. Decker, S. (1984). Effects of epidermal growth factor and 12-O-tetradecanoyl-phorbol-13-acetate on metabolism of the epidermal growth factor receptor in normal human fibroblasts. *Mol Cell. Biol.* **4**, 1718–1724.

102. Frantz, C., Stiles, C., and Scher, C. (1979). The tumor promoter 12-O-tetradecanoyl-phorbol-13-acetate enhances the proliferative response of Balb/c-3T3 cells to hormonal growth factors. *J. Cell. Physiol.* **100**, 413–424.

103. Whittenberger, B., Raben, D., Lieberman M., and Glaser, L. (1978). Inhibition of growth of 3T3 cells by extract of surface membranes. *Proc. Natl. Acad. Sci. U.S.A.* **75**, 5457–5461.

104. Heimark, R., and Schwartz, S. (1985). The role of membrane–membrane interactions in the regulation of endothelial cell growth. *J. Cell Biol* **100**, 1934–1940.

Cell-Associated Proteoglycans in Human Malignant Melanoma

John R. Harper and Ralph A. Reisfeld

Laboratory of Tumor Cell Biology, Scripps Clinic and Research Foundation, La Jolla, California 92037

I. EXTRACELLULAR MATRIX COMPOSITION DURING TUMORIGENESIS

The pathogenicity of cancer is largely determined by interactions mediated at the interface between transformed cells and the surrounding normal tissue. A major component in the interaction of a tumor cell with its environment is the basement membrane or extracellular matrix (ECM). The ECM elaborated by normal cells serves not only as a hydrated structural barrier through which cell nutrients diffuse, but also as a source of distinct signals regulating cellular differentiation and growth control. During the gradual, stepwise process of malignant transformation, premalignant cells continue to contribute both in composition and organization to the matrix, but no longer respond normally to its regulatory signals by cessation of proliferation and differentiation. Malignant progression is then characterized by alterations in cell surface composition and in secretion and organization of ECM components by the transformed cells. Moreover, the invasion and metastasis of malignant cells is accomplished, in part, by the secretion of enzymes by the tumor cells that degrade specific ECM components.

Alterations in the composition of basement membrane or ECM are believed to result from varying degrees of changes in biosynthesis and degradation of ECM molecules by tumor cells (Alitalo *et al.*, 1981; Liotta *et al.*, 1983; Liotta, 1984), whereas recent evidence has suggested that effector substances secreted by these cells may regulate the expression of ECM components by adjacent normal stromal cells such as fibroblasts (Iozzo, 1984a). Extensive studies into the changes in ECM composition that accompany neoplastic transformation have focused on the synthesis and degradation of glycosaminoglycans

BIOLOGY OF PROTEOGLYCANS

(GAG), such as hyaluronic acid and heparin, as well as heparan sulfate and chondroitin sulfate proteoglycans (for reviews, see Turley, 1984, and Iozzo, 1985). Collectively, these studies suggest that there does not seem to be a single aberration in the synthesis of such molecules that is shared by tumors of all histological types; however, in many of the cases examined, it appeared that the profile of GAG and/or proteoglycan synthesis may be altered compared to the corresponding normal tissue.

Biochemical and immunological approaches to the study of proteoglycan expression in normal and transformed tissues other than connective tissue have suggested that proteoglycans may vary in structure across histological tissue types, but share many characteristic structures and biosynthetic mechanisms with the corresponding molecules found in connective tissue. In this chapter, human malignant melanoma will be presented as an example of tumor tissue that expresses a cell-associated chondroitin sulfate proteoglycan whose basic structure and biosynthesis appears similar to that of cartilage proteoglycan. Along with certain characteristics shared with cartilage proteoglycans, melanoma-associated proteoglycan (MPG) has a relatively unique tissue distribution and some distinguishing antigenic determinants. Furthermore, as an illustration of the tissue-specific nature of this class of molecule, monoclonal antibodies that recognize determinants on the core protein of the MPG were isolated on the basis of their limited reactivity with normal human tissues and extensive reactivity with cultured human melanoma cells, as well as tumor tissue taken on biopsy. These antibodies have been useful as both functional and biochemical probes for establishing the human MPG as a model for the studying the physical characteristics, biosynthesis, and possible functional role of this macromolecule in the pathogenesis of melanoma.

II. Chondroitin Sulfate Proteoglycans Synthesized by Human Melanoma Cells: Production of Monoclonal Antibodies

Human malignant melanoma cells synthesize a cell-associated chondroitin sulfate proteoglycan that has a limited distribution of expression in normal human tissues. The discovery of this molecule was prompted by the findings of several laboratories involved in producing monoclonal antibodies against antigens expressed by melanoma cells and not by normal human tissues or tumors of other histological types (Harper et al., 1982). Throughout the course of such studies, murine monoclonal antibodies that react with antigenic determinants on the core glycoprotein of a chondroitin sulfate proteoglycan were isolated by using a variety of immunogens derived from cultured melanoma cells (Table I). For example, Morgan et al. (1981) isolated monoclonal antibody 9.2.27 by immunizing mice with a 4 M urea extract of melanoma cells that had been immunodepleted of immunodominant molecules, such as fibronectin and HLA antigens. An antibody referred to as 155.8 was later produced by immunizing mice with a crude membrane preparation obtained from a melanoma cell line (Harper et al., 1984a). Monoclonal antibodies that recognize MPG have also been isolated by immunizing mice with human astrocytoma cells, one of the few nonmelanoma cell types that express the molecule (Cairncross et al., 1982). Several anti-MPG monoclonal antibodies produced in different laboratories have been compared and found to react with different determinants on the same population of molecules with varying affinities (Table I).

Initial biochemical studies using these antibodies described the antigen as a complex containing a 250-kDa glycoprotein and a heterogeneous, high molecular weight component (Morgan et al., 1981). Bumol and Reisfeld (1982) reported later, using the monoclonal antibody 9.2.27, that not only was the high molecular weight component sulfated, but digestion of immunoprecipitated material with chondroitinase AC or ABC resulted in the loss of the high molecular weight component and increased amounts of the 250-kDa glycoprotein. Fur-

TABLE I

MONOCLONAL ANTIBODIES AGAINST MELANOMA-ASSOCIATED PROTEOGLYCAN (MPG)

Antibody (isotype)	Immunogen	Specificity	Reference
225.28S (IgG$_{2a}$)	Melanoma cells	MPG, C	Giacomini et al. (1983)
149.53 (IgG$_1$)	Melanoma cells	MPG, C	Giacomini et al. (1984)
763.74T (IgG$_1$)	Melanoma cells	MPG, C	Giacomini et al. (1984)
B5 (IgG$_{2a}$)	Melanoma cells	MPG, C	Real et al. (1985)
AO122 (IgG$_1$)	Astrocytoma cells	MPG, C	Cairncross et al. (1982)
48.7 (IgG$_1$)	Melanoma cells	MPG, C	Hellstrom et al. (1983)
F24.47 (IgG$_{2b}$)	Melanoma cells	MPG	H. M. Yang, personal communication
155.8 (IgG$_1$)	Melanoma membranes	MPG, C	Harper et al. (1984)
9.2.27 (IgG$_{2b}$)	4 M Urea melanoma extract	MPG, C	Morgan et al. (1981)
O$_1$-95-45 (IgG$_1$)	Melanoma cells	MPG, C	Herlyn et al. (1983)
691-19-19 (IgG$_1$)	Melanoma cells	MPG, C	Koprowski et al. (1978)

thermore, they found that the high molecular weight molecule was sensitive to the β-elimination reaction under dilute alkaline conditions, consistent with the presence of O-linked GAG. Based on these findings, the hypothesis was put forth that the sulfated, high molecular weight antigen immunoprecipitated by 9.2.27 was a chondroitin sulfate proteoglycan and the 250-kDa component was the free core glycoprotein without GAG chains attached. The molecular characterization of MPG has followed the approaches set forth by the extensive literature involving the study of chondroitin sulfate proteoglycans derived from cartilage and other connective tissue (Hascall, 1981).

A. Physicochemical Characteristics

Analysis of the MPG has been greatly facilitated by monoclonal antibodies 9.2.27 and 155.8, which were used for immunoprecipitation experiments. Figure 1 shows the molecular profile of MPG precipitated by 9.2.27 and 155.8, then analyzed by sodium dodecyl sulfate–polyacrylamide gel electrophoresis (SDS–PAGE) on a 5% gel. Whether immunoprecipitated by 9.2.27 or 155.8, MPG appears as a heterogeneous smear that enters the 5% resolving gel and incorporates $^{35}SO_4$ (lane B), in addition to a 250-kDa glycoprotein, whose migration increased slightly after digestion with neuraminidase (Bumol and Reisfeld, 1982).

Estimates of the molecular weight of the core protein obtained by SDS–PAGE vary slightly between laboratories from 250 to 260 kDa. Using gel exclusion chromatography on Sephacryl S-400 in nondenaturing conditions, Ross et al. (1983) estimated a molecular mass of 520 kDa for the native core protein and therefore suggested that the core protein exists as a noncovalent homodimer. It is not clear if this association is present naturally on the cell surface or if it only occurs during the isolation procedure. The molecular mass of mature MPG, containing a full complement of chondroitin sulfate chains, has been estimated at 420–1000 kDa, a value that would certainly vary depending on the number of GAG chains attached to the 250-kDA core protein (Ross et al., 1983; Garrigues et al. 1986). This would imply that the MPG core protein could contain from 3 to 12 GAG chains, using 60 kDa as an estimate for the size of chondroitin sulfate chains.

The GAG chains attached to the core protein are cleavable into disaccharide chains by chondroitinase AC and ABC (Fig. 1, lanes C and E), resulting in the release of core protein. When these assays are performed on $^{35}SO_4$-labeled cell lysates, the chondroitinase-released

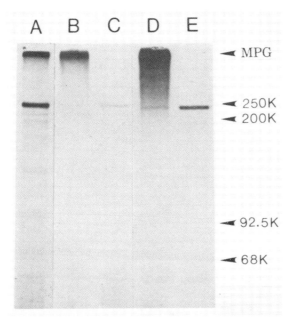

FIG. 1. Analysis of MPG and its core protein immunoprecipitated by antibodies 9.2.27 and 155.8. UCLA M21 melanoma cells were metabolically labeled with [³H]leucine (lane A) or ³⁵SO₄ to equilibrium and lysed with RIPA lysis buffer [100 mM Tris-Cl, pH 7.2, 0.15 M NaCl, 1% (w/v) deoxycholate, 1% (w/v) Nonidet- P-40, 0.1% sodium dodecyl sulfate (SDS), and 1% Trasylol (Sigma Chemical Co., St. Louis, Missouri)]. Lanes A–C are immunoprecipitated by antibody 155.8 and lanes D and E by 9.2.27. Lanes C and E represent immunoprecipitated molecules that were digested with chondroitinase ABC prior to electrophoretic separation on the 5% SDS–polyacrylamide gel. From Harper *et al.* (1984).

core protein appears to be sulfated. This residual sulfate may represent sulfated disaccharide stubs that remain protein bound following chondroitinase digestion (Heinegard and Hascall, 1974).

Identification of the GAG chains attached to the MPG core protein was performed by cellulose acetate electrophoresis (Harper *et al.*, 1984). Melanoma cells were labeled to equilibrium with [³H]glucosamine and MPG was isolated by immunoprecipitation with the 155.8 antibody. Following immunoprecipitation, GAG chains were released from the core protein via β-elimination by treatment of MPG with 0.05 N NaOH–1 M NaBH₄. Electrophoretic analysis of the liberated material on cellulose acetate in comparison to authentic GAG standards, showed that GAG chains linked to the MPG core protein were solely of chondroitin sulfate type; there was no evidence of dermatan or heparan sulfate chains.

Ross *et al.* (1983) analyzed carbohydrates released from purified MPG by using ion-exchange chromatography and also found a predominant chondroitin sulfate peak; however, they also found a small peak comigrating with heparan sulfate, which probably represents material that copurified with MPG from melanoma cell ghosts. The size of chondroitin sulfate chains associated with MPG appears to vary with a range from 30 to 60 kDa, depending on the type of molecular weight standards used in Sepharose CL-6B gel exclusion chromatography (Ross *et al.*, 1983; Garrigues *et al.*, 1986). The disaccharide composition of the chondroitin sulfate chains may vary according to the melanoma cell line used for the analysis.

Bhanvanandan (1981) reported that melanoma cells lacked the △Di-6S in chondroitin sulfate chains, whereas in a later report using 9.2.27-immunoprecipitated MPG, Bumol *et al.* (1984) found that both △Di-4S and △Di-6S disaccharides were present in chondroitinase ABC digests. It is not known whether this variation is an artifact of tissue culture or a naturally occurring diversity within this cell type.

Although the primary structure has yet to be determined, many biochemical properties of the MPG core protein have been reported. In analyzing the carbohydrates associated with MPG and its core protein, Ross and co-workers (1983) found that glycopeptides generated by pronase digestion of both components could be resolved into three size peaks. The fraction eluting in the void volume of a Sephadex G-50 column, present only in MPG digests, contained the chondroitin sulfate population. The fraction eluting next could not be eliminated by alkaline treatment and was common to both MPG and core protein, suggesting that it contained N-asparagine-linked oligosaccharides. Finally, a third small fraction released by alkali was found only in MPG digests, and believed to contain more O-linked oligosaccharide-containing glycopeptides; however, because of the limited amount of material available, this fraction could not be further analyzed. It appears from these studies that the MPG core protein contains only N-linked oligosaccharides, in contrast to chondroitin sulfate proteoglycan found in cartilage, which also contain substantial amounts of O-linked oligosaccharides (Lohmander *et al.*, 1980; Nilsson *et al.*, 1982).

The MPG core protein appears to be a sialoglycoprotein, since its migration in SDS–PAGE was increased after neuraminidase treatment (Bumol and Reisfeld, 1982). The isoelectric point of the MPG core protein ranges form 6.9 to 7.0, depending on whether or not the molecule has been treated with neuraminidase (Ross *et al.*, 1983). Table II contains the amino acid composition of the MPG core protein. This

TABLE II

AMINO ACID COMPOSITION OF THE MPG
CORE PROTEIN[a]

Amino acid	Residues/1000
Aspartic acid	75
Threonine	50
Serine	50
Glutamic acid	126
Proline	82
Glycine	97
Alanine	96
Valine	81
Methionine	8
Isoleucine	27
Leucine	128
Tyrosine	18
Phenylalanine	43
Histidine	30
Lysine	24
Arginine	65

[a] Data from Ross et al. (1983).

composition is one of a polar molecule (i.e., relatively high leucine but low lysine).

Rettig et al. (1986) have reported evidence suggesting that the gene coding for the MPG core protein resides on human chromosome 15. Discordancy analysis, using human–rodent cell hybrids containing different human chromosomes, of four monoclonal antibodies that recognize different epitopes on the MPG core protein found no discordant clones containing chromosome 15. This analysis allowed the chromosomal assignment of the core protein gene to human chromosome 15.

B. Cell Surface Localization

The arrangement of MPG and its core protein on the surface of cultured melanoma cells has been examined by several methods. Immunofluorescence analysis with the 155.8 monoclonal antibody revealed that MPG is distributed on sharply delineated microfilamentous structures on the apical and lateral surfaces of adherent melanoma cell (Fig. 2). These microspike-like structures can be seen forming

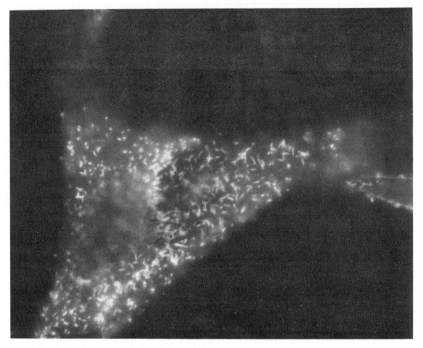

FIG. 2. Indirect immunofluorescence of Melur melanoma cells grown on coverslips with the 155.8 antibody X630. From Harper *et al.* (1984).

attachment sites with the solid substratum to which the cells are adhering. Based on this distribution, "substrate-attached material," such as that left behind following cell removal with chelating agents (i.e., EDTA; Culp *et al.*, 1979), were analyzed for the presence of MPG by immunofluorescence. While these adhesion plaques were rich in specific gangliosides, such as GD2 and GD3, MPG could not be detected (Cheresh *et al.*, 1984).

Distinct localization of MPG on the surface of melanoma cells has been established by Garrigues *et al.* (1986) using immunocytochemical techniques complemented by electron microscopy. These studies localized MPG and its core protein to specific clusters on the microspike structures, using the 48.7 and F24.47 antibodies as molecular probes. A statistical analysis of immunogold particle distribution revealed that there were 39.5 particles to a micron of microspike membrane compared to 0.35 and 2.9 particles per micron of bleb or basal cell surface, suggesting that MPG and its core protein were concentrated in specific coincidence on the microspikes. The microspike localization of MPG, as well as its presence at the points of cell attachment, would be consistent with a putative role of MPG in cell–cell and/or cell–substratum interactions.

C. Tissue Distribution

One of the richest sources of chondroitin sulfate proteoglycan in the body is cartilage and related connective tissue (Couchman et al., 1984). Several investigators using their respective anti-MPG monoclonal antibodies have studied the distribution of MPG and its core protein throughout the human body, as well as its expression in a variety of normal and neoplastic cultured cell lines.

Our initial studies showed that antibodies 9.2.27 and 155.8 fail to react with human cartilage, as determined by indirect immunoperoxidase staining; these tissue could be stained well with antibodies against fibronectin (J. R. Harper, T. F. Bumol, S. Levine, and R. A. Reisfeld, unpublished data). We obtained the same result after pretreating tissue with chondroitinase ABC, which partially ruled out the possibility that antibody binding was being sterically hindered by GAG chains or that GAG chains participate in the antigenic determinants. This was the first suggestion that the MPG core protein had unique structural determinants compared to the cartilage-derived core protein. In further screening of normal adult and fetal tissues, we and others found that the only tissues containing quantitatively comparable MPG were cells of benign nevi (Hellstrom et al., 1983). Adult and fetal melanocytes appear to express little if any MPG, suggesting that its expression may be somehow related to a particular proliferative state or stage of differentiation.

There appears to be some heterogeneity in the expression of MPG by primary and metastatic melanoma tumors. Even though extracts from a variety of normal tissues and other tumor types were negative of MPG expression, Hellstrom et al. (1983) found that 21 of 30 (70%) metastatic melanomas stained positive with antibody 48.7, while all primary tumors tested were MPG positive. One interesting case was found in which a section from a single nevus taken from a melanoma patient contained one nest of nevus cells staining strongly for MPG, while two others in the same nevus were clearly negative. The latter two nests of cells stained positive for the melanoma-associated antigen p97, suggesting that they may be premalignant lesions.

Ross et al. (1983) and Cairncross et al. (1982) have reported significant expression of MPG by malignant astrocytoma cells; in fact, while producing monoclonal antibodies against these tumor cells, they isolated one antibody, AO122, that recognizes the MPG core protein. This antibody reacted with 56% of astrocytomas and 100% of melanomas, but only slightly with fetal and adult brain and fetal skin fibroblasts. Real *et al.* (1985) have screened an extensive list of cell lines and tissues using antibody B5 and, with the exception of weak reactivity with adult keratinocytes and capillary endothelial cells, reported concurring results.

The extensive screening of fresh human tissues and cultured normal and tumor cells using a variety of monoclonal antibodies suggests that MPG has a very limited tissue distribution, quite distinct from that of cartilage-derived chondroitin sulfate proteoglycans. Its restricted expression by cells in a proliferative state, such as malignant melanoma and benign nevus cells, may imply a functional role in the process of cell proliferation. In this regard, resting cultured melanocytes, both *in vivo* and *in vitro*, do not express MPG, unless they are induced to proliferate by *in vitro* treatment with cholera toxin and the tumor promoter 12-*O*-tetradecanoylphorbol-13-acetate (TPA) (Real *et al.*, 1985). These agents induce cellular proliferation in some systems and differentiation in others (Yuspa *et al.*, 1982); thus the interpretation of these findings is complicated.

D. Functional Properties: Possible Role in Malignancy

Invasion and metastasis are processes that require tumor cells to express a complex array of molecular and biological phenotypes (Nicholson, 1982, 1984). Throughout the process of metastatic spread, ECM components, as well as molecules that regulate their expression and catalyze their degradation seem to be directly involved.

As tumor cells detach from the primary tumor and become invasive, they probably begin to secrete degradative enzymes, such as collagenases (Liotta, 1984), as well as increased amounts of GAG such as hyaluronic acid (Turley and Tretiak, 1985). After entering the circulatory or lymphatic systems and relocating in another tissue site via a blood-borne or lymphatic route, the metastatic tumor cell must provide a proper environment to support the establishment of colonies in the secondary site. This may occur by an induction of molecular changes in surrounding stromal cells (Iozzo, 1984a,b) or by a change in its own gene expression, resulting in the establishment of a suitable growth environment. Several investigators have been interested in the possible role of MPG in such processes, since it seems uniquely associated with specific types of tumors and proliferative cells.

Many studies, whose objective was to dissect the complex metastatic process and study the involvement of MPG, have focused on a variety of *in vitro* assays believed to simulate the various steps in the metastatic program. Human melanoma cells form proliferating colonies when suspended in soft agar culture medium. The ability to grow in an anchorage-independent manner is a unique characteristic of transformed cells required for colonization in a secondary organ site, the event that culminates a successful metastatic event (Cifone and Fidler, 1980). Monoclonal antibody 9.2.27 inhibits the colony-forming ability of cultured melanoma cells in soft agar by 65–70%, without mediating a direct cytotoxicity (Harper and Reisfeld, 1983). Since cells grown in soft agar are not attached and spread on a solid substratum, the antibody may be interfering with a cell–cell interaction or the interaction of MPG on the cell surface with the surrounding oligosaccharides in the agar medium that are important for anchorage-independent growth.

Due to the specificity with which 9.2.27 binds melanoma cells *in vitro* and *in vivo*, this antibody has potential as an effective diagnostic and therapeutic reagent against malignant melanoma. In correlation with *in vitro* data showing an inhibition of cell growth in soft agar by the 9.2.27 antibody, Bumol and co-workers (1983) reported that the 9.2.27 antibody had an effect on the growth of melanoma cells as tumors in nude mice. In these studies which were designed to test the cytotoxic effect of 9.2.27–diphtheria toxin (α-chain) conjugates *in vitro* and *in vivo*, Bumol *et al.* (1983) found that the antibody alone was noncytotoxic in liquid culture, while antibody–toxin conjugates were highly toxic. *In vivo*, the 9.2.27 antibody per se inhibited melanoma tumor growth in nude mice by as much as 65%, compared to 52% by the antibody–toxin conjugates. These results further support the idea that binding of antibody to MPG and its core protein may inhibit processes essential to tumor growth and development *in vivo*, possibly by interfering with cell–cell or cell–substratum interactions.

Bumol *et al.* (1984) examined the processes of cell adhesion and spreading on a solid substratum more directly, by studying the ability of melanoma cells to attach and spread on ECM elaborated by bovine aortic endothelial cells. Adhesion substrata, consisting of ECM, were prepared by removing confluent layers of endothelial cells using EDTA; then melanoma cells grown in serum-free medium were plated and allowed to adhere in presence or absence of the 9.2.27 antibody. Quantitatively, the effect of 9.2.27 on adhesion was relatively small: 27% inhibition compared to 9.5% inhibition by the control antibody (W6/32, anti-HLA A,B,C). The most significant effect of 9.2.27 was on the ability of cells to spread into the usual "fried-egg" morphology needed to form a tight interaction with the substratum. While the precise role played by MPG and its core protein in the biology of human melanoma cells is not yet clear, these data suggest that MPG may play an essential role related to the interactions between tumor cells and their immediate surroundings.

III. Biosynthesis and Intracellular Transport of Proteoglycans

A. Cartilage Proteoglycans

In order to understand fully the processes by which MPG and its core protein are synthesized, it is important to review the extensive data available on the biosynthesis and processing of cartilage-type chondroitin sulfate proteoglycans, a model that has provided a frame of reference for the study of MPG. Biosynthetic studies of cartilage-type proteoglycan have been performed predominantly using primary chondrocyte cultures derived from Swarm rat chondrosarcoma (Choi et al., 1971), a transplantable tumor that secretes large amounts of chondroitin sulfate proteoglycan resembling the embryonic type (Oegema et al., 1975, 1977).

During and subsequent to the translation of the core protein molecule, several types of posttranslational modifications occur that ultimately account for ~90% of the final mass of mature proteoglycan (Hascall, 1981). In the rat chondrosarcoma system, "high-mannose," N-asparagine-linked oligosaccharides appear to be added cotranslationally, as in the biosynthesis of other glycoproteins (Kimura et al., 1984). High-mannose, N-linked oligosaccharides are then processed to the complex type, probably at the rough endoplasmic reticulum border with the cis-Golgi (Kornfeld and Kornfeld, 1980). The newly synthesized core protein exists with a half-life of ~90 min in a pool within the cell, from which molecules are processed by the addition of GAG chains to form mature proteoglycans (Kimura et al., 1981).

Kinetic analysis of the incorporation of labeled carbohydrate precursors into O-linked oligosaccharides and GAG chains has shown that these two types of posttranslational modification probably occur almost simultaneously within vicinal regions of the Golgi compartment (Thonar et al., 1983). Viewing the total transport pathway of this secreted proteoglycan, 70% of its intracellular dwell time, prior to secretion, is spent in the rough endoplasmic reticulum, followed by the rapid addition of GAG chains and release into the extracellular medium (Fellini et al., 1984).

B. Melanoma-Associated Proteoglycan

The biosynthesis and intracellular transport of MPG have been studied extensively using pulse–chase experiments analyzed by immunoprecipitation with the 9.2.27 monoclonal antibody. This approach is particularly informative, since the antibody recognizes intracellular precursors of the core protein following the addition of N-linked oligosaccharides in their "high-mannose" form, the completed 250-kDa core protein, as well as the fully glycosylated proteoglycan.

At the 0 time point after a 5-min biosynthetic pulse with [35S]methionine, three polypeptides can be immunoprecipitated from lysates of melanoma cells (i.e., 210, 220, and 240 kDa), all of which are sensitive to digestion with endo-β-N-acetylglycosaminidase H (Endo H) (Bumol and Reisfeld, 1982). Between 10 and 20 min, the Endo H-resistant 250-kDa core protein appears, indicating that synthesis of the core protein and processing of the N-linked oligosaccharides occur rapidly. Once the mature core protein is observed in pulse–chase profiles, GAG are added almost simultaneously and intact MPG can be immunoprecipitated (Fig. 3).

Long-term pulse–chase studies suggest that the half-life of MPG core protein in its association with melanoma cells is ~15.6 hr (Bumol et al., 1984). In these experiments, MPG could be seen associated with cell lysates for as long as 72 hr. These dwell times exceeded those measured for proteoglycans secreted by Swarm rat chondrosarcoma chondrocytes, thus raising a question as to the distribution of MPG and its core protein on or within melanoma cells. Lactoperoxidase-catalyzed cell surface iodination of melanoma cells followed by immunoprecipitation of MPG and its core protein resulted in the labeling of both components (Hellstrom et al., 1983; Ross et al., 1983), and from this result it was hypothesized that the MPG core protein was present on the cell surface in two forms: as a free glycoprotein and with chondroitin sulfate chains attached.

Another approach used to determine the steady-state cellular location of MPG and its core protein was to apply a procedure that selectively immunoprecipitates metabolically labeled molecules from the surface of viable cells (Vitetta and Uhr, 1975 and Fitting and Kabat, 1982). Melanoma cells were first intrinsically labeled with [35S]methionine or [3H]leucine and then reacted with the 9.2.27 antibody in suspension. Since 9.2.27 recognizes intracellular core protein

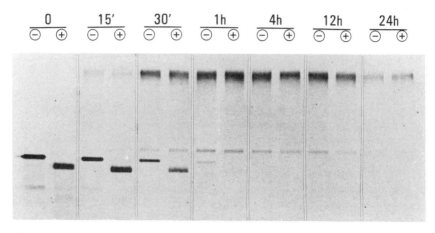

Fig. 3. Biosynthesis of MPG and its core protein by UCLA M21 melanoma cells analyzed by indirect immunoprecipitation with 9.2.27 followed by electrophoresis on a 5% polyacrylamide gel (SDS–PAGE). After a 5-min biosynthetic pulse with [³⁵S]-methionine, cells were incubated at 37°C in medium containing unlabeled methionine; aliquots were removed at the designated times for lysis in RIPA buffer, and immunoprecipitation of MPG and its precursors by 9.2.27 was performed on each sample. Each immunoprecipitated sample was then divided in half, such that one half was digested with Endo H (+), and the other sample incubated in enzyme dilution buffer alone (−).

precursors, the antibody-coated cells were then lysed in the presence of a vast excess of unlabeled cell extract, which prevented the exchange of labeled intracellular molecules with labeled surface components bound to the antibodies. The immunoprecipitation profiles from this type of experiment contained both MPG and its core protein in cell surface immunoprecipitated samples, but not the 240-kDa, Endo H-sensitive core protein precursor, which was only present in whole-cell lysates (Harper et al., 1986; Spiro et al., 1986). Since the unprocessed precursor serves as a control for intracellular molecules, it was therefore possible to conclude that after synthesis, the MPG core protein could, in fact, be transported directly to the surface of melanoma cells either as a glycoprotein or further modified by the addition of chondroitin sulfate chains within the Golgi compartment, prior to being transported to the cell surface.

The use of metabolic inhibitors that interfere with specific steps in the biosynthesis of proteoglycans has provided information that supplements the kinetic analyses as to the intracellular compartments serving as the sites of proteoglycan maturation. The effects of monensin on proteoglycan synthesis are multifaceted in various systems.

Early studies analyzing the effects of monensin on proteoglycan synthesis by chick sternal chondrocytes showed that proteoglycans secreted by cells treated with monensin were undersulfated (Tajiri et al., 1980). Furthermore, when the intracellular proteoglycan precursors were examined, it was found that monensin also induced an intracellular accumulation of underglycosylated, undersulfated core protein (Nishimoto et al., 1982). The complex effects of monensin on the proteoglycan-synthetic pathway in chick chondrocytes appears to involve, among other biosynthetic mechanisms, the enzymes and precursors responsible for the sulfation of GAG (Kajiwara and Tanzer, 1981); this activity is believed to be localized within the trans-Golgi compartment (Silbert and Freilich, 1980). In Swarm rat chondrosarcoma chondrocytes, synthesis and sulfation of chondroitin sulfate chains, but not hyaluronic acid, are inhibited by monensin (Mitchell and Hardingham, 1982; Stevens et al., 1985).

The synthesis of MPG by human melanoma cells is affected by monensin in a manner similar to that of proteoglycans in other cell systems. Monensin not only inhibits the assembly of the mature MPG molecule (Bumol and Reisfeld, 1982), but it interferes with the maturation of the 250-kDa core protein. While monensin does not appear to affect the synthesis of the initial precursors to the core protein (i.e., the 210, 220, and 240 kDa molecules), it does block the conversion of the 240-kDa molecule to the 250-kDa core protein, resulting in an Endo H-resistant, 245-kDa molecule (Bumol et al., 1984; Harper et al., 1986).

In addition to its effects on the initial glycosylation reactions in core protein synthesis, monesin inhibits the synthesis and sulfation of GAG in melanoma cells, as it does in chondrocytes (Harper et al., 1986). Melanoma cells, like chondrocytes, are capable of initiating and elongating chondroitin sulfate chains on an artificial acceptor such as p-nitrophenyl-β-D-xyloside. Likewise, the xyloside acceptor inhibits the synthesis of MPG, presumably by competing with the natural acceptor (e.g., core protein) for GAG-synthetic enzymes and carbohydrate donors (Harper et al., 1986; Bumol et al., 1984). In monensin-treated cells, both synthesis and sulfation of GAG chains on the xyloside acceptor is substantially inhibited (J. R. Harper, unpublished data), emphasizing the broad effects of this ionophore on cellular metabolism.

In order to understand further the MPG-biosynthetic pathway and the mechanism by which the core protein is transported through the various intracellular domains to the cell surface, other agents, whose action on cellular metabolism are more discrete, have been assessed for their effect on MPG synthesis. A reasonable candidate for such a reagent was NH_4Cl, which alters the intracellular proton distribution, resulting in the disruption of pH gradients established by certain organelles (Galloway et al., 1983). Ion gradients of this type are important in a variety of reactions within a cell, such as receptor-mediated endocytosis and targeting of molecules to the lysosomes (Distler et al., 1979; Natowicz et al., 1979; Tycko and Maxfield, 1982; Rome et al., 1979) and sorting of ACTH molecules to secretory granules (Moore et al., 1983). Furthermore, NH_4Cl shares with monensin inhibitory effects on receptor recycling.

Melanoma cells metabolically labeled in the presence of NH_4Cl synthesize an MPG core protein whose migration in SDS–PAGE is only slightly increased, if at all, suggesting that the addition and processing of N-linked oligosaccharides are not substantially affected (Harper et al., 1986, Spiro et al., 1986). A parallel comparison of core protein synthesized in the presence of NH_4Cl and monensin indicated that NH_4Cl allows further maturation of the core protein than does monensin (Fig. 4). Like monensin, NH_4Cl also inhibits the final step in MPG synthesis, namely the addition of GAG chains to the core protein. NH_4Cl does not appear to inhibit MPG assembly by interfering with enzymes and carbohydrate donors involved in GAG chain initiation, elongation, or sulfation, since NH_4Cl-treated cells retain their ability to synthesize chondroitin sulfate chains on a β-D-xyloside acceptor; xyloside-initiated chondroitin sulfate chains were composed predominantly of $\triangle$Di-4S disaccharides (Harper et al., 1986, Spiro et al., 1986). NH_4Cl did, however, inhibit the exocytosis of MPG and xyloside-initiated chondroitin sulfate chains, implying the need for a low pH-requiring vesicular transport mechanism in this reaction. Despite the inhibition of MPG synthesis, NH_4Cl does not affect the transport of the core protein to the surface of melanoma cells. This suggests that the addition of chondroitin sulfate chains to the core protein is neither a sorting signal nor a prerequisite for its cell surface expression.

The sequential order of steps in MPG synthesis affected by monensin and NH_4Cl was further established by reversibility and substitution experiments (Marnell et al., 1982). When cells were labeled with [³H]leucine in the presence of monensin, then switched to NH_4Cl, the 245-kDa molecule proceeded to form the 250-kDa molecule, suggesting that the NH_4Cl block was subsequent to that of monensin (Harper et al., 1986). Conversely, cells labeled in the presence of NH_4Cl, then switched to monensin, expressed the 250 kDa in the absence of any MPG synthesis. This was consistent with the fact that NH_4Cl allowed the core protein to reach a stage of maturation at which monensin

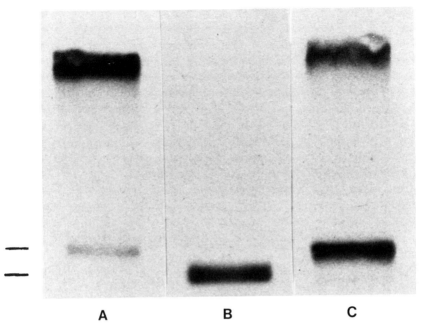

FIG. 4. Effect of monensin and NH_4Cl on the steady-state levels of MPG and core protein in UCLA M21 cells. Cells were labeled for 24 hr with [³H]leucine, lysed with RIPA buffer, then incubated with 9.2.27 bound to protein A–Sepharose. Immunoprecipitated proteins were resolved on a 5% polyacrylamide gel (SDS–PAGE). Cells represented in lane A were labeled in normal leucine-free medium; lanes B and C were labeled in the presence of 10^{-7} M monensin and 15 mM NH_4Cl, respectively. Bar markers indicate the migration of 250 kDa and 245 kDa.

could no longer affect its processing. Since monensin could independently inhibit the addition of GAG chains, a step downstream to the mechanism inhibited by NH_4Cl (i.e., MPG) was not synthesized. The final interpretation of these data was that monensin inhibits a step involved in core protein synthesis, as well as the addition of GAG chains, while NH_4Cl affects an intermediate step.

Diethylcarbamazine (DEC) is an antifilarial agent that has been shown to interfere with secretory reactions in mast cells from a variety of sources (Orange et al., 1971; Ishizaka et al., 1971; Razin et al., 1984), as well as the synthesis and exocytosis of proteoglycans produced by chondrocytes (Stevens et al., 1985). A detailed study of the mechanism by which DEC inhibited proteoglycan synthesis determined that the drug operates through the disruption of vesicular transport mechanisms within chondrocytes. Stevens et al. (1985) found that DEC not only blocked the transport of newly translated core protein from the endoplasmic reticulum to the Golgi compartment, where it undergoes further processing, but also interfered in the transport of proteoglycan from the Golgi to the cell surface. In ultrastructural studies, large vacuoles were seen in both pre- and post-Golgi regions of DEC-treated chondrocytes, providing further evidence for such blocks in intracellular transport.

Spiro and co-workers (1986) studied the effect of DEC on the synthesis and transport of MPG and its core protein, comparing this effect directly with that of NH_4Cl. DEC had little effect on the translation of the 240-kDa core protein precursor, but inhibited its conversion to the Endo H-resistant, 250-kDa molecule, an event that most probably localized in cis-Golgi vesicles. Moreover, by cell surface immunoprecipitation, it appeared that DEC, unlike NH_4Cl, also inhibited the transport of almost all proteoglycan products from the Golgi to the cell surface. As in chondrocytes, DEC induced the formation of large vesicles and dilation of both Golgi cisternae and portions of the endoplasmic reticulum, particularly when treated simultaneously with β-D-xyloside. This would be consistent with the earlier mentioned block in secretory and other transport mechanisms in mast cells and chondrocytes. Collectively, the biosynthetic studies using various metabolic inhibitors suggest that the synthesis and transport of MPG and its precursors involves a complex pathway of vesicular shuttling that is possibly mediated by low-pH steps and molecular signals yet to be identified.

IV. Conclusion

Proteoglycans of various types are present throughout the human body as part of the extracellular matrix (ECM). In connective tissue and in certain other organ systems, the matrix plays predominantly a structural role in maintaining the size and shape of the tissue it surrounds. The most obvious structural characteristic shared by proteoglycans is their modification by the addition of glycosaminoglycan (GAG) chains. As integral components of the ECM, proteoglycans, with their highly charged GAG chains, maintain the high degree of hydration necessary in providing flexibility of many structures, but may also serve as a signal that regulates growth and differentiation.

One prevailing question in the study of proteoglycans from various tissues is whether they share a common core protein, or at least certain structural domains within their protein core that are responsible for their biological function. This question can be extended to ask if malignant transformation results in changes in proteoglycan expression that is involved in the malignant phenotype.

Monoclonal antibodies that recognize the core protein of melanoma-type proteoglycan (MPG) have been isolated because of their specific reactivity with human melanoma cells when compared to normal tissues from throughout the human body. Antibodies from a variety of laboratories have been produced that react with several antigenic epitopes on the core protein molecule. Immunohistochemical examination on normal and neoplastic tissue has shown that this core protein is somewhat unique to tissues of neuroectodermal origin and is found predominantly in cells that are proliferating inappropriately, that is, malignant and benign tumors derived from melanocytes. Normal melanocytes, which divide very slowly in culture, do not produce significant amounts of MPG unless they are induced to proliferate rapidly by exposure to phorbol ester tumor promoters and other growth-stimulating agents.

First of all, the fact that antibodies against MPG and its core protein are highly specific strongly suggests that core protein structure may be tissue specific. An important experiment presently in progress is the direct comparison of MPG core protein with that of other core proteins by analysis of primary structure. Protein sequencing of purified material and recombinant DNA technology will soon reveal common domains, if they exist, that may lead to information relating to shared functional domains. From biosynthetic analysis, it appears that proteoglycan-producing cells, regardless of proteoglycan type or whether the molecules are on the cell surface or secreted, share similar intracellular compartments and enzymatic machinerx needed to carry out the appropriate posttranslational modification of a core protein. The precise molecular signature of the glycoprotein that designates it to be targeted to those compartments for GAG addition is still unknown.

The localization of MPG and its core protein to microspikes on the cell surface, along with several pieces of *in vitro* and *in vivo* functional data, seem to suggest that either or both of these structures have a functional role in regulating the growth and shape of melanoma cells. It is of particular interest that when cells are removed from substrata by EDTA, the MPG and core protein present in attachment sites seem to partition with cells, rather than remaining deposited in adhesion plaques left behind. This is in contrast to gangliosides, such as GD2 and GD3, which remain in adhesion plaques. The involvement of these molecules in cell adhesion and spreading is further suggested by the nature of factors which affect the expression of the MPG core protein.

Rettig *et al.* (1986) have shown that expression of the core protein gene, located on human chromosome 15, is induced in MPG-negative, chromosome 15-containing hybrids plated on endothelial cell-derived ECM. They have not yet determined which matrix component was responsible for the induction; however, these findings strongly correlate MPG expression with cell/cell and cell/substratum interactions. In the same study, these investigators have also reported that core protein gene expression was a function of the cellular differentiation state.

The question remains as to which form of the molecule (i.e., the free core protein or MPG) is the biologically active form. The ratio of MPG to free core protein present on the surface of melanoma cells can be modulated by disrupting an intracellular acidification mechanism using NH_4Cl (Harper *et al.*, 1985, 1986). The low-pH mechanism described in these studies seems to regulate the amount of core protein gaining access to the GAG-synthesizing compartment by either mediating the fusion of specific vesicles or preventing subtle glycosylation reactions that signal GAG chain initiation. *In vivo*, such a mechanism could allow cells to alter rapidly the amount of functional molecule in the extracellular domain in response to changing growth requirements without a need for changes in gene expression. In this regard, a metastatic melanoma cell has been isolated that expresses only the free MPG core protein on the cell surface, resembling cells treated with NH_4Cl (R. C. Spiro, D. A. Cheresh, and R. A. Reisfeld, unpublished data). The reason for this lack of GAG addition to the core protein is not presently known; however, this cell line will be a useful tool in determining the functional role of MPG and its core protein in the biology of melanoma cells.

REFERENCES

Alitalo, K., Keski-Oja, J., and Vaheri, A. (1981). *Int. J. Cancer* **27**, 755–761.
Bhanvanandan, V. P. (1981). *Biochemistry* **20**, 5595–5602.
Bumol, T. F., and Reisfeld, R. A. (1982). *Proc. Natl. Acad. Sci. U.S.A.* **79**, 1245–1249.
Bumol, T. F., Wang, Q. C., Reisfeld, R. A., and Kaplan, N. O. (1983). *Proc. Natl. Acad. Sci. U.S.A.* **80**, 529–533.
Bumol, T. F., Walker, L. E., and Reisfeld, R. A. (1984). *J. Biol. Chem.* **259**, 12733–12741.
Cairncross, J. G., Mattes, M. J., Beresford, H. R., Albino, A. P., Houghton, A. N., Lloyd, K. O., and Old, L. J. (1982). *Proc. Natl. Acad. Sci. U.S.A.* **79**, 5641–5645.
Cheresh, D. A., Harper, J. R., Schulz, G., and Reisfeld, R. A. (1984). *Proc. Natl. Acad. Sci. U.S.A.* **81**, 5767–5771.
Choi, H. U., Meyer, K., and Swarm, R. (1971). *Proc. Natl. Acad. Sci. U.S.A.* **68**, 877–879.
Cifone, M. A., and Fidler, I. J. (1980). *Proc. Natl. Acad. Sci. U.S.A.* **77**, 1039–1042.
Couchman, J. R., Caterson, B., Christner, J. E., and Baker, J. R. (1984). *Nature (London)* **307**, 650–652.

Culp, L. A., Murray, B. A., and Rollins, B. J. (1979). *J. Supramol. Struct.* **11**, 2161–2172.

Distler, J., Hieber, V., Sahagian, G., Schmickel, R., and Jourdian, G. W. (1979). *Proc. Natl. Acad. Sci. U.S.A.* **76**, 4235–4239.

Fellini, S. A., Hascall, V. C., and Kimura, J. H. (1984). *J. Biol. Chem.* **259**, 4634–4641.

Fitting, T., and Kabat, D. (1982). *J. Biol. Chem.* **257**, 14011–14014.

Galloway, C. J., Dean, G. E., Marsh, M., Rudnick, G., and Mellman, I. (1983). *Proc. Natl. Acad. Sci. U.S.A.* **80**, 3334–3338.

Garrigues, H. J., Lark, M. W., Lara, S., Hellstrom, I., Hellstrom, K. E., and Wight, T. N. (1986). *J. Cell Biol.*, **103**, 1699–1710.

Giacomini, P., Ng, A. K., Kantor, R. R. S., Natali, P. G., and Ferrone, S. (1983). *Cancer Res.* **43**, 3586–3590.

Giacomini, P., Veglia, F., Fei, P., Rehle, T., Natali, P. G., and Ferrone, S. (1984). *Cancer Res* **44**, 1281–1287.

Harper, J. R., and Reisfeld, R. A. (1983). *J. Natl. Cancer Inst.* **71**, 259–263.

Harper, J. R., Bumol, T. F., and Reisfeld, R. A. (1982). *Hybridoma* **1**, 423–432.

Harper, J. R., Bumol, T. F., and Reisfeld, R. A. (1984a). *J. Immunol.* **132**, 2096–2104.

Harper, J. R., Quaranta, V., and Reisfeld, R. A. (1984b). In "Extracellular Matrix: Structure and Function" (H. Reddi, ed.), pp. 367–372. Liss, New York.

Harper, J. R., Quaranta, V., and Reisfeld, R. A. (1986). *J. Biol. Chem.* **261**, 3600–3606.

Hascall, V. C. (1981). In "Biology of Carbohydrates," pp. 1–49. Wiley, New York.

Heinegard, D., and Hascall, V. C. (1974). *J. Biol. Chem.* **249**, 4250–4256.

Hellstrom, I., Garrigues, H. J., Cabasco, L., Mosely, G. H., Brown, J. P., and Hellstrom, K. E. (1983). *J. Immunol.* **130**, 1467–1472.

Herlyn, M., Steplewski, Z., Herlyn, D., Clark, W. H., Ross, A. H., Blaszczyk, M., Pak, K. Y., and Koprowski, H. (1983). *Cancer Invest.* **1**, 127–136.

Iozzo, R. V. (1984a). *Hum. Pathol.* **15**, 2–10.

Iozzo, R. V. (1984b). *J. Cell Biol.* **99**, 403–417.

Iozzo, R. V. (1985). *Lab. Invest.* **53**, 373–396.

Ishizaka, T., Ishizaka, K., Orange, R. P., and Austen, K. F. (1971). *J. Immunol.* **106**, 1267–1273.

Kajiwara, T., and Tanzer, M. L. (1981). *FEBS Lett.* **134**, 43–46.

Kimura, J. H., Caputo, C. B., and Hascall, V. C. (1981). *J. Biol. Chem.* **256**, 4368–4376.

Kimura, J. H., Lohmander, L. S., and Hascall, V. C. (1984). *J. Cell. Biochem.* **26**, 261–278.

Koprowski, H., Steplewski, Z., Herlyn, D., and Herlyn, M. (1978). *Proc. Natl. Acad. Sci. U.S.A.* **75**, 3405–3409.

Kornfeld, R., and Kornfeld, S. (1980). In "The Biochemistry of Glycoproteins and Proteoglycans," (W. J. Lennarz, ed.), pp. 1–27. Plenum, New York.

Liotta, L. A. (1984). *Am. J. Pathol.* **117**, 339–348.

Liotta, L. A., Rao, C. N., and Barsky, S. H. (1983). *Lab. Invest.* **49**, 636–649.

Lohmander, L. S., DeLuca, S., Nilsson, B., Hascall, V. C., Caputo, C. B., Kimura, J. H., and Heinegard, D. (1980). *J. Biol. Chem.* **255**, 6084–6091.

Marnell, M. H., Stockley, M., and Draper, P. K. (1982). *J. Cell Biol.* **93**, 57–62.

Mitchell, D., and Hardingham, T. (1982). *Biochem. J.* **202**, 249–254.

Moore, H. P., Grumbiner, B., and Kelly, R. B. (1983). *Nature (London)* **302**, 434–436.

Morgan, A. C., Galloway, D. R., and Reisfeld, R. A. (1981). *Hybridoma* **1**, 27–36.

Natowicz, M. R., Chi, M. M. Y., Lowry, O. H., and Sly, W. S. (1979). *Proc. Natl. Acad. Sci. U.S.A.* **76**, 4322–4326.

Nicholson, G. L. (1982). *Biochem. Biphys. Acta* **695**, 113–176.

Nicholson, G. L. (1984). *Exp. Cell Res.* **150**, 1–22.

Nilsson, B., DeLuca, S., Lohmander, S., and Hascall, V. C. (1982). *J. Biol. Chem.* **257,** 10920–10927.

Nishimoto, S. K., Kajiwara, T., and Tanzer, M. L. (1982). *J. Biol. Chem.* **257,** 10558– 10561.

Oegema, T. R., Hascall, V. C., and Dziewiatkowski, D. D. (1975). *J. Biol. Chem.* **250,** 6151–6159.

Oegema, T. R., Brown, M., and Dziewiatkowski, D. D. (1977). *J. Biol. Chem.* **252,** 6470– 6477.

Orange, R. P., Austen, W. G., and Austen, K. F. (1971). *J. Exp. Med.* **134,** 136–148.

Razin, E., Romeo, L. C., Krilis, S., Liu, F. T., Lewis, R. A., Corey, E. J., and Austen, K. F. (1984). *J. Immunol.* **133,** 938–945.

Real, F. X., Houghton, A. N., Albino, A. P., Cordon-Cardo, C., Melamed, M. R., Oettgen, H. F., and Old, L. J. (1985). *Cancer Res.* **45,** 4401–4411.

Rettig, W. J., Real, F. X., Spengler, B. A., Beidler, J. L., and Old, L. J. (1986). *Science* **231,** 1281–1284.

Rome, L. H., Weissman, B., and Neufeld, E. F. (1979). *Proc. Natl. Acad. Sci. U.S.A.* **76,** 2331–2334.

Ross, A. H., Cossu, G., Herlyn, M., Bell, J. R., Steplewski, Z., and Koprowski, H. (1983). *Arch. Biochem. Biophys.* **225,** 370–383.

Silbert, J. E., and Freilich, L. S. (1980). *Biochem. J.* **90,** 307–313.

Spiro, R. C., Parsons, W. G., Perry, S. K., Caufield, J. P., Hein, A., Reisfeld, R. A., Harper, J. R., Austen, K. F., and Stevens, R. L. (1986). *J. Biol. Chem.* **261,** 5121–5129.

Stevens, R. L., Parsons, W. G., Austen, K. F., Hein, A., and Caufield, J. P. (1985). *J. Biol. Chem.* **260,** 5777–5786.

Tajiri, K., Uchida, N., and Tanzer, M. L. (1980). *J. Biol. Chem.* **255,** 6036–6039.

Thonar, E. J.-M. A., Lohmander, S., Kimura, J. H., Fellni, S. A., Yanagishita, M., and Hascall, V. C. (1983). *J. Biol. Chem.* **258,** 11564–11570.

Turley, E. A. (1984). *Cancer Met. Rev.* **3,** 325–339.

Turley, E. A., and Tretiak, M. (1985). *Cancer Res.* **45,** 5098–5105.

Tycko, B., and Maxfield, F. R. (1982). *Cell* **28,** 643–651.

Vitetta, E. S., and Uhr, J. W. (1975). *J. Immunol.* **115,** 374–381.

Wilson, B. S., Imai, K., Natali, P. G., and Ferrone, S. (1981). *Int. J. Cancer* **28,** 293–300.

Yuspa, S. H., Ben, T., Hennings, H., and Lichti, U. (1982). *Cancer Res.* **42,** 2344–2349.

Intracellular Proteoglycans in Cells of the Immune System

Richard L. Stevens

Department of Medicine, Harvard Medical School, and Department of Rheumatology and Immunology, Brigham and Women's Hospital, Boston, Massachusetts 02115

I. Introduction

The first clear evidence for the existence of proteoglycans stored within the granules of a cell was obtained from studies of the skin mast cell. However, over the past decade it has become increasingly apparent that a number of cells which participate in immune and inflammatory responses, including mucosal mast cells, basophils, eosinophils, neutrophils, and natural killer (NK)* cells, also contain proteoglycans in their granules. The presence of a family of proteoglycans in cells that is distinct from plasma membrane proteoglycans or extracellular matrix proteoglycans suggests that these molecules may be important in the functions of such cells for tumor surveillance and host defense against bacterial, viral, fungal, and parasitic pathogens. This review will focus on the well-characterized proteoglycans of mast cells, basophils, and NK cells. The evidence for their localization within the secretory granule, the unique structural features of these proteoglycans, and their possible functions in the immune response will be addressed.

II. Localization and Characterization of Secretory Granule Proteoglycans in Different Cell Types

A. Heparin-Containing Mast Cells (HP-MC)

In 1878, Ehrlich discovered cells in connective tissues near glandular ducts, blood vessels, and nerves, as well as at sites of inflamma-

---

* Abbreviations used in this chapter are as follows: ChS-MC, chondroitin sulfate-containing mast cells; HPLC, high-performance liquid chromatography; HP-MC, heparin-containing mast cells; NK, natural killer; and RBL-1, rat basophilic leukemia 1.

tion and cancerous growth that possessed granules which became metachromatic when stained with cationic dyes, indicating the presence of acidic macromolecules in the granules of the cells. Subsequent isolation and characterization studies of these connective tissue cells that Ehrlich had designated as mast cells revealed that they contained a unique highly sulfated glycosaminoglycan which was termed heparin (Table I). While studies on the native heparin molecule were hampered by its location within a cell that is present in low concentration in connective tissue, advances in the isolation of sufficient quantities of a pure population of normal mast cells from the serosal cavity of rats (Yurt et al., 1977a; Holgate et al., 1980) and of an enriched population of mast cells from human lung (Paterson et al., 1976) have enabled in-depth studies to be carried out on this proteoglycan. The presence of small amounts of amino acids in some preparations of pig heparin led to the suggestion that heparin existed in its native state covalently bound to a protein (Lindahl et al., 1965). The findings that radiolabeled serine and glycine could be incorporated into the rat molecule, and that macromolecular rat heparin could be degraded to glycosaminoglycan-sized molecules using conditions known to hydrolyze the O-glycosidic bond between xylose and serine/threonine at the reducing terminus of glycosaminoglycan chains (Yurt et al., 1977a) indicated that mast cell-derived macromolecular heparin was a proteoglycan, rather than a complex carbohydrate polymer like hyaluronate or glycogen. Studies of macromolecular heparin isolated from pronase-treated rat skin (Robinson et al., 1978) and from purified rat serosal mast cells (Metcalfe et al., 1980a) revealed that this ~750,000 MW proteoglycan possessed a unique peptide core consisting almost entirely of serine and glycine. Because chondroitin sulfate glycosaminoglycans are linked through xylose to serine (Muir, 1958) at a serine–glycine sequence (Isemura and Ikenaka, 1975), it was hypothesized that the peptide core of rat HP-MC heparin proteoglycan was composed of altering serine and glycine amino acids (Robinson et al., 1978). Horner (1971) first demonstrated that macromolecular heparin from rat skin was distinct from all other proteoglycans in its resistance to pronase degradation. Yurt et al. (1977a) confirmed that the native molecule isolated from pure populations of rat serosal HP-MC was resistant to degradation not only by pronase but also by many other enzymes. To date, no protease has been found that will degrade heparin proteoglycan. This property of protease resistance appears to be a characteristic common to proteoglycans that reside inside the secretory granules of cells. It had been postulated that the acidic nature of the heparin chains and the high degree of substitution of

TABLE I

CHARACTERISTICS OF THE SECRETORY GRANULE PROTEOGLYCANS FROM SOME IMMUNE CELLS

	Cell type				
Property	Rat serosal HP-MC	Rat mucosal ChS-MC	RBL-1	Mouse bone marrow-derived ChS-MC	Human NK cell
Molecular weight	~750,000	~150,000	~150,000	~200,000	~200,000
GAG					
Molecular weight	~80,000	ND[a]	~12,000	~25,000	~50,000
Type	HP	ChS-diB	ChS-diB and HP	ChS-E	ChS-A
Major disaccharide	IdUA-2SO$_4$→GlcNSO$_4$-6SO$_4$	IdUA-2SO$_4$→GalNAc-4SO$_4$	IdUA-2SO$_4$→GalNAc-4SO$_4$	GlcUA→GalNAc-4,6-diSO$_4$	GlcUA→GalNAc-4SO$_4$
Peptide core					
Molecular weight	~20,000	ND	ND	~10,000	ND
Major amino acids	Ser, Gly	ND	Gly, Ser, Ala	Gly, Ser, Glu	ND
Protease resistance	Yes	Yes	Yes	Yes	Yes

[a] ND, Not determined.

these glycosaminoglycans along the peptide core conferred protease resistance upon this proteoglycan, but the lack of aromatic and basic amino acids is probably the reason why this molecule is resistant to degradation by enzymes like trypsin and chymotrypsin.

Although the early histochemical studies (Ehrlich, 1878) had suggested that skin-derived heparin proteoglycan probably resided in the mast cell's metachromatic secretory granules, definitive experiments were not performed until a number of years later. Jorpes *et al.* (1937) and then Benditt *et al.* (1956) demonstrated that heparin was associated with isolated mast cells. Using an X-ray dispersion technique, Hein and Caulfield (1982) demonstrated that the secretory granules were the only sites within the rat serosal HP-MC that possessed substantial amounts of sulfur-containing molecules. The parallel exocytosis of histamine and ^{35}S-labeled heparin proteoglycans upon immunologic activation of pure populations of [35]SO_4-labeled rat serosal HP-MC (Yurt *et al.*, 1977b) confirmed the secretory granule location of these proteoglycans. Since heparin cannot be detected in the skin of mice that are genetically deficient in mast cells (Nakamura *et al.*, 1981; Straus *et al.*, 1982), it appears that mast cells are the major source of heparin in the body.

Studies on the biosynthesis of heparin proteoglycans in normal cells have been hampered not only by a lack of a tissue source for large numbers of these cells but also by the inability to maintain these cells in culture. Nevertheless, by using cultured mouse mastocytoma cells Silbert and co-workers (1963, 1967) and Lindahl and co-workers (for reviews, see Lindahl and Höök, 1978; and Rodén, 1980) made considerable progress in determining the steps involved in the biosynthesis of the heparin glycosaminoglycan chain. Since the enzymatic steps involved in the biosynthesis of glycosaminoglycans are discussed elsewhere in this book, these steps will be summarized only briefly here. During the biosynthesis of heparin and chondroitin sulfate glycosaminoglycans, at least four glycosyltransferases located in the endoplasmic reticulum or the cis region of the Golgi produce the initial carbohydrate sequence of GlcUA(β1,3)$\rightarrow$Gal(β1,3)$\rightarrow$Gal(β1,4)$\rightarrow$Xyl (β1,3)$\rightarrow$serine at serine–glycine sequences along the peptide core. In the case of the heparin chain, alternating GlcNAc and GlcUA monosaccharides are added to produce the precursor chain. At least five enzymes then participate in its modification, resulting in the conversion of the majority of the GlcUA$\rightarrow$GlcNAc disaccharides to the characteristic major disaccharide of heparin, IdUA-2$SO_4\rightarrow$GlcNSO_4-6SO_4 (Fig. 1). The extent of polymer modification is often incomplete, and therefore the heparin chains can vary in their overall length, and the individual disaccharides within the chain can vary in the amount of

A   B

C   D

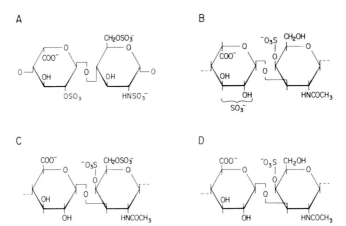

FIG. 1. Structure of the major disaccharide present in the glycosaminoglycan bound to (A) heparin proteoglycan from rat serosal HP-MC, (B) chondroitin sulfate diB proteoglycan from rat mucosal ChS-MC and RBL-1 cells, (C) chondroitin sulfate E proteoglycan from mouse bone marrow-derived ChS-MC, and (D) chondroitin sulfate A proteoglycan from human NK cells.

sulfation, position of sulfation, degree of acetylation, and extent of epimerization of glucuronic acid to iduronic acid. The molecular weights of the heparin glycosaminoglycans that are polymerized onto the peptide cores of rat skin heparin proteoglycans are 50,000–100,000 (Robinson *et al.*, 1978), but those bound to human lung heparin proteoglycans are considerably smaller (Metcalfe *et al.*, 1979). In the case of chondroitin sulfate, a different set of enzymes participate in the elongation of the tetrasaccharide initiation sequence and the subsequent modification of the precursor chain to yield a glycosaminoglycan with a repeating sulfated disaccharide containing GalNAc, rather than $GlcNSO_4$, and, alternatively, $\beta1,3 \rightarrow \beta1,4$ rather than $\alpha1,4 \rightarrow \alpha$ or $\beta1,4$-carbohydrate linkages. The factors that determine which hexosamine will be added during the polymerization reaction to result in the biosynthesis of a glycosaminoglycan that belongs to either the chondroitin sulfate family or the heparin family are not known. However, while isolated rat serosal mast cells synthesize only heparin proteoglycan when they are cultured immediately after their isolation (Stevens and Austen, 1982), upon treatment with an exogenous glycosaminoglycan acceptor such as *p*-nitrophenyl-β-D-xyloside prior to incubation with $^{35}SO_4$, the mast cells synthesize highly sulfated chondroitin sulfate E glycosaminoglycans (Stevens *et al.*, 1983). Although the rat serosal HP-MC produces both heparin and chondroitin sulfate glycosaminoglycans, the β-D-xyloside-treated HP-MC continues to polymerize only heparin chains onto the peptide core. Thus, the peptide core of the mast cell proteoglycan must play an important role in regulating which glycosaminoglycan will normally be synthesized.

Because cultured mouse mastocytoma cells possess an endoglycosidase that rapidly cleaves the heparin glycosaminoglycan chain from its peptide core (Ögren and Lindahl, 1975), it has been difficult to determine which protein translated by this tumor HP-MC is the peptide core for heparin proteoglycan. Thus, very little information has been obtained about the posttranslational modification, intracellular transport, sorting, and catabolism of the core peptide of heparin proteoglycan. With the recent discovery that nontransformed rat serosal HP-MC can be maintained at least 1 month when cultured in the presence of living fibroblasts (Levi-Schaffer et al., 1985), it is now possible to study the steps in the metabolism of the peptide core of this proteoglycan at either the molecular or the biochemical level.

Bourdon and co-workers (1985) obtained a cDNA clone from a rat yolk sac tumor poly(A)$^+$ RNA-derived cDNA library which encodes the peptide core of a plasma membrane chondroitin sulfate proteoglycan. The amino acid sequence deduced from the DNA sequence revealed a protein of 104 amino acids containing a region of 49 amino acids composed of alternating serine and glycine residues. Since the peptide cores of the secretory granule proteoglycans which have been studied to date are unusually rich in serine and glycine, it was postulated that this cDNA probe might recognize the mRNA for one or more of these proteoglycans, allowing for the determination of their amino acid sequences. In a preliminary experiment, total RNA was extracted from the rat serosal HP-MC. When analyzed by the Northern blot procedure using the rat yolk sac tumor cell cDNA proteoglycan probe, a mRNA transcript was detected from the rat serosal HP-MC which hybridized to the cDNA probe under conditions of high stringency (Tantravatti et al., 1986). This finding suggests that the peptide cores of the proteoglycans from the two cells are closely related in terms of their amino acid sequences. This finding also raises the interesting possibility that expression of a mast cell-like proteoglycan on the plasma membrane of a tumor cell may suppress the ability of the body's normal surveillance mechanisms to recognize the rat yolk sac tumor cell as transformed.

Rat HP-MC are long-lived in the connective tissue and appear to continuously synthesize heparin proteoglycan. Since the mast cells do not exocytose their proteoglycans unless they are activated, they must catabolize them so as not to exceed storage capacity. Although the regulation of catabolism of this secretory granule proteoglycan has not been well investigated, mast cells possess the full complement of acid hydrolases that are necessary for its catabolism (Schwartz and Austen, 1980). Extracellularly, the fate of heparin proteoglycan has also not been studied in great detail. Lloyd *et al.* (1966) found that when ^{35}S-labeled ox heparin glycosaminoglycan was administrated intra-peritoneally or intravenously to rats, ~50% of the glycosaminoglycan was degraded and eliminated into the urine in the first 12 hr. Probably more importantly, macrophages (Fabian *et al.*, 1978; Lindahl *et al.*, 1979), eosinophils (Mann, 1969), and fibroblasts (Rao *et al.*, 1983) avidly endocytose and degrade heparin proteoglycan when it is exocytosed from the mast cell.

B. Chondroitin Sulfate-Containing Mast Cells (ChS-MC)

By histochemical and ultrastructural criteria, Enerbäck (1966) identified a population of mast cells in the gastrointestinal mucosa of rats that was distinct from the skin HP-MC. An increase in the number of mucosal mast cells is seen in mice, rats, and humans infected with a variety of helminths, and the proliferation of these mast cells in rodents depends on a factor from T cells (Mayrhofer, 1979; Mayrhofer and Bazin, 1981). By comparing the absorbance spectra of granule meta-chromasia after toluidine blue staining, Tas and Berndsen (1977) concluded that rat mucosal mast cell granules contain little or no heparin, but rather a less-sulfated molecule. Studies of purified mucosal mast cells from rats infected with *Nippostrongylus brasiliensis* have revealed that they contain a protease-resistant proteoglycan (Table I) with a molecular weight of ~150,000, containing highly sulfated chondroitin sulfate glycosaminoglycans, and therefore these cells have been designated as ChS-MC (Stevens *et al.*, 1986). The chondroitin sulfate proteoglycan present in the rat mucosal ChS-MC is distinct from all other known mammalian proteoglycans because ~50% of its disaccharides are of the chondroitin sulfate diB structure (IdUA-2SO$_4$→GalNAc-4SO$_4$), and ~7% are of the chondroitin sulfate E structure (GlcUA→GalNAc-4,6-diSO$_4$) (Fig. 1). The presence of large amounts of disulfated disaccharides, as assessed by high-performance liquid chromatographic (HPLC) analysis (Seldin *et al.*, 1984) of chondroitinase digests (Saito *et al.*, 1968), has become a useful phenotypic marker for mast cell subclasses because matrix proteoglycans possessing chondroitin sulfate glycosaminoglycans are made up of predominantly monosulfated disaccharides.

Like the HP-MC, studies on the proteoglycans of ChS-MC have been hampered by the lack of a tissue source for large numbers of these cells. A number of investigators (Razin *et al.*, 1981; Schrader, 1981;

Tertian *et al.*, 1981, Galli *et al.*, 1982; Haig *et al.*, 1982; Sredni *et al.*, 1983; Ogawa *et al.*, 1983; Denburg *et al.*, 1983; Tadokoro *et al.*, 1983) have developed similar culture techniques to grow and differentiate ChS-MC. In these procedures, hematopoietic stem cells from sources such as mouse, rat, or human bone marrow or mouse fetal liver are cultured for 2–3 weeks in the presence of conditioned media derived from activated T lymphocytes. It is now apparent that a single factor produced by Lyt$^+$ T cells, termed interleukin 3 (Ihle *et al.*, 1982), regulates the growth and differentiation of mouse ChS-MC (Razin *et al.*, 1984a). *In vitro*-differentiated mouse ChS-MC derived from bone marrow, fetal liver, or lymph node progenitors, cultured in concanavalin A splenocyte-conditioned medium, WEHI-3-conditioned medium, or purified interluekin 3 (Razin *et al.*, 1982, 1983, 1984a,b) do not contain heparin proteoglycans but rather contain chondroitin sulfate proteoglycans that possess 30–50% of their disaccharides as GlcUA→GalNAc-4,6-diSO$_4$ (Table I and Fig. 1). The structure of this disaccharide is similar to that present in squid chondroitin sulfate E glycosaminoglycan (Suzuki *et al.*, 1968), and therefore the mouse ChS-MC-derived molecule was termed chondroitin sulfate E proteoglycan. Because *in vitro*-differentiated rat mucosal ChS-MC contain some chondroitin sulfate E in their proteoglycans, it is clear that *in vitro*-differentiated mouse ChS-MC are closely related to *in vivo*-differentiated rat mucosal ChS-MC. However, there appear to be species differences, since the rat-derived cells possess predominantly chondroitin sulfate diB and the mouse-derived cells possess predominantly chondroitin sulfate E.

It was concluded that chondroitin sulfate E proteoglycans reside in the cell's metachromatic granules rather than on the plasma membrane because (a) incubation of whole $^{35}SO_4$-labeled cells with chondroitinase ABC failed to degrade the ^{35}S-labeled proteoglycan (Stevens, unpublished findings); (b) transmission electron microscopy and autoradiography of $^{35}SO_4$-labeled cells revealed the presence of ^{35}S-labeled molecules in the cell's granules (Galli *et al.*, 1982); (c) X-ray dispersion analysis demonstrated sulfur in the secretory granules (Stevens *et al.*, 1983); and (d) proteoglycans were exocytosed in soluble form when the cells were sensitized with IgE and activated with antigen (Razin *et al.*, 1983). Recently, these proteoglycans were extracted from ~6 × 10^9 cultured ChS-MC using zwittergent and 4 *M* guanidine hydrochloride and purified to apparent homogeneity using density gradient centrifugation, ion-exchange chromatography, and gel filtration chromatography (Stevens *et al.*, 1985). The ~200,000 MW proteoglycans were found to consist of approximately seven 25,000 MW chondroitin sulfate side chains attached to peptide cores which are ~10,000 in MW. The purified mouse ChS-MC proteoglycans cannot be degraded by collagenase, clostripain, trypsin, chymotrypsin, elastase, chymopapain, V8 protease, proteinase K, or pronase, as assessed by gel filtration chromatography and by SDS–PAGE. When analyzed for neutral hexose content by gas–liquid chromatography, the proteoglycans contained ~2% mannose, fucose, galactose, and other sugars by weight, indicating that oligosaccharides might also be linked to the peptide core. Amino acid analysis of the core peptide of the intact proteoglycans revealed that glycine, serine, and glutamic acid/glutamine accounted for 70% of the total amino acids, and were present in a molar ratio of 4.3:1.6:1.0. Thus, the mouse ChS-MC chondroitin sulfate E proteoglycans, like the rat serosal HP-MC heparin proteoglycans, are markedly protease resistant, have highly sulfated glycosaminoglycans, and contain a peptide core that is rich in serine and glycine. Although the peptide core of mouse heparin proteoglycan has not been sequenced, the higher glutamic acid content of the mouse chondroitin sulfate E proteoglycan compared to rat heparin proteoglycan suggests that if the peptide cores for the two proteoglycans are derived from the same gene, a glutamic-rich region is removed in its processing. Using the rat yolk sac proteoglycan cDNA probe, it should be possible to determine the extent of homology of the peptide cores of these proteoglycans.

C. Basophils and Rat Basophilic Leukemia (RBL) Cells

Because the peripheral blood basophil contains metachromatic granules, it has been concluded that this cell also contains an intragranular proteoglycan. Studies on guinea pig basophils demonstrated a cell-associated chondroitin sulfate proteoglycan (Orenstein et al., 1978). In order to study the relationship between mast cells and basophils, a leukemia composed of basophilic cells was generated a number of years ago in rats with the chemical carcinogen β-chloroethylamine (Eccleston et al., 1973) and was subsequently established as the RBL-1 cell line. Because the RBL-1 cell and the rat mucosal ChS-MC have similar staining properties and the identical intragranular neutral protease, it was hypothesized that the RBL-1 line is a transformed ChS-MC which has lost its T-cell growth factor dependence (Seldin et al., 1985a); its relationship to circulating basophils is presently unknown. This transformed basophil/ChS-MC has become an interesting model cell for investigating regulation of proteoglycan synthesis because the cell produces a proteoglycan which is a hybrid of two different types of glycosaminoglycans (Table I) (Metcalfe et al., 1980b; Seldin *et al.*, 1985b). RBL-1 cells contain in their granules ~0.5 $\mu g/10^6$ cells of a protease-resistant proteoglycan with a MW of 90,000–150,000, with <10 glycosaminoglycan side chains which are ~12,000 in MW. Like the rat mucosal ChS-MC, the majority of the glycosaminoglycans bound to the RBL-1 proteoglycan peptide core are chondroitin sulfate rich in disaccharides of the diB structure. However, in addition to the chondroitin sulfate diB, approximately one-third of the bound glycosaminoglycans are similar to heparin in that they are susceptible to nitrous acid degradation. Amino acid compositional analysis has revealed that the peptide core of the RBL-1 proteoglycan is homologous to, but not identical to, the peptide cores of the rat serosal HP-MC-derived heparin proteoglycan and the mouse bone marrow ChS-MC-derived chondroitin sulfate E proteoglycan, probably due to differential posttranslational degradation of the peptide cores. Further investigation of this cell may give additional insights as to how cells regulate the type of glycosaminoglycan to be polymerized onto different peptide cores. Although the findings in the RBL-1 cell may represent loss of normal biosynthetic control due to the transformed nature of this cell, a chondroitin sulfate–heparan sulfate hybrid proteoglycan has also been recently described on the cell surface and basement membrane of mouse mammary epithelial cells (David and Van den Berghe, 1985).

D. *Natural Killer Cells*

NK activity is mediated by a heterogeneous population of peripheral blood mononuclear cells which are morphologically classified as large granular lymphocytes (Herberman and Ortaldo, 1981). Mouse cloned NK cells (Bland *et al.*, 1984; Dvorak *et al.*, 1983), and human peripheral blood mononuclear cells (Levitt and Ho, 1983) have been found to possess cell-associated proteoglycans bearing chondroitin sulfate glycosaminoglycan side chains. Transmission electron microscopy and autoradiography localized ^{35}S-labeled molecules to the cytoplasmic granules of one clone of $^{35}SO_4$-labeled mouse NK cells, suggesting that these proteoglycans are present in the secretory granules (Dvorak *et al.*, 1983). More recent characterization studies on cloned human NK cells have revealed that they contain a chondroitin sulfate proteoglycan in their secretory granules (Table I) (MacDermott *et al.*, 1985). Like the mast cell proteoglycans, this proteoglycan is also resistant to degradation by pronase. However, unlike the rat ChS-MC proteoglycans, the NK-cell proteoglycans possess glycosaminoglycan chains which are ~50,000 MW. These chondroitin sulfate side chains are also not as highly sulfated as those present in ChS-MC. HPLC analysis of the chondroitinase ABC digests revealed only $\triangle$Di-4S, a disaccharide derived from chondroitin sulfate A (GlcUA→GalNAc-4SO$_4$). Recent studies on the fate of this proteoglycan during the cytotoxicity reaction have shown that when the human cloned NK cell comes in contact with a tumor target cell sensitive to lysis, the NK cell exocytoses its chondroitin sulfate proteoglycan. Using six separate human NK-cell clones with different target specificities and seven different tumor cells, a 0.92 correlation coefficient was obtained for the percentage of maximal proteoglycan release vs the percentage cytotoxicity against a given target (Fig. 2) (Schmidt et al., 1985). In addition, when cloned NK cells were preincubated with monoclonal antibodies that inhibit the cytotoxicity reaction by blocking cellular recognition determinants on either the target cell or the effector cell, exocytosis of proteoglycan was markedly inhibited. Thus, release of secretory granule proteoglycans from the human NK cells occurs in the process of killing tumor cell targets. In these studies the presence of the target cells made it difficult to study the metabolism of the NK-

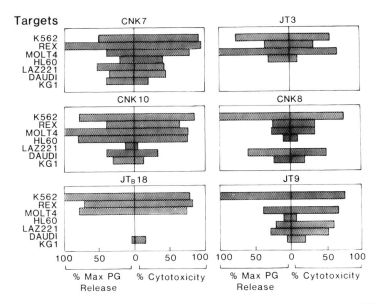

FIG. 2. Comparison of cytotoxicity and proteoglycan (PG) release from human NK-cell clones (CNK7, JT3, CNK10, CNK8, JT$_B$18, and JT9) interacting with seven target tumor cell lines (K562, REX, MOLT4, HL60, LAZ221, DAUDI, and KG1). The release of proteoglycan is expressed as a percentage of the maximal proteoglycan release obtained with the optimal NK–target combination. From Schmidt et al. (1985).

cell proteoglycan and its possible association with other granule-bound molecules. In order to address the role of NK cell-derived proteogly-cans in the cytotoxicity reaction, monoclonal antibodies which trigger exocytosis of intracellular proteoglycans in the absence of target cells were sought. Since incubation with monoclonal antibodies directed against the T3 and NKTa determinants on the plasma membrane of NK cells will also induce exocytosis of the NK-cell proteoglycans (Schmidt *et al.*, 1985), it is now possible to study the biology of these proteoglycans in the absence of the tumor target cell.

III. Speculated Functions of Intracellular Proteoglycans in Immunity and Inflammation

That substantial amounts of proteoglycans are in the secretory granules of a variety of cells that participate in immune and inflam-matory reactions suggests that these polysulfated macromolecules may play an important role in the host response to diverse pathogens. Further evidence for the importance of these proteoglycans has come from the finding that similar proteoglycans are present in the secreto-ry granules of turtle and human basophils (R. L. Stevens and D. C. Seldin, unpublished findings). Probably the basophil has continued to synthesize these proteoglycans through evolution because they are necessary for certain immunological functions. Proteoglycans are un-usual in the extent of their posttranslational modification, but secreto-ry granule proteoglycans such as heparin are exceptionally complex in structure, requiring an even more complicated enzymatic machinery for their biosynthesis than matrix or plasma membrane proteoglycans. This is further circumstantial evidence of their importance. Because of the advances that have been made in culture techniques to grow, dif-ferentiate, and maintain these cells and in biochemical techniques to isolate and characterize proteoglycans, it is now possible to carry out studies on secretory granule proteoglycans that were previously im-possible. The ability to clone some of these cells has eliminated the possibility that these cell-associated proteoglycans might be from a contaminating cell. Despite the fact that to date only a few of these intracellular proteoglycans have been characterized in any detail, enough is now known about their general properties to permit some speculation as to what their functions might be.

The mast cell and RBL-1 cell secretory granule proteoglycans are unusual in the extent of their sulfation. In both rats and mice, heparin proteoglycan has ~4000 sulfate residues and ~2000 carboxylic acid residues per molecule, making it the most acidic macromolecule in the body. The cells that synthesize these highly acidic proteoglycans are designed to respond quickly to immunological challenges. They contain preformed molecules, termed "mediators," which are stored in their secretory granules. In the case of the HP-MC, ChS-MC, and the basophil, the principal components of the secretory granules are serine proteases (Lagunoff and Pritzl, 1976; Woodbury and Neurath, 1978; Everitt and Neurath, 1980; Orenstein *et al.*, 1981; Schwartz *et al.*, 1981a; DuBuske *et al.*, 1984), vasoactive amines, and proteoglycans. NK cells (Petty *et al.*, 1984) and cytotoxic T lymphocytes (Pasternak and Eisen, 1985) also possess serine proteases, although their intracellular localization has not been established. Since both the proteases and the amines of mast cells are positively charged at physiological pH, the proteoglycans probably act like an ion-exchange resin to bind and concentrate these positively charged molecules in the cell's secretory granule. Once bound, the proteoglycan–protease–amine complexes may serve to inactivate reversibly the intragranular proteases in order to prevent nonspecific enzymatic activity and unwarranted destruction of the cell and its microenvironment. It has been shown that upon exocytosis, proteases of the rat serosal HP-MC remain bound to heparin proteoglycan (Schwartz *et al.*, 1981b), and when bound their enzymatic activity is substantially reduced (Yurt and Austen, 1977). Since serine proteases also remain tightly associated with chondroitin sulfate E proteoglycan when exocytosed from the mouse bone marrow-derived ChS-MC (W. E. Serafin *et al.*, 1986), this may be a common mechanism for controlling the diffusion of the low molecular weight proteases into the surrounding connective tissue.

Different immune cells may synthesize proteoglycans with glycosaminoglycans that vary in their degree and position of sulfation and their stereochemistry in order to increase the affinity of binding of proteoglycan with the other molecules present in the granule. Although all mast cells contain large amounts of serine neutral proteases, it is clear that the enzyme in the rat ChS-MC is different from that in the rat HP-MC. Even in closely related species like the mouse and the rat, the enzymes in the respective ChS-MC are not identical as assessed by differential reactivity to polyclonal antibodies. Thus, the different types of glycosaminoglycans may have evolved to increase the affinity of the proteoglycan–protease interaction.

Proteoglycans possess steric-exclusion properties which enable them to occupy different solvent volumes and different conformational states in varied environmental conditions (Meyer *et al.*, 1971). This physical property, which is dependent on molecular weight as well as charge (Comper and Preston, 1974), suggests a stabilizing effect of proteoglycans inside the granule. Just as DNA collapses into tight, thermally stable globular structures when complexed with histones and other basic molecules (Jordan et al., 1972; Lerman, 1971), negatively charged secretory granule proteoglycans may also form a more thermodynamically stable structure when complexed with positively charged proteases and amines. Indeed, some proteoglycan-containing cells, such as the human lung HP-MC (Caulfield et al., 1980) and the human eosinophil (Miller et al., 1966), possess secretory granules that have their contents packaged in distinctive highly ordered crystalline arrays which can be distinguished at the electron-microsopic level.

Because of their ability to attract cations, matrix proteoglycans function to regulate the osmotic pressure of connective tissue (Hedbys and Dohlman, 1963; Comper and Laurent, 1978). Likewise, secretory granule proteoglycans may function to regulate the osmotic pressure of the granule. It has been demonstrated morphologically (Caulfield et al., 1980) that before the human lung HP-MC exocytose proteoglycans and other preformed mediators, a flux of water into the granules occurs, resulting in a dramatic swelling of the granules. The granules then fuse with one another and with the plasma membrane. The difference in osmotic pressure between the granule and the extracellular milieu may be the driving force for efficient exocytosis of mediators from deep within the cell.

Secretory granule proteoglycans may act as mediators themselves. Although caution must be exercised when interpreting experiments in which these highly acidic proteoglycans are examined for their effect on a particular biological system in vitro, evidence has been obtained that mast cell proteoglycans can themselves regulate both the alternative complement pathway (Weiler et al., 1978; Wilson et al., 1984) and the Hageman factor-dependent contact activation system (Hojima et al., 1984). In the case of the effects of secretory granule proteoglycans on the alternative complement pathway, the type of uronic acid that makes up the glycosaminoglycan chain is actually more important than the degree or position of sulfation. Whether or not HP-MC-derived heparin proteoglycan functions in vivo as an anticoagulant is unproven, but is has been used extensively clinically to prevent coagulation of blood because of its ability to bind to antithrombin III. This anticoagulant effect of heparin is dependent on a very specific structure within the heparin chain (Lindahl et al., 1984).

In addition to the general functions ascribed above to secretory granule proteoglycans, the NK cell-derived molecule may protect the NK cell from being lysed by the cytotoxic molecules in the granule by preventing the deposition of these molecules on the surface of the NK cell. The proteoglycans might also participate in the assembly of the lytic subunits or act as insertion molecules in the final ring-forming structure on the target cell.

IV. Summary

Proteoglycan research on cells that participate in immune responses has progressed from the early novel finding that heparin proteoglycans are present in the secretory granules of the HP-MC to the recent findings that many secretory granule-containing cells possess chondroitin sulfate proteoglycans in their granules. In the case of the RBL-1 cell, both heparin and chondroitin sulfate diB glycosaminoglycans reside in the same granule and appear to be covalently bound to a common peptide core. Characterization studies of secretory granule proteoglycans have revealed that they possess a number of similar features. They have peptide cores which are very resistant to proteolytic degradation and are substituted with glycosaminoglycans that tend to be highly sulfated. Although the secretory granule proteoglycans vary in molecular weight from ~750,000 (rat serosal HP-MC) to 150,000 (RBL-1 cell), this variation has been found primarily to be the result of differences in the length of the bound glycosaminoglycans. Each secretory granule-containing cell type so far studied, however, appears to possess glycosaminoglycans that differ in such parameters as the type of hexosamine, location of sulfation, degree of sulfation, or extent of epimerization of the uronic acid. Since the type of glycosaminoglycan present in the secretory granule proteoglycan has been found to be a characteristic of that cell, the structure of the cell-associated proteoglycan is one of the properties now used phenotypically to distinguish immune cells. The preliminary finding that oligosaccharides are present on mouse bone marrow ChS-MC-derived chondroitin sulfate E proteoglycans raises the possibility that secretory granule proteoglycans may also vary in the type of low molecular weight glycans present on their peptide cores. The amino acid compositional analyses of heparin proteoglycans from the rat serosal HP-MC, chondroitin sulfate E proteoglycans from the mouse bone marrow-derived ChS-MC, and chondroitin sulfate diB proteoglycans from the RBL-1 cell, indicate their peptide cores are homologous to one another. It is not yet known if the minor differences reflect a species variation or are the result of different proteolytic degradation of a common peptide core. The availability of the cDNA probe from the rat yolk sac cell will be invaluable for answering such questions. The presence of proteoglycans in the secretory granules of cells which are distinct from those proteoglycans that reside in the extracellular matrix or on the plasma membrane suggests that the secretory granule proteoglycans have evolved to carry out important functions in the immune system. Which of the postulated functions of secretory granule proteoglycans prove to be important will be an exciting area of future investigation, because of the crucial role that these cells and their mediators have in our defense against different pathogens and tumor cell proliferation.

ACKNOWLEDGMENTS

This work was supported by a grant (AM35984, AI23483) from the National Institutes of Health.

REFERENCES

Benditt, E. P., Arase, M., and Roeper, M. E. (1956). Histamine and heparin in isolated rat mast cells. *J. Histochem. Cytochem.* **4,** 419–420.

Bland, C. E., Rosenthal, K. L., Pluznik, D. H., Dennert, G., Hengartner, H., Bienenstock, J., and Metcalfe, D. D. (1984). Glycosaminoglycan profiles in cloned granulated lymphocytes with natural killer function and in cultured mast cells: Their potential use as biochemical markers. *J. Immunol.* **132,** 1937–1942.

Bourdon, M. A., Oldberg, Ä., Pierschbacher, M., and Ruoslahti, E. (1985). Molecular cloning and sequence analysis of a chondroitin sulfate proteoglycan cDNA. *Proc. Natl. Acad. Sci. U.S.A.* **82,** 1321–1325.

Caulfield, J. P., Lewis, R. A., Hein, A., and Austen, K. F. (1980). Secretion in dissociated human pulmonary mast cells. Evidence for solubilization of granule contents before discharge. *J. Cell Biol.* **85,** 299–312.

Comper, W. D., and Laurent, T. C. (1978). Physiological function of connective tissue polysaccharides. *Physiol. Rev.* **58,** 255–315.

Comper, W. D., and Preston, B. N. (1974). Model connective tissue systems. A study of polyion-mobile ion and of excluded-volume interactions of proteoglycans. *Biochem. J.* **143,** 1–9.

David, G., and Van den Berghe, H. (1985). Heparan sulfate–chondroitin sulfate hybrid proteoglycan of the cell surface and basement membrane of mouse mammary epithelial cells. *J. Biol. Chem.* **260,** 11067–11074.

Denberg, J. A., Richardson, M., Telizyn, S., and Bienstock, J. (1983). Basophil/mast cell precursors in human peripheral blood. *Blood* **61,** 775–780.

DuBuske, L., Austen, K. F., Czop, J., and Stevens, R. L. (1984). Granule-associated serine neutral proteases of the mouse bone marrow-derived mast cell that degrade fibronectin: Their increase after sodium butyrate treatment of the cells. *J. Immunol.* **133,** 1535–1541.

Dvorak, A. M., Galli, S. J., Marcum, J. A., Nabel, G., Der Simonian, H., Goldin, J., Monahan, R. A., Pyne, K., Cantor, H., Rosenberg, R. D., and Dvorak, H. F. (1983). Cloned mouse cells with natural killer function and cloned suppressor T cells express ultrastructural and biochemical features not shared by cloned inducer T cells. *J. Exp. Med.* **157,** 843–861.

Eccleston, E., Leonard, B. J., Lowe, J. S., and Welford, H. J. (1973). Basophilic leukaemia in the albino rat and a demonstration of the basopoietin. *Nature (London) New Biol.* **244**, 73–76.

Ehrlich, P. (1878). Beiträge zur Theorie und Praxis der histologischen Färbung. Doctoral thesis, University of Leipzig, East Germany.

Enerbäck, L. (1966). Mast cells in rat gastrointestinal mucosa. II. Dye binding and metachromatic properties. *Acta Pathol. Microbiol. Scand.* **66**, 303–312.

Everitt, M. T., and Neurath, H. (1980). Rat peritoneal mast cell carboxypeptidase: Localization, purification and enzymatic properties. *FEBS Lett.* **110**, 292–296.

Fabian, I., Bleiberg, I., and Aronson, M. (1978). Increased uptake and desulphation of heparin by mouse peritoneal macrophages. *Biochim. Biophys. Acta* **544**, 69–76.

Galli, S. J., Dvorak, A. M., Marcum, J. A., Ishizaka, T., Nabel, G., Der Simonian, H., Pyne, K., Goldin, J. M., Rosenberg, R. D., Cantor, H., and Dvorak, H. F. (1982). Mast cell clones: A model for the analysis of cellular maturation. *J. Cell. Biol.* **95**, 435–444.

Haig, D. M., McKee, T. A., Jarrett, E. E. E., Woodbury, R., and Miller, H. R. P. (1982). Generation of mucosal mast cells is stimulated *in vitro* by factors derived from T cells of helminth-infected rats. *Nature (London)* **300**, 188–190.

Hedbys, B. O., and Dohlman, C. H. (1963). A new method for the determination of the swelling pressure of the corneal stroma *in vitro*. *Exp. Eye Res.* **2**, 122–129.

Hein, A., and Caulfield, J. P. (1982). Sulfur and calcium in rat peritoneal mast cell granules. *J. Cell Biol.* **95**, 397a.

Herberman, R. B., and Ortaldo, J. R. (1981). Natural killer cells: Their role in defenses against disease. *Science* **214**, 24–30.

Hojima, Y., Cochrane, C. G., Wiggins, R. C., Austen, K. F., and Stevens, R. L. (1984). *In vitro* activation of the contact (Hageman factor) system of plasma by heparin and chondroitin sulfate E. *Blood* **63**, 1453–1459.

Holgate, S. T., Lewis, R. A., and Austen, K. F. (1980). 3′,5′-Cyclic adenosine monophosphate-dependent protein kinase of the rat serosal mast cell and its immunologic activation. *J. Immunol.* **124**, 2093–2099.

Horner, A. A. (1971). Macromolecular heparin from rat skin. Isolation, characterization, and depolymerization with ascorbate. *J. Biol. Chem.* **246**, 231–239.

Ihle, J. N., Rebar, L., Keller, J., Lee, J. C., and Hapel, A. J. (1982). Interleukin 3: Possible roles in the regulation of lymphocyte differentiation and growth. *Immunol. Rev.* **63**, 5–31.

Isemura, M., and Ikenaka, T. (1975). β-Elimination and sulfite addition reaction of chondroitin sulfate peptidoglycan and the peptide structure of the linkage region. *Biochim. Biophys. Acta* **411**, 11–21.

Jordan, C. F., Lerman, L. S., and Venable, J. H. (1972). Structure and circular dichroism of DNA in concentrated polymer solutions. *Nature (London) New Biol.* **236**, 67–70.

Jorpes, E., Holmgren, H., and Wilander, O. (1973). Über das Vorkommen von Heparin in den Gefabwanden und in den Augen. *Z. Mikrosk. Anat. Forsch.* **42**, 279–302.

Lagunoff, D., and Pritzl, P. (1976). Characterization of rat mast cell granule proteins. *Arch. Biochem. Biophys.* **173**, 554–563.

Lerman, L. S. (1971). A transition to a compact form of DNA in polymer solutions. *Proc. Natl. Acad. Sci. U.S.A.* **68**, 1886–1890.

Levi-Schaffer, F., Austen, K. F., Caulfield, J. P., Hein, A., Bloes, W. F., and Stevens, R. L. (1985). Fibroblasts maintain the phenotype and viability of the rat heparin-containing mast cell *in vitro*. *J. Immunol.* **135**, 3454–3462.

Levitt, D., and Ho, P.-L. (1983). Induction of chondroitin sulfate proteoglycan synthesis and secretion in lymphocytes and monocytes. *J. Cell Biol.* **97**, 351–358.

Lindahl, U., and Höök, M. (1978). Glycosaminoglycans and their binding to biological macromolecules. *Annu. Rev. Biochem.* **47**, 385–417.

Lindahl, U., Cifonelli, J. A., Lindahl, B., and Rodén, L. (1965). The role of serine in the linkage of heparin to protein. *J. Biol. Chem.* **240**, 2817–2820.

Lindahl, U., Pertoft, H., and Seljelid, R. (1979). Uptake and degradation of mast cell granules by mouse peritoneal macrophages. *Biochem. J.* **182**, 189–193.

Lindahl, U., Thunberg, L., Backstrom, G., Riesenfeld, J., Nordling, K., and Björk, I. (1984). Extension and structural variability of the antithrombin-binding sequence in heparin. *J. Biol. Chem.* **259**, 12368–12376.

Lloyd, A. G., Embery, G., Wusteman, F. S., and Dodson, K. S. (1966). The metabolic fate of ^{35}S-labeled heparin and related compounds. *Biochem. J.* **98**, 33P–34P.

MacDermott, R. P., Schmidt, R. E., Caulfield, J. P., Hein, A., Bartley, G. T., Ritz, J., Schlossman, S. F., Austen, K. F., and Stevens, R. L. (1985). Proteoglycans in cell-mediated cytotoxicity. Identification, localization, and exocytosis of a chondroitin sulfate proteoglycan from human cloned natural killer cells during target cell lysis. *J. Exp. Med.* **162**, 1771–1778.

Mann, P. R. (1969). An electron-microscope study of the relations between mast cells and eosinophil leukocytes. *J. Pathol.* **98**, 183–186.

Mayrhofer, G. (1979). The nature of the thymus dependency of mucosal mast cells. II. The effect of thymectomy and of depleting recirculating lymphocytes on the response to *Nippostrongylus brasiliensis*. *Cell. Immunol.* **47**, 312–322.

Mayrhofer, G., and Bazin, H. (1981). Nature of the thymus dependency of mucosal mast cells. III. Mucosal mast cells in nude mice and nude rats, in B rats and in a child with the Di George syndrome. *Int. Arch. Allergy Appl. Immunol.* **64**, 320–331.

Metcalfe, D. D., Lewis, R. A., Silbert, J. E., Rosenberg, R. D., Wasserman, S. I., and Austen, K. F. (1979). Isolation and characterization of heparin from human lung. *J. Clin. Invest.* **64**, 1537–1543.

Metcalfe, D. D., Smith, J. A., Austen, K. F., and Silbert, J. E. (1980a). Polydispersity of rat mast cell heparin. Implications for proteoglycan assembly. *J. Biol. Chem.* **255**, 11753–11758.

Metcalfe, D. D., Wasserman, S. I., and Austen, K. F. (1980b). Isolation and characterization of sulfated mucopolysaccharides from rat leukaemic (RBL-1) basophils. *Biochem. J.* **185**, 367–372.

Meyer, F. A., Comper, W. D., and Preston, B. N. (1971). Model connective tissue systems. A physical study of gelatin gels containing proteoglycans. *Biopolymers* **10**, 1351–1364.

Miller, F., DeHarven, E., and Palade, G. E. (1966). The structure of eosinophil leukocyte granules in rodents and in man. *J. Cell Biol.* **31**, 349–362.

Muir, H. (1958). The nature of the link between protein and carbohydrate of a chondro-itin sulfate complex from hyaline cartilage. *Biochem. J.* **69**, 195–204.

Nakamura, N., Kojima, J., Okamoto, S., and Kitamura, Y. (1981). Absence of heparin in glycosaminoglycan fractions isolated from the skin of genetically mast cell-depleted W/W^v mice. *Biochem. Int.* **3**, 449–456.

Ogawa, M., Nakahata, T., Leary, A. G., Sterk, A. R., Ishizaka, K., and Ishizaka, T. (1983). Suspension culture of human mast cells/basophils from umbilical cord blood mononuclear cells. *Proc. Natl. Acad. Sci. U.S.A.* **80**, 4494–4498.

Ögren, S., and Lindahl, U. (1975). Cleavage of macromolecular heparin by an enzyme from mouse mastocytoma. *J. Biol. Chem.* **250**, 2690–2697.

Orenstein, N. S., Galli, S. J., Dvorak, A. M., Silbert, J. E., and Dvorak, H. F. (1978). Sulfated glycosaminoglycans of guinea pig basophilic leukocytes. *J. Immunol.* **121**, 586–592.

Orenstein, N. S., Galli, S. J., Dvorak, A. M., and Dvorak, H. F. (1981). Glycosaminogly-cans and proteases of guinea pig basophilic leukocytes. *In* "Biochemistry of the Acute Allergic Reactions" (E. L. Becker, A. S. Simon, and K. F. Austen, eds.), pp. 131–143. Liss, New York.

Pasternak, M. S., and Eisen, H. N. (1985). A novel serine esterase expressed by cytotoxic T lymphocytes. *Nature (London)* **314**, 743–745.

Paterson, N. A. M., Wasserman, S. I., Said, J. W., and Austen, K. F. (1976). Release of chemical mediators from partially purified human lung mast cells. *J. Immunol.* **117**, 1356–1362.

Petty, H. R., Hermann, W., Dereski, W., Frey, T., and McConnell, H. M. (1984). Activata-ble esterase activity of murine natural killer cell-YAC tumour cell conjugates. *J. Cell Sci.* **72**, 1–13.

Rao, P. V. S., Friedman, M. M., Atkins, F. M., and Metcalfe, D. D. (1983). Phagocytosis of mast cell granules by cultured fibroblasts. *J. Immunol.* **130**, 341–349.

Razin, E., Cordon-Cardo, C., and Good, R. A. (1981). Growth of a pure population of mouse mast cells *in vitro* with conditioned medium derived from concanavalin A-stimulated splenocytes. *Proc. Natl. Acad. Sci. U.S.A.* **78**, 2559–2561.

Razin, E., Stevens, R. L., Akiyama, F., Schmid, K., and Austen, K. F. (1982). Culture from mouse bone marrow of a subclass of mast cells possessing a distinct chondroitin sulfate proteoglycan with glycosaminoglycans rich in N-acetylgalactosamine-4,6-di-sulfate. *J. Biol. Chem.* **257**, 7229–7236.

Razin, E., Mencia-Huerta, J.-M., Stevens, R. L., Lewis, R. A., Liu, F.-T., Corey, E. J., and Austen, K. F. (1983). IgE-mediated release of leukotriene C_4, chondroitin sulfate E proteoglycan, β-hexosaminidase, and histamine from cultured bone marrow-derived mouse mast cells. *J. Exp. Med.* **157**, 189–201.

Razin, E., Ihle, J. N., Seldin, D., Mencia-Huerta, J.-M., Katz, H. R., LeBlanc, P. A., Hein, A., Caulfield, J. P., Austen, K. F., and Stevens, R. L. (1984a). Interleukin 3: A differ-entiation and growth factor for the mouse mast cell that contains chondroitin sulfate E proteoglycan. *J. Immunol.* **132**, 1479–1486.

Razin, E., Stevens, R. L., Austen, K. F., Caulfield, J. P., Hein, A., Lui, F.-T., Clabby, M., Nabel, G., Cantor, H., and Friedman, S. (1984b). Cloned mouse mast cells derived from immunized lymph node cells and from foetal liver cells exhibit characteristics of bone marrow-derived mast cells containing chondroitin sulfate E proteoglycan. *Im-munology* **52**, 563–575.

Robinson, H. C., Horner, A. A., Höök, M., Ögren, S., and Lindahl, U. (1978). A pro-teoglycan form of heparin and its degradation to single-chain molecules. *J. Biol. Chem.* **253**, 6687–6693.

Rodén, L. (1980). Structure and metabolism of connective tissue proteoglycans. *In* "The Biochemistry of Glycoproteins and Proteoglycans" (W. J. Lennarz, ed.), pp. 267–371. Plenum, New York.

Saito, H., Yamagata, T., and Suzuki, S. (1968). Enzymatic methods for the determina-tion of small quantities of isomeric chondroitin sulfates. *J. Biol. Chem.* **243**, 1536–1542.

Schmidt, R. E., MacDermott, R. P., Bartley, G., Bertovich, M., Amato, D. A., Austen, K. F., Schlossman, S. F., Stevens, R. L., and Ritz, J. (1985). Specific release of pro-teoglycans from human natural killer cells during target cell lysis. *Nature (London)* **318**, 289–291.

Schrader, J. W. (1981). The *in vitro* production and cloning of the P cell, a bone marrow-derived null cell that expresses H-2 and Ia-antigens, has mast cell-like granules, and is regulated by a factor released by activated T cells. *J. Immunol.* **126**, 452–458.

Schwartz, L. B., and Austen, K. F. (1980). Enzymes of the mast cell granule. *J. Invest. Dermatol.* **74,** 349–353.

Schwartz, L. B., Lewis, R. A., and Austen, K. F. (1981a). Tryptase from human pulmonary mast cells. Purification and characterization. *J. Biol. Chem.* **256,** 11939–11943.

Schwartz, L. B., Riedel, C., Caulfied, J. P., Wasserman, S. I., and Austen, K. F. (1981b). Cell association of complexes of chymase, heparin proteoglycan, and protein after degranulation by rat mast cells. *J. Immunol.* **126,** 2071–2078.

Seldin, D. C., Seno, N., Austen, K. F., and Stevens, R. L. (1984). Analysis of polysulfated chondroitin disaccharides by high performance liquid chromatography. *Anal. Biochem.* **141,** 291–300.

Seldin, D. C., Adelman, S., Austen, K. F., Stevens, R. L., Hein, A., Caulfield, J. P., and Woodbury, R. G. (1985a). Homology of the rat basophilic leukemia cell and the rat mucosal mast cell. *Proc. Natl. Acad. Sci. U.S.A.* **82,** 3871–3875.

Seldin, D. C., Austen, K. F., and Stevens, R. L. (1985b). Purification and characterization of protease-resistant secretory granule proteoglycans containing chondroitin sulfate di-B and heparin-like glycosaminoglycans from rat basophilic leukemia cells. *J. Biol. Chem.* **260,** 1131–1139.

Serafin, W. E., Katz, H. R., Austen, K. F., and Stevens, R. L. (1986) Complexes of heparin proteoglycans, chondroitin sulfate E proteoglycans, and [^{3}H]diisopropyl flurophosphate-binding proteins are exocytosed from activated mouse bone marrow-derived mast cells. *J. Biol. Chem.* **261,** 15017–15021.

Silbert, J. E. (1963). Incorporation of ^{14}C and ^{3}H from nucleotide sugars into a polysaccharide in the presence of a cell free preparation from mouse mast cell tumors. *J. Biol. Chem.* **238,** 3542–3546.

Silbert, J. E. (1967). Biosynthesis of heparin. III. Formation of a sulfated glycosaminoglycan with a microsomal preparation from mast cell tumors. *J. Biol. Chem.* **242,** 5146–5152.

Sredni, B., Friedman, M. M., Bland, C. E., and Metcalfe, D. D. (1983). Ultrastructural, biochemical, and functional characteristics of histamine-containing cells cloned from mouse bone marrow: tentative identification as mucosal mast cells. *J. Immunol.* **131,** 915–922.

Stevens, R. L., and Austen, K. F. (1982). Effect of *p*-nitrophenyl-β-D-xyloside on proteoglycan and glycosaminoglycan biosynthesis in rat serosal mast cell cultures. *J. Biol. Chem.* **257,** 253–259.

Stevens, R. L., Razin, E., Austen, K. F., Hein, A., Caulfield, J. P., Seno, N., Schmid, K., and Akiyama, F. (1983). Synthesis of chondroitin sulfate E glycosaminoglycan onto *p*-nitrophenyl-β-D-xyloside and its localization to the secretory granules of rat serosal mast cells and mouse bone marrow-derived mast cells. *J. Biol. Chem.* **258,** 5977–5984.

Stevens, R. L., Otsu, K., and Austen, K. F. (1985). Purification and analysis of the core protein of the protease-resistant intracellular chondroitin sulfate E proteoglycan from the interleukin 3-dependent mouse mast cell. *J. Biol. Chem.,* **260,** 14194–14200.

Stevens, R. L., Katz, H. R., Seldin, D. C., and Austen, K. F. (1986). Biochemical characteristics distinguish subclasses of mammalian mast cells. *In* "Mast Cell Heterogeneity" (A. D. Befus, J. Bienenstock, and J. Denburg, eds.). Raven, New York.

Straus, A. H., Nader, H. B., and Dietrich, C. P. (1982). Absence of heparin or heparin-like compounds in mast-cell free tissues and animals. *Biochim. Biophys. Acta* **717,** 478–485.

Suzuki, S., Saito, H., Yamagata, T., Anno, K., Seno, N., Kawai, Y., and Furuhashi, T. (1968). Formation of three types of disulfated disaccharides from chondroitin sulfates by chondroitinase digestion. *J. Biol. Chem.* **243,** 1543–1550.

Tadokoro, K., Stadler, B. M., and DeWeck, A. L. (1983). Factor dependent *in vitro* growth of human normal bone marrow-derived basophil-like cells. *J. Exp. Med.* **158,** 857–871.

Tas, J., and Berndsen, R. G. (1977). Does heparin occur in mucosal mast cells of the rat small intestine? *J. Histochem. Cytochem.* **25,** 1058–1062.

Tantravahi, R. V., Stevens, R. L., Austen, K. F., and Weis, J. H. (1986). A single gene in mast cells encodes the core peptides of heparin and chondroitin sulfate proteoglycans. *Proc. Natl. Acad. Sci. (USA)* **83,** 9207–9210.

Tertian, G., Yung, Y.-P., Guy-Grand, D., and Moore, M. A. S. (1981). Long term *in vitro* culture of murine mast cells. I. Description of a growth factor-dependent culture technique. *J. Immunol.* **127,** 788–794.

Weiler, J. M., Yurt, R. W., Fearon, D. T., and Austen, K. F. (1978). Modulation of the formation of the amplification convertase of complement, C3b,Bb, by native and commercial heparin. *J. Exp. Med.* **147,** 409–421.

Wilson, J. G., Fearon, D. T., Stevens, R. L., Seno, N., and Austen, K. F. (1984). Inhibiton of the function of activated properdin by squid chondroitin sulfate E glycosaminoglycan and murine bone marrow-derived mast cell chondroitin sulfate E proteoglycan. *J. Immunol.* **132,** 3058–3063.

Woodbury, R. G., and Neurath, H. (1978). Purification of an atypical mast cell protease and its levels in developing rats. *Biochemistry* **17,** 4298–4304.

Yurt, R. W., and Austen, K. F. (1977). Preparative purification of the rat mast cell chymase. Characterization and interaction with granule components. *J. Exp. Med.* **146,** 1405–1419.

Yurt, R. W., Leid, R. W., Jr., Austen, K. F., and Silbert, J. E. (1977a). Native heparin from rat peritoneal mast cells. *J. Biol. Chem.* **252,** 518–521.

Yurt, R. W., Leid, R. W., Jr., Spragg, J., and Austen, K. F. (1977b). Immunologic release of heparin from purified rat peritoneal mast cells. *J. Immunol.* **118,** 1201–1207.

Index